*Masters and Johnson
on Sex and Human Loving*

Masters and Johnson on Sex and Human Loving

WILLIAM H. MASTERS
VIRGINIA E. JOHNSON
ROBERT C. KOLODNY

LITTLE, BROWN AND COMPANY

BOSTON   TORONTO   LONDON

A substantial portion of the material in this book has been previously published in the Little, Brown and Company college textbook *Human Sexuality* by Masters, Johnson, and Kolodny.

Acknowledgments for permission to reprint material are on page 622.

Library of Congress Cataloging-in-Publication Data

Masters, William H.
  Masters and Johnson on sex and human loving.

  Rev. ed. of: Human sexuality. 2nd ed. © 1985
  Bibliography: p.
  Includes index.
  1. Sex.   2. Sex (Psychology)   3. Sex (Biology)
4. Sexual disorders.   I. Johnson, Virginia E.
II. Kolodny, Robert C.   III. Masters, William H.
Human sexuality.   IV. Title.   V. Title: On sex and
human loving. [DNLM: 1. Sex.   2. Sex Behavior.
3. Sex
Disorders. HQ 21 M423ha]
HQ21.M46158   1986   306.7   85-23950
ISBN 0-316-54998-3 (hc)
ISBN 0-316-50160-3 (pb)

10   9   8   7   6   5   4

RRD VA

*Published simultaneously in Canada
by Little, Brown & Company (Canada) Limited*

PRINTED IN THE UNITED STATES OF AMERICA

Contents

Masters and Johnson
on Sex and Human Loving

CHAPTER ONE

Perspectives on Sexuality

EVERY PERSON has sexual feelings, attitudes, and beliefs, but everyone's experience of sexuality is unique because it is processed through an intensely personal perspective. This perspective comes from both private, personal experience and public, social sources. It is impossible to understand human sexuality without recognizing its multidimensional nature.

Sexuality has fascinated people in all walks of life from ancient times until the present. Sexual themes have been common in art and literature. Religions, philosophies, and legal systems — all concerned with shaping human behavior — have typically tried to establish sexual values and sexual taboos. At various times in history, illness, creativity, aggression, emotional disorders, and the rise and fall of cultures have all been "explained" as the result of too much or too little sexual activity or unusual sexual practices or thoughts.

While keeping in mind the private, public, and historical sources of our sexual heritage, we can broaden and deepen our understanding by studying sexuality from biological, psychosocial, behavioral, clinical, and cultural perspectives. In examining sexuality from these varied viewpoints, however, we must be careful not to forget that learning about sexuality, in all its forms, is really learning about people and the complexities of human nature.

Learning accurate information about sexuality can help prevent sexual problems and help us be better sex educators for our children. Becoming well informed about sex can also help us deal more effectively with certain types of problems if they occur in our lives (e.g., infertility, sexual dysfunctions, sexually transmitted diseases, sexual harassment). Even more important, studying sexuality can help us become more sensitive and aware in our interpersonal relationships, thus contributing to the growth of intimacy and sexual satisfaction in our lives.

Unfortunately, it is also true that these results do not happen automatically. There is no guarantee that careful study of this text will make finding (or keeping) sexual partners any easier, nor that it will lead to sexual bliss. Instead, we believe that learning about sexuality in an objective fashion will enable our readers to examine important sexual issues — some intensely personal, some social, some moral — and emerge with deeper insight into themselves and others. We also believe that sexual knowledge can lead to reasoned, responsible interpersonal sexual behavior and can help people make important personal decisions about sex. In short, learning about sexuality is an invaluable preparation for living.

DIMENSIONS OF SEXUALITY: SOME DEFINITIONS

One would certainly think that there could be no doubt about what is to be understood by the term "sexual." First and foremost, of course, it means the "improper," that which must not be mentioned. (Freud, 1943, p. 266)

Sex is not a mere physiological transaction to the primitive South Sea Islander any more than it is to us; it implies love and love-making; it becomes the nucleus of such venerable institutions as marriage and the family; it pervades art and it produces its spells and its magic. It dominates in fact almost every aspect of culture. *Sex*, in its widest meaning . . . is rather a sociological and cultural force than a mere bodily relation of two individuals. (Malinowski, 1929, p. xxiii)

"Francie, you bloody fucker," I used to say, "you've got the morals of a clam." "But you like me, don't you?" she'd answer. "Men like to fuck, and so do women. It doesn't harm anybody and it doesn't mean you have to love everyone you fuck, does it?" (Miller, 1961, p. 262)

What is sexuality? As shown by the quotes above, there is no simple answer to this question. Freud saw sex as a powerful psychological and biological force, while Malinowski emphasized its sociological and cultural dimensions. Henry Miller used frank portrayals of sex in his novels to make a philosophical statement about the human condition. In everyday life, the word "sex" is often used to mean male or female (biological gender) or to refer to physical activity involving the genitals ("having sex"). The word "sexuality" generally has a broader meaning since it refers to all aspects of being sexual. Sexuality means a dimension of personality instead of referring to a person's capacity for erotic response alone.

Unfortunately, our language for talking about sex and sexuality is very limited.* We may distinguish between sex acts (such as masturbation, kissing, or sexual intercourse) and sexual behavior (which includes not only specific sex acts but being flirtatious, dressing in certain ways, reading *Playboy,* or dating) without having yet scratched the surface of sexuality. We may describe different types of sex as *procreational* (for having children), *recreational* (for having fun, with no other goal), or *relational* (for sharing with a cared-for person), and find that our categories are still too few. While we cannot fully answer the question "What is sexuality?" in this chapter, we can briefly introduce the dimensions of sexuality that are the subject of this book.

A CASE PROFILE

David and Lynn sat anxiously in the waiting room at a sex therapy clinic. Although feeling embarrassed and unsure, they were determined to seek a solution to the sexual problems that had troubled their relationship for the past three months. Although they had been living together for almost two years and had planned to marry after their graduation from college, the dissatisfactions that now rocked their lives had thrown these future plans into doubt.

Once inside the clinic, they told their story in a straightforward fashion. They had met three years ago, at age eighteen, during their freshman year at school. Romance blossomed as they dis-

* Wardell Pomeroy, a leading sexologist, likes to ask an audience for a four-letter word ending in the letter *k* that means "intercourse" — and then pauses knowingly before he says, "Talk."

covered many shared interests and easily developed an intimate sexual relationship. Neither David nor Lynn was a virgin when they met, and they felt a strong sexual attraction to each other. Their first shared lovemaking experience was passionate and sensuous. As their relationship matured, their sexual interaction continued to be a major source of pleasure. Living together was a natural outgrowth of these feelings, and it had been fun — until recently.

Trouble first appeared during Christmas vacation when they visited Lynn's parents in Boston. David was upset because he and Lynn were not allowed to share the same bedroom. Lynn was put off by her parents' apparent coolness toward David. Their only sexual opportunity (on a Sunday morning while Lynn's parents were at church) was hurried and felt mechanical. They were both relieved to return to school in time for a big New Year's Eve party with some friends.

The party lasted until 4:00 A.M. with great quantities of champagne consumed by all. Once back at their apartment, David and Lynn tried to make love but David was unable to get an erection. They laughed it off and fell asleep, happy to be "home."

The next morning David had a terrible hangover. He took some aspirin, ate a quick breakfast, and invited Lynn into their bedroom. She wasn't very enthusiastic, having a slight hangover herself, but didn't object. Once again David was unable to get an erection. Although Lynn was very understanding and supportive, David worried about his sexual performance (or lack of it) all day. He decided he needed some rest and relaxation before trying again, so he went to bed early that evening without any romantic overtures.

David awakened the next day feeling rested and refreshed and immediately turned to embrace Lynn.

Despite feeling good, he found himself having only a partial erection, and even that disappeared when they tried to have intercourse. From that point on, David was plagued by difficulties getting or keeping an erection, and Lynn — despite initial attempts at helping him — was getting increasingly upset. Whereas their relationship *had* been relaxed and comfortable, now they were becoming short-tempered and abrupt. They talked about splitting up but believed they still loved each other and that they could — with some expert help — overcome this problem.

This real-life example, drawn from our files, allows us to introduce the various perspectives on sexuality that we will examine in greater detail later in this book. By looking at David and Lynn's situation, we can see the importance of the different dimensions of sexuality that interact in all of our lives.

The Biological Dimension

David's problem with erections first occurred after he had consumed a lot of champagne. This is not very surprising, since alcohol is a depressant to the nervous system. Because the nervous system normally transmits physical sensations to the brain and activates our sexual reflexes, too much alcohol can block *anyone's* sexual response.

The biological dimension of sexuality is far more inclusive than this, however. Biological factors largely control sexual development from conception until birth and our ability to reproduce after puberty. The biological side of sexuality also affects our sexual desire, our sexual functioning, and (indirectly) our sexual satisfaction. Biological forces are even thought to influence certain sex differences in behavior, such as the tendency of males to act more aggressively than females. And sexual turn-ons, no matter what their source, produce specific biological events: the pulse quickens, the sexual organs respond, and sensations of warmth or tingling spread through our bodies.

The Psychosocial Dimension

David and Lynn reacted to their situation in different ways. David became anxious and goal-oriented and lost his self-confidence, while Lynn, who started out being supportive and understanding, became irritated and aloof. Clearly, the nature of their relationship changed in response to the stress of their sexual problem. David and Lynn even began to doubt whether they were in love and wanted to marry, although while visiting Lynn's parents they were convinced this plan was "right."

These responses illustrate the psychosocial dimension of sexuality, which includes psychological factors (emotions, thoughts, and personalities) in combination with social elements (how people interact). In this case, David's concern about his first sexual "failure" led him into further difficulties even when the original biological "cause" — too much alcohol — was removed from the situation. His anxieties led him into trying too hard to make sex work, with the result being exactly the opposite of what he and Lynn wanted.

The psychosocial side of sexuality is important because it sheds light not only on many sexual problems but also on how we develop as sexual beings. From infancy, a person's *gender identity* (the personal sense of being male or female) is primarily shaped by psychosocial forces. Our early sexual attitudes — which often stay with us into adulthood — are based largely on what parents, peers, and teachers tell us or show us about the meanings and purposes of sex. Our sexuality is also social in that it is regulated by society through laws, taboos, and family and peer group pressures that seek to persuade us to follow certain paths of sexual behavior.

The Behavioral Dimension

Talking with David and Lynn separately, we learned that the pattern of their sexual interaction changed considerably during the three months of their problem. The frequency of attempted lovemaking fell drastically, while in the past it had been four or five times per week. David masturbated several times a week (which he had not done for several years) after finding that he could easily get erections this way. On the other hand, Lynn masturbated only once, since she felt guilty about this activity. She also shied away from initiating sexual activity or even acting romantic with David because she thought this would put extra pressure on him.

These aspects of David and Lynn's situation reflect the behavioral dimension of sexuality. Sexual behavior is a product of both biological and psychosocial forces, yet studying it in its own right can be enlightening. The behavioral perspective allows us to learn not only *what* people do but to understand more about *how* and *why* they do it. For example, David may have used masturbation to boost his self-confidence by "proving" to himself that he could still get erections. Lynn's withdrawal from initiating physical intimacy may have been well intended, but David may have interpreted it as rejection.

In discussing this topic, we should avoid judging other people's sexual behavior by our own values and experiences. Too often, people have a tendency to think about sexuality in terms of "normal" versus "abnormal." "Normal" is frequently defined as what we ourselves do and feel comfortable about, while the "abnormal" is what others do that seems different or odd to us. Trying to decide what is normal for others is not only a thankless task but one ordinarily doomed to failure because our objectivity is clouded by our values and experiences.

The Clinical Dimension

David and Lynn entered our sex therapy program and resolved their difficulties within two weeks. Not only did their sexual interaction return to its previously pleasurable state, but both felt that the therapy experience improved their relationship in other ways. As Lynn commented to us: "Overcoming the sexual problem was great, but we've also learned so much about ourselves. Our communication is about 1,000 percent better now, and we really feel like we have a solid relationship that can cope with any kind of problem that comes up."

Although sex is a natural function, many types of obstacles can lessen the pleasure or spontaneity of our sexual encounters. Physical problems such as illness, injury, or drugs can alter our sexual response patterns or knock them out completely. Feelings such as anxiety, guilt, embarrassment, or depression and conflicts in our personal relationships can also hamper our sexuality. The clinical perspective of sexuality examines the solutions to these and other problems that prevent people from reaching a state of sexual health and happiness.

Greatly improved results have been obtained in the treatment of a wide variety of sexual difficulties in the last two decades. Two key changes have contributed to this success: a better understanding of the multidimensional nature of sexuality and the development of a new discipline, called sexology, devoted to the study of sex. Doctors, psychologists, nurses, counselors, and other professionials trained in sexology can integrate this knowledge with training in sex counseling or sex therapy to help a high percentage of their patients.

The Cultural Dimension

David and Lynn's lives, like all of ours, reflect the input of the culture in which they live. For example, Lynn's parents refused to let them sleep in the same bedroom although they knew David and Lynn were living together. As another example, Lynn's sense of guilt toward masturbation stemmed largely from her religious upbringing. And David's anxiety over his sexual difficulties was partly a reaction to the prevailing American notion that men should be instantly erect at the first moment of a sexual encounter.

Our own cultural attitudes toward sexuality are far from universal. In some societies, a man's special obligations to guests or friends are discharged by an invitation to have sexual relations with his wife. Ford and Beach (1951, p. 49) listed eight cultural groups in which kissing was unknown, pointing out: "When the Thonga first saw Europeans kissing they laughed, expressing this sentiment: 'Look at them — they eat each other's saliva and dirt.' " While these cultural differences may shock or amuse us, they can also help us understand that our viewpoint is not shared by all people in all places.

Sexual topics are often controversial and value-laden, but the controversy is often relative to time, place, and circumstance. What is labeled as "moral" or "right" varies from culture to culture, from century to century. Many of the moral issues pertaining to sex relate to certain religious traditions, but religion has no monopoly on morality. People who have no closely held religious creed are just as likely to be moral as those whose values are tied to a religious position. *There is no sexual value system that is right for everyone and no single moral code that is indisputably correct and universally applicable.*

In America, messages about sexual behavior that prevailed in the first half of this century now appear to be changing. Three trends deserve particular mention. The first is a loosening of gender role stereotypes. *Gender role* is the public expression of gender identity — that is, how an individual asserts his or her maleness or femaleness in social settings. Traditionally, women and girls were cast as sexually passive and unresponsive creatures, while men were seen as virile sexual aggressors. According to this view, the male was expected to be the sexual initiator and expert, and the female who was aggressive or enjoyed sex too much was frowned upon. This notion has now been replaced for many people by a concept of mutual participation and satisfaction. A second trend is the greater degree of openness about sexuality. All forms of the media from television to cinema to the printed word reflect this change, and, as a result, sex has become less shameful and mysterious. The third trend is the growing acceptance of relational and recreational sex as opposed to reproductive sex. This shift, which has been especially evident in the past twenty years, is due partly to improved contraceptive techniques and concern for overpopulation. The emergence of a positive sex philosophy is also tightly intertwined with the sexual emancipation of women and greater societal openness toward sex.

HISTORICAL PERSPECTIVES ON SEXUALITY

A major obstacle to understanding our own sexuality is realizing we are prisoners of past societal attitudes toward sex. (Bullough, 1976, p. xi)

To understand the present, it is helpful to begin by examining the past. In certain respects, we are bound by a sexual legacy passed on from generation to generation, but in other ways modern views of sex and sexuality differ drastically from past patterns.

Although written history goes back almost 5,000 years, only limited information is available describing sexual behavior and attitudes in various societies prior to 1000 B.C. Clearly, a prominent taboo against incest had already been established, and women were considered as property, with sexual and reproductive value. Men were free to have many sexual partners, prostitution was widespread, and sex was accepted as a straightforward fact of life.

With the advent of Judaism, an interesting interplay of sexual attitudes began to emerge. In the first five books of the Old Testament, the primary source of Jewish laws, there are rules about sexual conduct: adultery is forbidden in the Ten Commandments (Exodus 20:14), for example, and homosexual acts are strongly condemned (Leviticus 18:22, Leviticus 21:13). At the same time, sex is recognized as a creative and pleasurable force, as depicted in the Song of Songs. Sex was neither considered inherently evil nor restricted to procreative purposes alone.

In ancient Greece, however, there was tolerance and even enthusiasm regarding male homosexuality in certain forms. Homosexual relations between an adult man and adolescent boy past the age of puberty were commonplace, usually occurring in an educational relationship where the man was reponsible for the boy's moral and intellectual development. At the same time, exclusive homosexuality and homosexual acts between adults were frowned upon, and homosexual contact between adults and boys under the age of puberty was illegal. There was a strong emphasis on marriage and family yet women were second-class citizens, if they could be considered citizens at all: "In Athens, women had no more political or legal rights than slaves; throughout their lives they were subject to the absolute authority of their male next-of-kin. . . . As everywhere else in the first millennium B.C., women were chattels, even if some of them were independent-minded ones. To the Greeks, a woman (regardless of age or marital status) was *gyne,* whose linguistic meaning is 'bearer of children' " (Tannahill, 1980, pp. 94–95).

As Christianity developed in its early forms, there was an intermingling of Greek and Jewish attitudes toward sexuality. In contrast to Judaism, which did not distinguish physical from spiritual love, Christian theology borrowed from the Greek and separated *eros,* or "carnal love," from *agape,* a "spiritual, nonphysical love." The Hellenistic era in Greece (beginning in 323 B.C.) was marked by a denial of worldly pleasures in favor of developing the purely spiritual. Along with the New Testament portrayal of the imminent end of the world, this led to Christianity's placing a high ideal on celibacy, although St. Paul allowed that while "It is good for a man not to touch a woman . . . it is better to marry than to burn" (I Corinthians 7:1–12).

By the end of the fourth century A.D., despite small groups of Christians whose views of sexuality were less rigid and constrained, the Church's negative attitudes toward sex were dramatically presented in the writings of St. Augustine, a religious leader whose background included a vivid and varied set of erotic experiences before he renounced worldly ways. Augustine confessed in stark terms, "I muddied the stream of friendship with the filth of lewdness and clouded its clear waters with hell's black river of lust" (*Confessions,* Book III:i). He believed that sexual lust came from the downfall of Adam and Eve in the Garden of Eden and that this sinfulness was transmitted to children by the inherent lust that separated humanity from God. Thus, sex was strongly condemned in all forms, although Augustine and his contemporaries apparently felt that marital procreative sex was less evil than other types.

Elsewhere in the world, sexual thinking varied remarkably from that just described. In particular, Islamic, Hindu, and ancient Oriental sexual attitudes were considerably more positive. The historian Vern Bullough states that "almost anything in the sexual field received approval from some segment of the Hindu society" and that in China "sex was not something to be feared, nor was it regarded as sinful, but rather, it was an act of worship" and even a path toward immortality (Bullough, 1976, pp. 275 and 310). The *Kama Sutra,* compiled at about the same time Augustine was writing his *Confessions,* is a detailed Indian sex manual; in ancient China and Japan, similar manuals were abundant and glorified sexual pleasure and variety. These divergent patterns continued, although our focus for now will remain with the history of sex in the Western world.

The early Christian traditions regarding sexuality became more firmly entrenched in Europe during the twelfth and thirteenth centuries as the Church assumed greater power. Theology often became synonymous with common law, and there was a generally oppressive "official" attitude toward sex except for the purpose of procreation. There was, however, a certain hypocrisy between professed Church policies and actual practices: "religious houses themselves were often hotbeds of sexuality" (Taylor, 1954, p. 19).

During this era, a new style of living emerged among the upper classes that brought about a drastic separation between actual practice and religious teachings. This style, called courtly love, introduced a new code of acceptable behavior in which women (at least high-ranking women) were elevated to an immaculate plane and romanticism, secrecy, and valor were celebrated in song, poetry, and literature. Pure love was seen as incompatible with the temptations of the flesh, and sometimes this concept was tested by lovers lying together in bed naked to see if they could prove the fullness of their love by refraining from sexual intercourse. Needless to say, it is unlikely that courtly love was always the unconsummated romantic ideal portrayed in story and verse.

Not too long after the era of courtly love began, chastity belts made their appearance. These devices allowed husbands to lock up their wives just as they would protect their money; while they may have been originally designed to prevent rape, they also served to guard "property":

> The belt of medieval times was usually constructed on a metal framework that stretched between the woman's legs from front to back. It had two small, rigid apertures that allowed for waste elimination but effectively prevented penetration, and once it was locked over the hips the jealous husband could take away the key. (Tannahill, 1980, p. 276)

The rebirth of humanism and the arts that subsequently engulfed Europe in the sixteenth and seventeenth centuries was accompanied by a loosening of sexual restrictions as well as less adherence to the formulas of courtly love. The Protestant Reformation, led by Martin Luther, John Calvin, and others, generally advocated less negative attitudes toward sexual matters than the Catholic Church did. For example, although Luther was hardly liberal in his sexual attitudes, he thought that sex was not inherently sinful and that chastity and celibacy were not signs of virtue. At the same time, Europe was caught in a massive epidemic of syphilis — possibly imported from the Americas — that might have worked to limit sexual freedom.

When we speak of the attitudes of a historical era, we must keep in mind that there was variance among different countries, levels of society, and religious groups. Although evidence can be cited to show a rather broad tolerance toward sexuality in England and France in the 1700s, the Puritan ethic reigned in colonial America. Sex outside marriage was condemned and family solidarity was exalted; those giving in to the passions of adultery or premarital sex, if discovered, were flogged, put in pillories or stocks, or forced to make public confessions. Nathaniel Hawthorne's *Scarlet Letter* presents an account of colonial times that is notable primarily for its understatement.

In America, the Puritan ethic was carried over into the nineteenth century with a curious schism. As American frontiers expanded and as large cities took on a more cosmopolitan flair, there was a corresponding loosening of notions about sexual propriety, and prostitution became commonplace. This new development was met by the formation in the 1820s and 1830s of several groups whose primary mission was to combat the social evils of prostitution and rescue the "fallen women" who plied this trade. Despite the organized resistance of such groups as the American Society for the Prevention of Licentiousness and Vice and the Promotion of Morality and the American Society for Promoting the Observance of the Seventh Commandment, prostitution flourished. During a three-year period in the 1840s, the government prosecuted 351 brothels in Massachusetts alone, and by the eve of the Civil War, a guidebook listing fashionable brothels in big cities described 106 establishments in New York, 57 in Philadelphia, and dozens of others in Baltimore, Boston, Chicago, and Washington, DC (Pivar, 1973).

By the mid-1800s, as the Victorian era began, reserve and prudery emerged once again in Europe, although this time less connected to religious edict. The spirit of Victorianism was sexual repression and a strong sense of modesty necessitated by the presumed purity and innocence of women and children. Taylor points out, "So delicate did the sensibilities of the Victorians become, so easily were their thoughts turned to sexual matters, that the most innocent actions were taboo in case they might lead to lurid imaginings. It became indelicate to offer a lady a *leg* of chicken . . . ," and clothing styles, showing not even a glimpse of ankle or bare neck, mirrored this conservatism (Taylor, 1954, pp. 214–215). The prudishness of this period is astonishing to us today: in some Victorian homes, piano legs were covered with crinolines, and books by authors of opposite sexes were not shelved side by side unless the authors were married to each other.

In America, although the influence of Victorianism was strongly felt, crosscurrents sent the mainstream of moral thinking into a dizzying spin. For example, in 1870 the St. Louis City Council found a loophole in state law that allowed it to legalize

FEIFFER

SEXTERROR #2

EVERYTHING I HAVE LEARNED ABOUT SEX:

ITS DIRTY. ITS CLEAN.

ITS FORBIDDEN. ITS HEALTHY.

NICE GIRLS DON'T DO IT. EVERYONE DOES IT.

IT GIVES V.D. V.D. IS CURABLE.

IT GIVES HERPES. HERPES IS INCURABLE.

ITS FUN. IT GIVES PREGNANCY. PREGNANCY IS CURABLE. ABORTION IS A CHOICE. NO, ITS A SIN.

TO LEARN EVERYTHING ABOUT SEX IS TO KNOW NOTHING ABOUT SEX.

TAKE ME.

prostitution, causing an uproar across the nation. Groups were again formed to combat sexual immorality and managed to find allies in other organizations dedicated to the cause of temperance (abolishing the sale of alcoholic beverages). This movement achieved several legislative successes. In 1886, for example, twenty-five states fixed the age of consent at ten (thus permitting child prostitution to flourish), but by 1895, only five states retained this low age, and eight states had raised the age of consent to eighteen.

Although the mainstream of Victorianism was antisexual — pornography was first banned by law in this era — there was another side to the times. A sexual "underground" of pornographic writings and pictures was widely read, as Stephen Marcus describes in a book called *The Other Victorians.* Prostitution was common in Europe, and in the 1860s it was legalized and regulated by an act of the British Parliament. Furthermore, Victorian prudery in sexual behavior and attitudes was not standard for all social classes. The middle and lower classes did not practice the sexual pretensions of the upper class. Indeed, it was the abject poverty of the lower classes that forced many young women into prostitution, and the middle classes — despite the ideal of the docile, sexless Victorian lady — not only had sexual feelings and desires, but acted on them in much the same way women do today. Victorian women had (and enjoyed) marital sex and occasionally had torrid love affairs, as is seen in a number of diaries that detailed the number and quality of their orgasms. In fact, a female sex survey conducted by a woman named Clelia Duel Mosher in 1892 has recently come to light, providing additional evidence that viewing the Victorian period as strictly antisexual is incorrect. In addition, an interesting viewpoint has been advanced about female sexuality in Victorian times:

> Although it is obvious that many Victorians suffered from sexual repression, it appears on closer observation that those women who contributed to the concept of prudishness were far closer to today's feminists than most are willing to admit. . . . The Victorian woman sought to achieve a sort of sexual freedom by denying her sexuality . . . in an effort to keep from being considered or treated as a sex object. Her prudery was a mask that conveniently hid her more "radical" effort to achieve freedom of person. (Haller and Haller, 1977, p. xii)

Science and medicine reflected the antisexualism of the era thoroughly. Masturbation was variously branded as a source of damage to the brain and nervous system and a cause of insanity and a wide range of other illnesses. Women were thought to have little or no capacity for sexual response and were viewed as inferior to men both physically and intellectually. In 1878, the prestigious *British Medical Journal* printed a series of letters in which a number of physicians offered evidence supporting the idea that the touch of a menstruating woman would spoil hams. And even as eminent a scientist as Charles Darwin, the father of the theory of evolution, wrote in his *Descent of Man and Selection in Relation to Sex* (1871) that "Man is more courageous, pugnacious, and energetic than woman, and has a more inventive genius," and that "the average of mental power in man must be above that of women."

In the latter part of the nineteenth century Richard von Krafft-Ebing, a German psychiatrist, undertook a detailed classification of sexual disorders. The impact of his *Psychopathia Sexualis* (1886), which went through twelve editions, was profound and influenced subsequent public attitudes and medical and legal practice for more than three-quarters of a century. There were positive and negative aspects to this influence: on the one hand, Krafft-Ebing advocated sympathetic medical concern for the so-called sexual perversions and reform in laws dealing with sex criminals, while on the other hand his book seemed to lump sex, crime, and violence together. Much of his attention was devoted to aspects of sexuality he considered abnormal, such as sadomasochism (sexual arousal from inflicting or experiencing pain), homosexuality, fetishism (sexual arousal by an object rather than a person), and bestiality (sexual contact with animals). Because he frequently used lurid examples (sexual murders, cannibalism, and intercourse with the dead, to name just a few) which he presented in the same pages with less frightening sexual variations, many readers were left with a general loathing for almost all forms of sexual conduct. Nevertheless, Krafft-Ebing is often considered the founder of modern sexology.

By the turn of the new century, sexuality began to be investigated in a more objective manner. Although Victorian attitudes still prevailed in many circles, the work of serious scientists such as Albert Moll, Magnus Hirschfeld, Iwan Bloch, and Havelock Ellis combined with the dynamic theories of Freud to initiate a striking reversal in thinking about sex.

Sigmund Freud (1856–1939) was a Viennese physician who, more successfully than any figure before or since, demonstrated the central importance of sexuality to human existence. Today Freud's genius is recognized as partly a matter of original discovery and partly a reflection of his ability to synthesize emerging ideas into a cohesive and persuasive theoretical framework. Freud believed that sexuality was both the primary force in the motivation of all human behavior and the principal cause of all forms of *neurosis,* a mild form of mental disorder in which anxiety is prominent and coping skills are distorted although a sense of reality is maintained. He clearly described the existence of sexuality in infants and children, expanding views expressed by other sexologists between 1880 and 1905, and formulated a detailed theory of psychosexual development.

Freud devised many innovative concepts related to sexuality. The best known, the *Oedipal complex,* refers to an inevitable sexual attraction of the young male child to his mother accompanied by an ambivalent mixture of love, hate, fear, and rivalry toward his father. Freud also believed that boys were concerned about the possible loss of their penis as a terrible form of punishment (*castration anxiety*) and that girls felt a sense of inadequacy and jealousy at not having a penis (*penis envy*). Freud saw these situations as operating primarily at the unconscious level — a level of the personality deeper than conscious awareness. From the rich theoretical tapestry of his thought, Freud wove a clinical method called psychoanalysis for assessing and treating the unconscious conflicts that lead to psychological problems. Although many modern sexologists disagree with Freud's formulations, as we will discuss in subsequent chapters, psychoanalysis remains a widely used method of treatment today.

At about the same time, an English physician named Havelock Ellis (1859–1939) was publishing a six-volume series called *Studies in the Psychology of Sex* (1897–1910). Ellis anticipated much that Freud later wrote about childhood sexuality and had remarkably modern views in certain areas. For example, he recognized the common occurrence of masturbation in both sexes at all ages, took exception to the Victorian idea that "good" women had no sexual desire, and emphasized the psychological rather than physical causes of many sexual problems. His writings also focused on the varied nature of human sexual behavior and provided an important balancing influence to Krafft-Ebing's view of sexual variations as diseases.

Needless to say, the views of Freud and Ellis were at first seen as rather heretical, and even the medical community resisted giving them serious consideration in the early part of the century. However, as their ideas crossed the Atlantic they slowly began to have some impact. Coincidentally or not, by the end of World War I, massive social changes were emerging in both Europe and America that differed drastically from Victorian practices. Influenced by increasing social and economic freedom for women and the availability of the automobile, sexual attitudes became increasingly less inhibited in the Jazz Age and were accompanied by corresponding changes in fashion, dance, and literature. Women had become involved professionally in the sexual revolution that was brewing. Margaret Sanger was a leader of the birth control movement in America. Katharine Davis conducted a survey of the sex lives of 2,200 women, which was published initially as a series of scientific articles between 1922 and 1927 and then as a book. An Englishwoman, Marie

Stopes, wrote an explicit marriage manual that sold well on both sides of the Atlantic.* By 1926, when a gynecologist named Theodore van de Velde published *Ideal Marriage,* providing specific details about a wide range of sexual techniques and endorsing such practices as oral-genital sex, his book became an instant international best-seller.

The Roaring Twenties came to a sudden end in 1929 with the stock market crash. In the Great Depression that followed, concern for sustenance, shelter, and survival seemed to take precedence over sex. In the 1940s, the world was quickly at war again, and the postwar era brought instant notoriety to another sexologist who was to leave an indelible mark on scientific history.

Alfred C. Kinsey (1894–1956), a zoologist at Indiana University, had been asked to participate in teaching a noncredit college course on marriage in the summer of 1938. Struck by the lack of scientific data about human sexual behavior, he used this opportunity to administer questionnaires to some of his students for the purpose of gathering information about their sexual histories. Soon thereafter, Kinsey decided that personal interviewing was a more promising technique for obtaining such case history material since it permitted greater flexibility and detail, and he embarked on a course of action that eventually led to interviews with thousands of men and women across the country. Joined by his coauthors and colleagues, Wardell Pomeroy and Clyde Martin, Kinsey published the monumental *Sexual Behavior in the Human Male* on January 5, 1948. Five years later, with Paul Gebhard, they published the companion volume, *Sexual Behavior in the Human Female.*

The Kinsey reports were based on extensive face-to-face interviews with 12,000 people from all segments of the population, and the findings were often startling. For instance, 37 percent of American men were reported to have had at least one homosexual experience to the point of orgasm after the age of puberty; 40 percent of husbands had been unfaithful to their

---

* It is interesting to note that Stopes, who had obtained a doctorate and was an accomplished scientific researcher, was very much a victim of Victorian prudishness about sex. Six months after her marriage to another scientist, Dr. Reginald Ruggles Gates, she "began to feel instinctively that something was lacking," and went to the British Museum to try to discover what it was. Finding out that her marriage had not been consummated, she successfully sued for divorce and later undertook the writing of her book to help others avoid such problems (Harrison, 1977).

wives; and 62 percent of the women studied had tried masturbation.

The publication of *Sexual Behavior in the Human Male* instantly
catapulted the Kinsey research into the public eye. By mid-
March more than 100,000 copies had been sold, and the book
remained on the best-seller list for twenty-seven weeks.

Although Kinsey and his colleagues attempted without moral
or medical value judgments to describe how people behave sexually, their work was severely criticized on methodological and
moral grounds. Prestigious *Life* magazine called it "an assault on
the family as a basic unit of society, a negation of moral law, and
a celebration of licentiousness." Margaret Mead criticized Kinsey
for dealing with sex "as an impersonal, meaningless act" (*New
York Times*, March 31, 1948), a charge that was echoed by many
critics, including one professor from Columbia University who
stated that "there should be a law against doing research dealing
exclusively with sex" (*New York Times*, April 1, 1948). However,
the Kinsey report was also praised as having "done for sex what
Columbus did for geography."

All in all, the reception of Kinsey's first volume was fairly
positive, but the same cannot be said for his second book, *Sexual
Behavior in the Human Female* (1953). Many newspapers de-
nounced this report in editorials and refused to give it coverage
in their news columns. For example, the *Times* of New Philadelphia, Ohio, justified this decision by saying, "We believe it would
be offensive to a large portion of our readers" (August 20,
1953). Church leaders and educators called Kinsey's findings
amoral, antifamily, and even tainted with communism.

Kinsey died in 1956, embittered and disillusioned, but the
impact of his energetic investigations was to be strongly apparent in the years ahead. In addition to the cultural and scientific
legacy he left behind, he and his colleagues formed the Institute
for Sex Research at Indiana University, which continues as a
major center under the leadership of June Reinisch.

In the aftermath of Kinsey's studies, there was an era marked
by quite a bit of sexual confusion. Premarital sex became more
commonplace than it had been before, although it seems to have
been restricted mainly to engaged couples. Popular descriptions
of sex began to appear in books (such as the then-steamy *Peyton
Place*) and movies (mostly imported from overseas), and even

popular music began to present sexual themes. In the mid-1950s, Professor Pitirim Sorokin of Harvard, horrified by what he heard and saw, sourly noted that the "sexualization" of music made it "naked, seductive . . . lusty and perverse," with performers whose "bleating is underscored by their gyrations, contortions, and bodily rhythms all too clear in sexual innuendo and undisguised meaning." (Please note if you will, that Professor Sorokin offered his critical comments even before Elvis Presley burst into national prominence. We wonder what he would say today on seeing — and hearing — Prince, Boy George, and Madonna.)

At the same time, the 1950s were a time when females were expected to be glamorous but brainless creatures — something along the lines portrayed by Marilyn Monroe in her movies — whose primary ambitions should be directed toward marriage and motherhood. *Harper's Magazine* (January, 1950) noted, "If an American girl wears plain, unadorned eyeglasses, instead of highly colored and fancifully shaped specs, she might just as well be dead, for all the dating it will get her." And *See* magazine (January, 1950) solemnly advised readers: "It is quite legitimate for a girl to use falsies and not to tell her husband about them before marriage."

Albert Ellis (1959, p. 227) succinctly summarized the prevailing mores of the time this way: "The fundamental law underlying all our sex, love, and marriage attitudes can be stated with absolute and appalling clarity in two simple statements: (a) If it's FUN you mustn't do it; (b) If it's DUTY you must."

In the early 1960s, several factors influenced the start of a sexual revolution that was more visible than any America had previously seen. Many factors contributed to this revolution: (1) the availability of birth control pills; (2) the protest movement among adolescents and young adults; (3) the reemergence of feminism in modern form; and (4) greater openness in discussions and displays of sex. While it isn't possible to render any final historical judgment on the relative importance of each of these factors in fueling the sexual revolution, it appears certain that each had a strong influence.

The pill made premarital sex considerably safer and permitted millions to think of sex as relational or recreational rather than procreative, as we have already noted. Indeed, the availability of the pill provided a sense of freedom for many women and probably contributed more to changing sexual behavior than has generally been imagined. The protest movement among the young, which began with the civil rights movement and expanded with the growing disillusionment with the Vietnam war, led teens and young adults to challenge their parents' generation ("the establishment") in every way imaginable. This challenge was expressed not only in the younger generation's clothing, long hair, and music, but also in their recreational drug use and their support of sexual freedom ("Make love, not war").

With their consciousness raised at many levels to political and social injustices, young adults in the sixties also embraced the women's movement with enthusiasm. Since the pill had given women a new degree of control over their sexual destinies, it is not surprising that female sexuality was increasingly accepted as a natural fact of life.

In society at large, the initial reactions to the sexual revolution were mixed. While some sought to join the movement enthusiastically, many others seemed to regard it as a passing phase that would eventually fade away. And it's probably safe to say that a sizable segment of the population watched this upheaval with great distaste and alarm, concerned that the moral fabric of American society was disintegrating before their eyes. Nevertheless, sexuality became more talked about, shown, and studied, and the sixties saw the advent of "topless" bars, nudity in Broadway shows (first with *Hair,* later with *Oh! Calcutta!*), and the publication of a revolutionary study of human sexual function.

Kinsey and his collaborators had investigated the nature of human sexuality by interviews designed to find out how, when, and how often people behaved sexually. Since then, sex research has expanded in various directions in an attempt to answer questions that had not previously been resolved. Among the first and most significant departures from Kinsey's methods were those used by William H. Masters and Virginia E. Johnson, a physician and a behavioral scientist at Washington University Medical School in St. Louis.

Masters and Johnson believed that to understand the complexities of human sexuality, people must understand sexual anatomy and physiology as well as psychological and sociological data. Unsatisfied with the relevance to humans of information gathered by studies of sexual response in animals, Masters and Johnson decided that only a direct approach to the problem would be illuminating. A laboratory investigation began in 1954 to observe and record the physical details of human sexual arousal. By 1965 more than 10,000 episodes of sexual activity by 382 women and 312 men had been observed, and the report that followed, *Human Sexual Response* (Masters and Johnson, 1966), drew rapid public attention. Although some health care professionals quickly grasped the importance of these findings, others were shocked by the methods employed. Amid the accusations of "too mechanistic an approach" and the cries of moral outrage, relatively few people recognized that the physiological information was not an end point but was instead a foundation on which a treatment method for people with sexual problems could be based.

In the seventies and eighties, the new openness about sexuality was readily apparent. In 1970, Masters and Johnson published *Human Sexual Inadequacy,* a landmark book that described a startlingly new approach to the treatment of sexual problems that had previously required lengthy treatments without very high rates of success. With a two-week treatment program and only a 20 percent failure rate, this work was soon to give rise to an entire new profession — sex therapy — with the eventual proliferation of thousands of sex clinics across the country before the end of the decade, and the development of other therapy approaches by doctors such as Helen Kaplan and Jack Annon.

Other less technical books about sex were published by the dozens, with Alex Comfort's *Joy of Sex* (1972) probably being the most accomplished and certainly the most successful (with sales of over nine million copies). Television became a notable force in the sexual revolution, too, as a number of programs tackled previously taboo sexual themes. Not to be outdone, movies became more sexually explicit, and, in the early days of the home video market, pornographic films were the single best-selling category.

A number of other trends occurred in this time period that affected the ways Americans viewed sexuality: (1) the practice of nonmarital cohabitation — living together — began to assume increasing importance as a stage preceding marriage; (2) the legalization of abortion by the U.S. Supreme Court in 1973 made it possible to obtain safe abortions, but also provoked considerable controversy about the morality of this practice; (3) the 1974 decision by the American Psychiatric Association to remove homosexuality from classification as a mental disorder set the stage for advances to be made in the gay rights movement; (4) a growing awareness of the significance of all forms of sexual victimization — in part an outgrowth of the women's movement and in part a result of the work of scientists and scholars who effectively showed that rape is a crime of violence rather than a crime of passion led to major legislative changes aimed at modernizing procedures for trying rape cases, as well as the rapid growth in rape crisis centers across the country; (5) the appearance of new reproductive technologies encompassing the birth of the world's first "test tube baby" in 1978 (with more than 1,000 babies conceived by similar means now alive) has now proceeded to even more startling techniques, such as embryo transfer methods and the controversial "surrogate mother" practice.

The late 1970s and early 1980s were also a time for a backlash against what some perceived as overly permissive, even immoral sexual practices. The Moral Majority sought to block sex education in public schools and campaigned against "promiscuous" sexual behavior, which seemed to include anything other than marital sex. The "Right to Life" movement challenged the legality of abortion and unsuccessfully tried to pass a constitutional amendment that would have banned abortion under all circumstances. In 1983, the Reagan administration tried to implement a policy requiring notice to the parents of teenagers requesting contraceptives; this proposal, which became derisively known as the "Squeal Rule," fortunately never got off the ground.

Particularly alarming to some was the appearance in the late 1970s and early 1980s of seemingly new epidemics of sexually transmitted diseases (STDs): genital herpes, primarily among heterosexuals, and AIDS (acquired immune deficiency syndrome), and since they seemed to be linked incontrovertibly to promiscuous sexual behavior, some observers suggested that they were a form of punishment from God for sexual transgressions.

We cannot know, of course, if the changes and trends we see as significant today will have any lasting impact on our sexual behavior over time. Nor can we be certain that a century from now, historians won't label our era with a single word (like "Victorian") and reduce the many complexities of our sexual attitudes to a single notion. The only thing we can be sure of is that our attitudes and behaviors will continue to change — what directions those changes will take, however, is impossible to predict with any accuracy.

CHAPTER TWO

Sexual Anatomy

Karen is a twenty-year-old college junior who generally avoided dating because she felt her breasts were too small. She wrote: "I hate looking in a mirror or wearing a bathing suit because I see how flat I am. I would be *mortified* to let a guy touch or see my breasts."

Brad is an athletic seventeen-year-old who quit his school's basketball team because his breasts were large. He told us that his teammates kidded him mercilessly in the locker room and showers about when he was going to get a bra. He was afraid he might "turn into a woman."

A married couple in their mid-twenties who were sex therapy patients said they frequently used stimulation of the clitoris as part of making love. When asked to identify the clitoris during a physical exam, the husband pointed to a large freckle on the lower part of his wife's labia majora.

A class of eighty college sophomores was given a brief test on sexual anatomy the first day of their course on sexuality. The average student had more mistakes than right answers.

AS THESE EXAMPLES SHOW, many of us have inaccurate information or negative feelings about our sexual anatomy. This should not be surprising for a variety of reasons: we are taught to keep our sex organs covered by clothing; we are scolded or punished for touching our "private parts"; we are not likely to be told the correct terminology to describe our sexual anatomy; we are discouraged from conversations or questions about sex; and the sexual images we are exposed to in movies and magazines are likely to present almost unattainable standards to measure ourselves against. It is no wonder that our sexual anatomy can be a source of anxiety, shame, guilt, mystery, and curiosity, as well as a source of pleasure.

The mixed feelings we have about our sexual body parts are mirrored in the words we use to talk about them: some words are "clean" and "proper," while others are "dirty" or "impolite." These differences are a result of how we interpret words, not an innate property of the words themselves. Consider, for example:

> In Nigeria, the moral taboos of sex were taught by missionaries and administrators who used only clean words. These were the words that became taboo. The dirty words used as part of the vernacular of sailors, traders, and the like, became part of Nigerian vernacular English, with no taboo attached. In consequence, today it is as forbidden to say sexual intercourse, penis and vagina on Nigerian television as it is to say fuck, cock, and cunt on the national networks in the United States. In Nigeria, the latter terms are considered normal and respectable. (Money, 1980, pp. 50–51)

In this book we will not use slang terms about sex because they convey a negative message to some people.

People of all ages have common concerns about sexual anatomy. What is the normal size of the penis? Is something wrong if one breast is smaller than the other? Does circumcision lessen sexual pleasure? Do large breasts indicate a passionate woman? Is it abnormal if one testicle hangs lower than the other? Where and what is the clitoris? The answers to such questions begin with learning about sexual anatomy. The basis for understanding the way our bodies function sexually will be discussed in the next chapter.

FEMALE SEXUAL ANATOMY

We are encouraged to feel as if our bodies are not ours. Our "figure" is for a (potential) mate to admire. Our breasts are for "the man in our lives" to fondle during lovemaking, for our babies to suckle, for our doctors to examine. The same kind of "hands-off" message is even stronger for our vaginas. (The Boston Women's Health Book Collective, 1976, p. 24)

Anyone who has been around young children knows that baby girls play with their genitals just as they touch and explore all parts of their body. Although this activity seems pleasurable and interesting, most girls are quickly taught that it's "not nice" or "dirty," a prohibition that is probably reinforced during toilet training when the two- or three-year-old girl is urged to "wipe carefully" and "be clean." The sex-negative tone of these early childhood messages is consistently reinforced for most girls as they grow up, commonly creating anxieties and inhibitions about sex in general and their sexual anatomy in particular. These difficulties are compounded by many people's perception of the female sex organs as unattractive and unclean.

Menstruation is one source of such negative attitudes: menstrual periods are sometimes called "the curse," menstrual flow is contained by "sanitary" napkins (which suggests an underlying condition of uncleanliness), sex during menstruation is often avoided by men and women because it may be messy, and in some societies, there are strong taboos surrounding menstruating women that isolate them so they will not contaminate food, plants, or people. In our cosmetic-conscious society of perfumes, deodorants, and after-shave lotions, women have been told that their vaginal odors are unpleasant and should be hidden. As a result, "feminine hygiene deodorant sprays" were widely used until it became apparent that they frequently caused vaginal irritation and itching.

Many women have not taken a direct look at their own genitals or cannot accurately name and identify the parts of their sexual anatomy. (Sex organs in the pelvic region — in females, the outer sexual structures and the vagina, and in males, the penis, scrotum, and testes — are customarily called the genitals.) While we cannot imagine a person unable to distinguish between eyes, nose, mouth, and chin, many men and women have no idea of the location of the female urethra, clitoris, or hymen.

The Vulva

The external sex organs of the female, called the *vulva* (meaning "covering"), consist of the mons, the labia, the clitoris, and the perineum (Figure 1). Although the vagina has an external opening (the *introitus,* or entrance), it is principally an internal organ and will be discussed separately.

The *mons veneris* (Latin: mound of Venus; Venus was the Roman goddess of love) is the area over the pubic bone that consists of a cushion of fatty tissue covered by skin and pubic hair. Since this region has numerous nerve endings, touch and/ or pressure here may lead to sexual arousal. Many women find that stimulation of the mons area can be as pleasurable as direct clitoral touch.

The *outer lips* (labia majora) are folds of skin covering a large amount of fat tissue and a thin layer of smooth muscle. Pubic hair grows on the sides of the outer lips, and sweat glands, oil glands, and nerve endings are liberally distributed in them. In the sexually unstimulated state, the outer lips usually are folded together in the midline, providing mechanical protection for the urethral (urinary) opening and the vaginal entrance.

The *inner lips* (labia minora) are like curving petals. They have a core of spongy tissue rich in small blood vessels and without fat cells. The skin covering the inner lips is hairless but has many sensory nerve endings. The inner lips meet just above the clitoris, forming a fold of skin called the *clitoral hood* (see Figure 1). This portion of the inner lips is sometimes referred to as the female foreskin.

The labia are an important source of sexual sensations for most women since their many nerve endings serve as sensory receptors. When the skin of the labia is infected, sexual intercourse may be painful, and itching or burning may occur.

Women's external genitals may vary greatly in appearance. There are differences in the size, shape, and color of the labia, in the color, texture, amount, and distribution of pubic hair, and in the appearance of the clitoris, vaginal opening, and hymen. Sexual anatomy varies just as much as facial anatomy differs from one person to another.

Bartholin's glands lie within the labia minora and are connected to small ducts that open on the inner surface of the labia next to the vaginal opening. Although once thought to play a major role in the production of vaginal lubrication, it is now clear that the few drops of secretion usually produced by these glands during

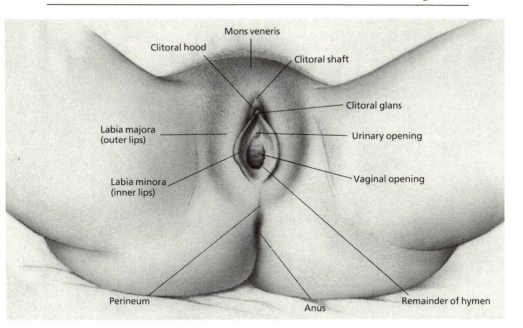

Mons veneris
Clitoral hood
Clitoral shaft
Clitoral glans
Labia majora (outer lips)
Urinary opening
Labia minora (inner lips)
Vaginal opening
Perineum
Anus
Remainder of hymen

FIGURE 1. The Vulva

sexual arousal are not important contributants to vaginal lubrication, although they may slightly moisten the labia.

The *clitoris,* one of the most sensitive areas of a female's genitals, is located just beneath the point where the top of the inner lips meet. The only directly visible part of the clitoris is the head or *clitoral glans,* which looks like a small, shiny button. This head can be seen by gently pushing up the skin, or *clitoral hood,* that covers it. The clitoral hood also hides the *clitoral shaft,* the spongy tissue that branches internally like an inverted V into two longer parts or *crura.* The crura lead to the bony pelvis (see Figure 2). The clitoris is richly endowed with nerve endings, which make it highly sensitive to touch, pressure, and temperature. It is unique because it is the only organ in either sex whose only known function is to focus and accumulate sexual sensations and erotic pleasure.

The clitoris is often regarded as a miniature penis, but this notion is sexist and incorrect. The clitoris has no reproductive or urinary function and does not usually lengthen like the penis when stimulated, although it does become engorged. The clito-

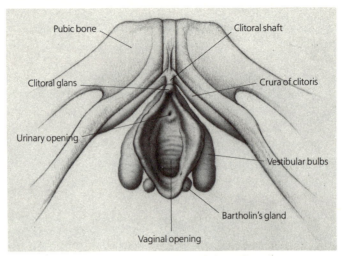

Pubic bone · Clitoral shaft · Clitoral glans · Crura of clitoris · Urinary opening · Vestibular bulbs · Bartholin's gland · Vaginal opening

FIGURE 2.   Anatomy of the Clitoris

ris and the penis, however, are derived embryologically from the same tissues.

The size and appearance of the clitoris vary considerably among women, but there is no evidence that a larger clitoris provides more intense sexual arousal. Contrary to the opinion of some physicians, masturbation rarely causes enlargement of this organ.

Clitoral circumcision — surgical removal of the clitoral hood — has been said to improve female sexual responsivity by exposing the clitoral glans to more direct stimulation.* We believe, however, that this procedure is rarely useful since it has two major drawbacks: (1) the clitoral glans is often exquisitely sensitive to direct touch, to the point of pain or irritation (in this sense, the clitoral hood serves a protective function),† and (2) during intercourse the thrusting of the penis indirectly stimu-

* Some tribes in Africa and South America practice surgical removal of the clitoris (clitoridectomy) as a ritual rite of puberty. In fact, more than one-half of all young girls in Egypt are still undergoing this painful procedure today. Although this practice is sometimes called "clitoral circumcision," it is not at all the same. Clitoridectomy does not destroy the capacity for sexual arousal or orgasm, but it certainly does not help it at all. When a resolution was introduced at the Sixth World Congress of Sexology in 1983 to oppose this procedure formally, it was defeated because many delegates were concerned that they had no business meddling in the long-standing customs of other cultures.

† In masturbation most women stroke areas around the clitoral glans but avoid its direct stimulation for this very reason. Apparently, the advocates of clitoral circumcision (usually men, oddly enough) have overlooked this finding.

lates the clitoris by moving the inner lips of the vagina, causing the clitoral hood to rub back and forth across the clitoral glans. A less drastic procedure than circumcision is advocated by several sexologists to improve some women's sexual responsiveness. A probe is used to loosen adhesions or thickened secretions (*smegma*) between the clitoral hood and clitoral glans. In more than thirty years of practice, we have seen very few cases that required such an approach and remain skeptical about the use of this procedure on a routine basis.

The *perineum* is the hairless area of skin between the bottom of the labia and the anus (the opening for evacuation of the bowels). This region is often sensitive to touch, pressure, and temperature and may be a source of sexual arousal.

The opening of the vagina is covered by a thin tissue membrane called the *hymen*. The hymen, which has no known function, typically has perforations in it that allow menstrual flow to pass from the body at puberty. The hymen usually stretches across some but not all of the vaginal opening and may vary in shape, size, and thickness, as Figure 3 reveals.

Historically, it has been important for a woman to have an intact hymen at the time of marriage as proof of her virginity. In some societies, a bride who does not have an intact hymen is returned to her parents, subjected to public ridicule, physically punished, or even put to death. In modern Japan and Italy, plastic surgeons are kept busy by reconstructing the hymens of many engaged women to create "neovirginity" for those who wish to conceal their sexual histories from their future husbands.

Contrary to the fears of some females, a doctor cannot usually tell if they are virgins by conducting a pelvic examination. The presence or absence of an intact hymen is not an accurate indication of prior sexual behavior. The hymen may be broken or stretched at an early age by various exercises or by inserting fingers or objects in the vagina. Some females are born with only a partial hymen or none at all. In addition, intercourse does not always tear the hymen; instead, it may simply stretch it. Under most circumstances, the first intercourse experience for a girl or women is not painful or marked by a great deal of bleeding. The excitement of the moment is usually enough so that the pressure on the hymen is barely noticed.

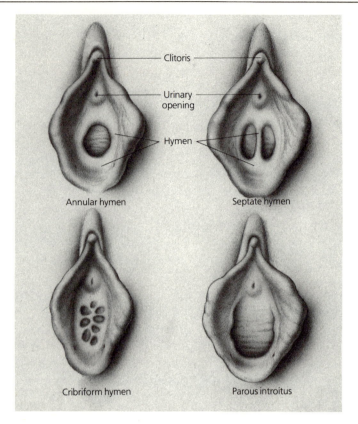

FIGURE 3.   Variations in the Hymen

The annular *hymen forms a ring around the vaginal opening; the* septate *hymen has one or more bands of tissue that cross the diameter of the vaginal opening; and the* cribriform *hymen stretches completely across the vaginal opening but has many small perforations. The* parous introitus *is the opening of the vagina in a woman who has previously given birth to a child by vaginal delivery; only small remnants of the tissue of the hymen are visible.*

The Vagina and Other Internal Sex Organs

The *vagina* is a muscular internal organ that tilts upward at a 45° angle diagonally pointed toward the small of the back (Figure 4). In the sexually unstimulated state, the vagina's walls are collapsed. In a woman who has never had a child, the back wall of the vagina averages 8 centimeters (about 3 inches) in length, while the front wall is approximately 6 centimeters (2½ inches) long.

The vagina functions as a potential space that like a balloon can change shape and size. It can contract and expand, accommodate the passage of a baby during childbirth, or adjust in size to fit snugly around a finger.*

Many people wonder about the relationship between vaginal size and sexual gratification. Since the vagina adjusts equally well to a large or small penis, it is unusual for size differences between male and female sex organs to lead to sexual difficulties. Following childbirth, the vagina usually enlarges moderately and loses elasticity. Exercises to strengthen the muscles supporting the vagina are thought by some authorities to improve this condition and foster sexual responsiveness.†

The inside of the vagina is lined with a surface similar to the lining inside the mouth. This *mucosa* is the source of vaginal lubrication. There are no secretory glands in the vagina, but there is a rich supply of blood vessels. The vagina has relatively few sensory nerve endings except near its opening. As a result, the inner two-thirds of the vagina are relatively insensitive to touch or pain.

Recently, there have been claims that a region in the front wall of the vagina midway between the pubic bone and the cervix has a special sensitivity to erotic stimulation. Called the *G spot* (or Grafenberg spot, for the German physician who first suggested its presence in 1950), it has been described as a mass of tissue about the size of a small bean in the unstimulated state. When stimulated the tissue purportedly swells to the size of a dime or larger.

---

* Despite its ability to contract, the human vagina cannot "clamp down" on the penis during intercourse and make physical separation impossible. In dogs, there is a type of intravaginal "locking," but it occurs primarily because of expansion of the head of the penis.

† The "Kegel exercises" are done by contracting the pelvic muscles that support the vagina (most notably, the *pubococcygeus* and *bulbocavernosus* muscles). These same muscles are used when a woman stops the flow of urine or tightens the vagina against an inserted object such as a tampon, a finger, or an erect penis. The muscles are contracted firmly for one or two seconds and then released; this is repeated in a series of ten contractions several times a day for maximum results. In addition to strengthening muscular contractions, these exercises can improve a woman's sense of self-awareness. Whether they really improve sexual responsivity is less certain at present.

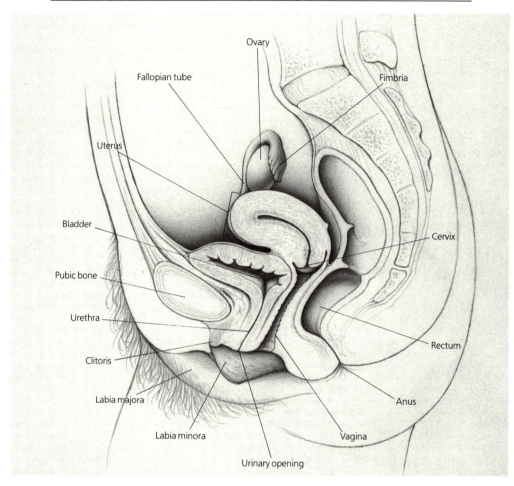

FIGURE 4.   Internal Side View of the Female Reproductive System

Ladas, Whipple, and Perry, authors of a book about the "G spot," state that examinations of more than 400 women identified the "G spot" in each one; they explain that it has generally been overlooked in the past because "in its sexually unstimulated state, it is relatively small and difficult to locate, especially since you can't see it." This explanation does not fit the findings of a subsequent research project in which Whipple herself participated in which the "G spot" could be found in only four out of eleven women, nor does it coincide with our studies at the Masters & Johnson Institute, where less than 10 percent of a sample of over 100 women who were carefully examined had an area of heightened sensitivity in the front wall of the vagina or a tissue mass that fit the various descriptions of this area. Another recent study, by Alzate and Londono, also was unable to find evidence supporting the existence of the "G spot," although many of the women studied showed signs of erotic sensitivity in the front wall of the vagina. Thus, at the present time it seems that additional research is needed to establish whether the "G spot" exists as a distinct anatomic structure or whether, as the well-known sex therapist Helen Kaplan says, "the knowledge that many women have erotically sensitive areas in their vaginas which contribute to pleasure and orgasm is not new or controversial."

The *cervix* is the bottom part of the uterus that protrudes into the vagina. Through the vagina, the cervix of a women who has never been pregnant looks like a smooth pink button with a rounded face and a small central hole. At the mouth of the cervix (the *cervical os*), sperm cells enter the uterus and menstrual flow passes into the vagina. The *endocervical canal* (a thin tubelike communication between the mouth of the cervix and the cavity of the uterus) contains many secretory glands that produce mucus. The consistency of cervical mucus varies during different phases of the menstrual cycle in response to changing hormonal stimulation: just before or at the time of ovulation (when the egg is released from the ovary), cervical secretions become thin and watery; at other times, these secretions are thick and form a mucus plug that blocks the entrance to the cervix.

The cervix has no surface nerve endings so it experiences little in the way of sexual feelings. If the cervix is removed surgically, there is no loss of sexual responsivity.

The *uterus* (womb) is a hollow muscular organ shaped like an inverted pear somewhat flattened from front to back. It is about 7.5 centimeters (3 inches) long and 5 centimeters (2 inches) wide. Anatomically, the uterus consists of several parts (Figure 5). The inside lining of the uterus (the *endometrium*) and the muscular component of the uterus (the *myometrium*) have separate and distinct functions. The inner lining changes during the menstrual cycle and is where a fertilized egg implants at the beginning of a pregnancy. The muscular wall facilitates labor and delivery. Both aspects of uterine function are regulated by

chemicals called hormones, which also play a part in the growth
of the uterus during pregnancy.

The uterus is held loosely in place in the pelvic cavity by six
ligaments. The angle of the uterus in relation to the vagina varies from woman to woman; ordinarily, it is relatively perpendic-
ular to the axis of the vaginal canal, but in about 25 percent, the
uterus is tipped backward and in approximately 10 percent, it is
tilted farther forward. If the uterus is rigidly fixed in position by
scar tissue or inflammation, it may be a source of pain during
sexual activity, requiring surgical correction.

The *Fallopian tubes,* or oviducts, begin at the uterus and extend
about 10 centimeters (4 inches) laterally (Figure 5). The far ends
of the Fallopian tubes are funnel-shaped and terminate in long
fingerlike extensions called *fimbria,* which hover near the ovaries.
The inside lining of the Fallopian tubes consists of long, thin
folds of tissue covered by hairlike *cilia.* The Fallopian tubes pick
up eggs produced and released by the nearby ovary and then
serve as the meeting ground for egg and sperm.

The *ovaries,* or female gonads, are paired structures located
on each side of the uterus. About the size of unshelled almonds
(about 3 x 2 x 1.5 centimeters, or 1.2 x 0.8 x 0.6 inches), they
are held in place by connective tissue that attaches to the broad
ligament of the uterus. The ovaries have two separate functions:
manufacturing hormones (most notably, estrogen and progesterone) and producing and releasing eggs.

Before a baby girl is born, development of future eggs begins
in her just-forming ovaries. About halfway through her mother's pregnancy, the girl's ovaries contain 6 or 7 million future
eggs, most of which degenerate before birth. About 400,000
immature eggs are present in the newborn girl, and no new eggs
are formed after this time. During childhood, continued degen-
eration reduces the number of eggs still further. The immature
eggs are surrounded by a thin capsule of tissue forming a *follicle.*

When puberty arrives and girls begin to have menstrual cycles
each cycle is marked by a process of maturation in which some
immature eggs divide twice, splitting their genetic material in
half. Through this process, called *meiosis,* each young egg divides
into four cells, only one of which is a mature egg (*ovum*). A
mature egg is about 0.135 millimeters ($\frac{1}{175}$ inch) in diameter
and is surrounded by a zone of jellylike material called the *zona*

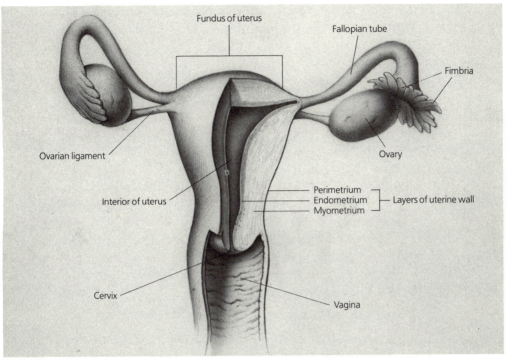

FIGURE 5.   Front View of the Internal Female Reproductive System
The uterus and vagina in this figure are shown partially cut away.

pellucida. A human egg is just barely visible, appearing as a speck smaller than the period at the end of this sentence. The other three cells, called *polar bodies,* have no known function, and eventually degenerate.

Although a number of different follicles begin growing in each cycle, usually only one develops to the point where it moves to the surface of the ovary and ruptures, releasing the egg in a process called *ovulation.* For every follicle that ovulates, about a thousand undergo various degrees of growth and then degenerate. Fewer than 400 follicles are usually involved in ovulation during the female's reproductive years.

The Breasts

Although the breasts are not reproductive organs, they are clearly part of the sexual anatomy. In American society, the female breasts have a special erotic allure and symbolize sexuality, femininity, and attractiveness. Prominent attention is devoted to the breasts in clothing styles, men's magazines, advertising, television, and cinema. This attitude is not universal by any means, and in some cultures, little or no erotic importance is attached to the breasts. For example, in Japan women traditionally bound their breasts to make them inconspicuous. Today, however, the westernization process has brought about changes in Japan and the breasts have become rather fully eroticized.

As the big-breasted female has become an almost universal sex symbol — the image used to promote everything from car sales to X-rated films — men and women have been bombarded on a daily basis with the not very subtle suggestion that a woman with large breasts has a definite sexual advantage. This has led to a number of harmful misconceptions. For example, men and women alike have come to believe that the larger a woman's breasts are, the more sexually excitable she is or can become. Another fallacy, still firmly subscribed to by many men, holds that the relatively flat-chested woman is less able to respond sexually and actually has little, if any, interest in sex.

The fact is that there is absolutely no evidence to suggest that breast size bears any relation to a woman's level of sexual interest, to her capacity for sexual response, or to the ease with which she attains orgasm. Actually, many women experience very little sexual sensation when their breasts are fondled or caressed, and this is as true of those with large breasts as it is of those with small ones. Furthermore, the woman who does become sexually excited when her breasts are stimulated does so regardless of their size.

For all their erotic significance, breasts are actually just modified sweat glands. The female breasts undergo changes in size and shape during puberty, gradually becoming conical or hemispherical with the left breast usually slightly larger than the right. Each breast contains fifteen to twenty subdivided *lobes* of glandular tissue arranged in a grapelike cluster, with each lobe drained by a duct opening on the surface of the nipple (Figure 6). The glandular lobes are surrounded by fatty and fibrous tissue, giving a soft consistency to the breast.

The *nipple* is located at the tip of the breast and mostly consists

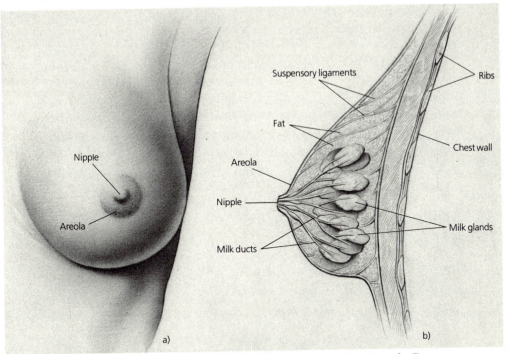

Suspensory ligaments

Ribs

Fat

Chest wall

Nipple

Areola

Nipple

Areola

Milk glands

Milk ducts

a)   b)

FIGURE 6.   External and Internal Anatomy of the Female Breast

of smooth muscle fibers and a network of nerve endings that make it highly sensitive to touch and temperature. The dark wrinkled skin of the nipple extends 1 or 2 centimeters onto the surface of the breast to form the *areola,* a circular area of dark skin with many nerve fibers and with muscle fibers that cause the nipple to stiffen and become erect.

The sexual sensitivity of the breast, areola, and nipple do not depend upon breast size or shape. Personal preference, learned habit, and biology all contribute to their responsiveness. Nevertheless, the American male's fascination with female breasts leads many women who consider themselves "flat-chested" or "underdeveloped" to seek to improve their sexual attractiveness and self-esteem by the use of exercises, lotions, or mechanical devices such as suction machines to enlarge the breasts. These methods, though widely advertised, do not work. For this reason, so-called breast augmentation surgery has become popular. In the past, liquid silicone was injected directly into the breasts to increase their size, but this technique proved to be highly unsatisfactory since it led to many medical complications. Today, soft thin plastic pouches filled with silicone gel are implanted through a simple breast incision to increase breast size while retaining a natural-appearing and soft-feeling breast.

Conversely, some women are troubled by breasts that are too large. This condition, called *mammary hyperplasia* or *macromastia*, can be treated by reduction mammaplasty, a fairly simple operation to reduce breast size and weight. Other common breast problems include *inverted nipples* (the nipples are pushed inward), a harmless anatomical variation that does not interfere with nursing, and *extra nipples*, which are minor errors of development that have no adverse health consequences but may be a source of embarrassment.

MALE SEXUAL ANATOMY

> It is not much of an exaggeration to say that penises in fantasyland come in only three sizes — large, gigantic, and so big you can barely get them through the doorway. . . .
>
> Accepting your own merely human penis can be difficult. You know it is somewhat unpredictable and, even when functioning at its best, looks and feels more like a human penis than a battering ram or a mountain of stone. But you do have one small advantage. You are alive and can enjoy yourself whereas the supermen of the model with the gigantic erections are unreal and feel nothing. (Zilbergeld, 1978, pp. 23 and 26)

The male sex organs are more visible and accessible than the female sex organs. Unlike the clitoris or vagina, the penis is involved directly in the process of urination so that boys become accustomed to touching and handling their penises at a relatively early age. The sexual aspects of the male organ are hard for a boy to miss. He learns about them by watching, touching, and playing with his penis as it becomes erect (a pleasurable experience) or by hearing stories and jokes that graphically portray the sexual and reproductive purposes of the penis. Despite such exposure, many males are not fully informed about the details of the anatomy and function of their sex organs.

The Penis

The penis is an external organ that consists primarily of three parallel cylinders of spongy tissue bound in thick membrane sheaths (Figure 7). The cylindrical body on the underside of the penis is called the *spongy body* (*corpus spongiosum*). The *urethra* (a tube that carries urine or semen) runs through the middle of the spongy body and exits at the tip of the penis via the *urinary opening* (*urethral meatus*). When the penis is erect, the spongy body on the underside looks and feels like a straight ridge. The other two cylinders, called the *cavernous bodies* (*corpora cavernosa*), are positioned side-by-side above the spongy body. All three consist of irregular spongelike tissue dotted with small blood vessels. The tissue swells with blood during sexual arousal, causing the penis to become erect.

Internally, beyond the point where the penis attaches to the body, the cavernous bodies branch apart to form tips (*crura*) that are firmly attached to the pelvic bones. The penis has numerous blood vessels, both inside and apart from the cylindrical bodies; a pattern of veins is often visible on the outer skin of the erect penis. The penis also has many nerves, making it highly sensitive to touch, pressure, and temperature.

The tip of the penis, the *glans* or head, consists entirely of corpus spongiosum. This region has a higher concentration of sensory nerve endings than the shaft of the penis and is thus particularly sensitive to physical stimulation. Two other areas particularly sensitive to touch are the rim of tissue that separates the glans from the shaft of the penis (the *coronal ridge*) and the small triangular region on the underside of the penis where a thin strip of skin (the *frenulum*) attaches to the glans. Many males find that direct stimulation of the glans may become painful or irritating and prefer to masturbate by rubbing or stroking the penile shaft.

The skin that covers the penis is freely movable and forms the *foreskin,* or prepuce, at the glans. Inflammation or infection of the foreskin or glans may cause pain during sexual activity. Sometimes the foreskin sticks to the underlying glans when *smegma,* a naturally occurring substance of cheesy consistency made up of oily secretions, dead skin cells, dirt particles, sweat, and bacteria, is not regularly washed away from underneath the

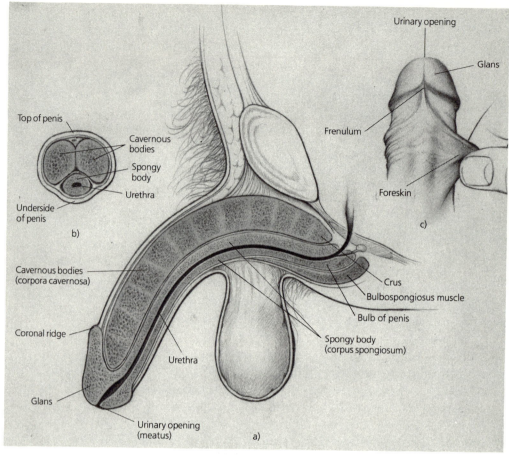

FIGURE 7.   Anatomy of the Penis
(a) Internal side view of the penis. (b) A cross-section of the penis. (c) A view of the underside of the penis showing the location of the frenulum.

foreskin. This type of problem occurs only in uncircumcised men and is one argument in favor of routine circumcision.

Circumcision is the surgical removal of the foreskin. As a result of this minor operation, usually done shortly after birth, the glans of the penis is fully exposed. Circumcision is sometimes a religious practice, as in Islam or in Judaism, where it symbolizes the covenant with God made by Abraham. In the United States, it is often done routinely for nonreligious reasons, whereas it is less common in Canada and Europe.

The advantages of circumcision are primarily related to hygiene and health: smegma does not collect, the glans of the penis is easier to clean, conditions of inflammation or infection are less likely to occur, and cancer of the penis is less frequent. Although the rate of cancer of the cervix is considerably lower in the spouses of circumcised men, it is not certain that this is a cause–effect relationship. Opponents of routine circumcision see no clear reason for this operation and suggest that removing the skin protecting the glans weakens the region's sexual sensitivity since it constantly rubs directly against clothing. Others believe that circumcision increases the risk of premature ejaculation (this is probably not true, since the foreskin of the erect uncircumcised penis retracts, exposing the glans, and researchers have not found a difference in the rates of premature ejaculation in circumcised versus uncircumcised men). We are not aware of any believable evidence demonstrating that circumcision affects male sexual function one way or the other. In any event, uncircumcised men who practice routine hygenic care are unlikely to be at any major health disadvantage.

In 1982, plastic surgeon Donald Greer and his co-workers reported on a small number of men who were so dissatisfied with having been circumcised as infants that they underwent a complicated series of operations to reconstruct the foreskin. While the results were reported as uniformly pleasing to these men, it was also noted that the reconstructed foreskin (which was taken from the scrotum) had a noticeable difference in skin texture, color, and contour from the skin on the shaft of the penis. The series of operations takes up to a year to complete.

The appearance of the penis varies considerably from one male to another. These variations are due to differences in color, size, shape, and the status of the foreskin (circumcised or uncircumcised). Some examples of different male genitals are shown in Figure 8.

Concerns about penile size are common in males of all ages. Although the size of the nonerect penis differs widely from one male to another (the average length is approximately 9.5 centimeters or just under 4 inches), in adulthood this variation is less apparent in the erect state. Erection can be thought of as "the great equalizer" since men with a penis that is smaller when flaccid (nonerect) usually have a larger percentage volume increase during erection than men who have a larger flaccid penis.

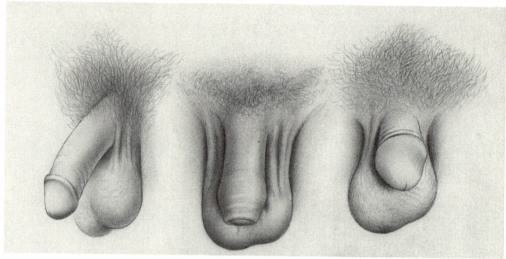

FIGURE 8.   Some Variations in the Appearance of the Male Genitals
The middle drawing shows an uncircumcised penis.

Such interest in penis size has several different components. First, it shows a concern for being "normal" — the same as everyone else. Second, it is related to a wish to be sexually adequate. Our society generally believes that "biggest is best," and the notion that a "big" penis will provide more sexual satisfaction to a woman is widespread. Actually, penile size has little physiological effect for the woman (although it may have positive or negative psychological significance) since the vagina accommodates its size equally well to an erect penis that is relatively smaller or larger in circumference. The length of the penis, which determines the depth of vaginal penetration, is also relatively unimportant, since the inner portion of the vagina and the cervix have few sensory nerve endings. Third, there is often an element of status-seeking in wishing to have a large penis. Finally, some males feel that a larger penis would make them more sexually attractive. Most of these points apply to both heterosexual and homosexual males.

In art and in the media (particularly erotic books, male magazines, and movies) there is a tendency to portray male genital size in "bigger than life" dimensions. This distortion reflects the triumph of anxious perception versus reality, literary and cinematic license (the use of particular camera angles or close-up

"Doris, I thought you told me size wasn't important."

shots, for example), and the deliberate selection of male subjects whose genital proportions are decidedly larger than average. Male readers should remember, too, that there is a visual difference between the view you get of your own genitals (they appear shortened because of your viewing angle) and the view you get of someone else's penis size in the locker room or on the movie screen.

Recently, a team of Canadian researchers studied the psychological impact of penis size on sexual arousal. They found that reading erotic passages that differed only in the description of the size of the penis produced no differences in the levels of arousal of male or female undergraduates. Thus, they concluded that "penis size may be as unimportant on a psychological level as it appears to be on a physical level."

There *is* a rare medical condition called *micropenis* in which the penis is formed properly but is miniature in size. This condition is marked by a penis length of less than 2 centimeters (approximately ¾ inch), and sometimes is due to a treatable deficiency of testosterone. In other circumstances, there is no means of increasing penis size by drugs, creams, gadgets, hypnosis, or hormones although there are advertisements for such "treatments" that exploit the myth that bigger is necessarily better.

Men who are preoccupied or extremely anxious about the size of their penis appear to be more likely to develop sexual difficulties than other men. These difficulties range from the avoidance of potentially sexual relationships because of embarrassment or worry to difficulty in obtaining or maintaining an erection due to poor self-confidence, tension, and anxiety. Fortunately, this type of problem can usually be overcome by brief sex counseling or therapy.

The Scrotum

The *scrotum* is a thin loose sac of skin underneath the penis that is sparsely covered with pubic hair and contains the testicles (testes). The scrotum has a layer of muscle fibers that contract involuntarily as a result of sexual stimulation, exercise, or exposure to cold, causing the testes to be drawn up against the body. In hot weather, the scrotum relaxes and allows the testes to hang more freely away from the body. These reflexes of the scrotum help to maintain a stable temperature in the testes, an important function because sperm production (occurring in the testes) is impaired by heat or cold. In response to cold, the scrotum lifts the testes closer to the body to provide the warmer environment. In hotter conditions, the scrotum loosens, thereby moving the testes away from the body and providing a larger skin surface for the dissipation of heat. Tightening of the scrotum with sexual arousal or physical exercise may be a protective reflex that lessens the risk of injury to the testes.

The Testes

The *testes* (the male gonads) are paired structures usually contained in the scrotum (Figure 9a). The two testes are about equal in size, averaging 5 x 2 x 3 centimeters (2 x 0.8 x 1.2 inches) in adults, although one testicle generally hangs lower than the other. Most often, the left testis is lower than the right one, but in left-handed men the reverse is usually true. There is no significance attached to the relative height of the testes within the scrotal sac, but if one testis is considerably larger or smaller than the other there could be a medical problem and a doctor should be seen.

The testes are highly sensitive to pressure or touch. Some men find that light caressing or stroking of the scrotum or gentle squeezing of the testes during sexual activity is arousing, but many others are uncomfortable with touching in this region.

The testes have two separate functions: hormone and sperm production. The cells that manufacture hormones — most importantly, *testosterone*, which controls male sexual development and plays an important part in sexual interest and function — are called *Leydig's cells*. Sperm production occurs in the *seminiferous tubules*, tightly coiled tubes of microscopic size that collectively measure almost 500 meters (more than a quarter mile) in length. The entire process of sperm production takes seventy days. Unlike the female, who creates no new eggs after birth, the male produces sperm from puberty on, manufacturing billions of sperm annually.

A mature sperm is considerably smaller than the size of a human egg, being about 0.06 millimeter ($\frac{1}{500}$ inch) in length and thousands of times smaller than the egg in volume. Sperm are only visible with the aid of a microscope, which shows that they consist of three pieces: a head, a midpiece, and a tail. The head of the sperm contains genetic material (*chromosomes*) and a chemical reservoir (the *acrosome*). The midpiece contains an energy system that allows the sperm to swim by lashing its long tail back and forth.

The Epididymis and Vas Deferens

The seminiferous tubules (the tubes where sperm are produced) empty into the *epididymis,* a highly coiled tubing network folded against the back surface of each testis (Figure 9b). Sperm cells generally spend several weeks traveling slowly through the epididymis as they reach full maturation. From here, sperm are carried into the *vas deferens,* a long tube (approximately 40 centimeters, or 16 inches) a pair of which leave the scrotum and curve alongside and behind the bladder. Both the right and left vas deferens are cut when a vasectomy is done.

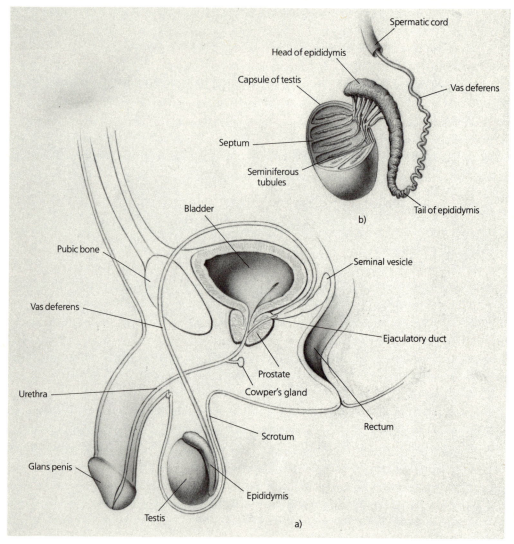

FIGURE 9.   Internal Side View of the Male Reproductive System

The Prostate and Accessory Organs

The *prostate gland,* normally about the size of a chestnut, consists of a muscular and a glandular portion. The prostate is located directly below the bladder and surrounds the urethra (the tube through which urine passes) as it exits from the bladder. The relationship of the prostate to the urethra is like a large bead (the prostate) on a string (the urethra). Because the *rectum* (the lowest part of the bowels) is directly behind the prostate, the prostate can be examined by a physician during a rectal examination. This is important because the prostate can become infected or cancerous.

The prostate produces clear fluid that makes up about 30 percent of *seminal fluid,* the liquid that is expelled from the penis during ejaculation. The other 70 percent of seminal fluid comes from the *seminal vesicles.* These two small structures lie against the back portion of the base of the bladder and join with the ends of the vas deferens to form the *ejaculatory ducts.* These ducts in turn join the urethra, thereby creating a continuous tubing system that leads to the end of the penis.

On the average, there are 3 to 5 milliliters (5 milliliters = about one teaspoonful) of semen (seminal fluid plus sperm) per ejaculate. Although the concentration of sperm is highly variable, depending in part on the frequency of ejaculation, a count of 40 to 120 million per milliliter is considered normal. This means that there may be 120 to 600 million sperm in a single ejaculate.*

Seminal fluid ranges in color from whitish to tones of yellow or gray and has a creamy, sticky texture. Right after ejaculation, seminal fluid is rather thick but then it liquefies quickly. It consists of water, mucus, and a large number of chemical substances that include sugar (providing an energy source for sperm), bases (for neutralizing the acidity of the male urethra and the female vagina), and prostaglandins (hormones that cause contractions in the uterus and Fallopian tubes, possibly aiding upward transit of sperm).

Cowper's glands are two pea-sized structures connected to the urethra just below the prostate gland. They produce a few drops of fluid, which sometimes appear at the tip of the penis during sexual arousal but before ejaculation. Some men never notice this pre-ejaculatory fluid, while others may produce a teaspoonful or more of this slippery secretion. Although this pre-

---

* After a vasectomy, although there are no longer sperm in the ejaculate, the amount of fluid in the ejaculate remains the same.

ejaculatory fluid may buffer the acidity of the urethra, there is
no certainty about its function. Occasionally, the pre-ejaculatory
fluid contains a small number of live sperm cells, accounting for
at least some of the "failures" as a method of birth control of
withdrawal of the penis before ejaculation.

The Breasts

The male breasts have nipples and areolas but have little underlying glandular tissue or fatty padding. The male nipple and
areola seem to be less sensitive to touch and pressure than the
same structures in adult females. Nevertheless, some males find
that having their breasts or nipples stroked or licked is sexually
arousing. Others do not notice any erotic pleasure from such
practices.

Sometimes one or both of a male's breasts may become enlarged. This condition, called *gynecomastia,* occurs in 40 to 60
percent of boys during puberty but usually disappears within a
year or two. In adulthood, it may be caused by alcoholism, liver
disease, thyroid disease, drug ingestion, or certain forms of cancer. When gynecomastia is so severe that it creates major psychological problems, it can be corrected by relatively simple surgery.

The male breasts can also become enlarged if a man takes
estrogen over a period of time. As we will discuss in chapter 8,
most male-to-female transsexuals undergo such treatment. We
have also seen a case in which a man unwittingly took birth
control pills for several months, causing the same result.

OTHER EROGENOUS ANATOMY

Many parts of the body besides those involved in reproduction
are potential sources of sexual arousal in both sexes. Surprisingly, the largest sensory organ for both females and males is
the skin itself. The insides of the thighs, the neck, and the perineum are often sources of sexual pleasure. In our genitally oriented society, where sex is often thought of as synonymous with
intercourse, it is easy to overlook the importance of touching
and body-to-body contact as a form of intimacy and gratification.
Stroking, caressing, and massage can be forms of nonverbal

communication, sensual pleasure, or invitations to further sexual activity.

Some people are well aware of the erotic sensations they can experience from touch, while others pay little attention to this component of their sexual arousal. However, there are wide differences from person to person in such matters: for some, the skin outside the genital region has relatively little sexual input or may actually dampen sexual feelings (what would happen to your level of arousal if a touch felt persistently ticklish or irritating?); at the other extreme, some women can be aroused to the point of orgasm by having the small of their back rubbed without any other stimulus. (However, the likelihood of being or encountering a female capable of reaching orgasm by back-rubbing alone is less than one in a million.)

The mouth, including the lips and tongue, is an area of high erotic potential. Kissing is one practice that uses the sensitivity of this region in a sexually stimulating fashion. In addition to the sensory signals activated by kissing, it is also an act of intimacy that can symbolize passion and penetration (think of the form of kissing called "French kissing" or "soul kissing," in which one partner's tongue enters the other's mouth). Oral-genital contact — stimulation of one person's genitals in a licking or sucking fashion by the partner's lips or tongue — is another common form of sexual stimulation.

The anus, rectum, and buttocks are also potentially erogenous areas. The anus is highly sensitive to touch and the insertion of a finger, object, or penis in the anus and rectum is part of some people's sexual activity. Anal intercourse is often thought to be primarily an act of male homosexuals. However, numerically speaking, far more heterosexual couples engage in this activity than homosexuals, and many homosexual men have not had experience with this type of sexual behavior.

The buttocks are regarded in some cultures as symbolic of female sexuality in much the sense that our society regards the female breasts. The buttocks are bulky groups of muscles covered by fat and skin, with a relatively sparse distribution of nerves sensitive to touch. The underlying muscles are important in the mechanical process of pelvic thrusting during sexual intercourse. As a target for spanking, the buttocks are sometimes provocative for those of both sexes who find this activity erotically arousing. As a visible part of the anatomy, the buttocks (especially when displayed in tight jeans, swim trunks, bikinis, or similar apparel) commonly serve as a form of sexual enticement.

Many other parts of the body can also have erotic allure. For instance, hair can be sensual or sexual: some women are turned on by their partner's hairy chest and some lovers like to stroke each other's hair. Well-developed muscles make males more attractive to some females, whereas others are less impressed or actually turned off by this "he-man" appearance. Nibbling an earlobe, caressing the face, and touching fingertips can all be part of a sexual encounter and all may be a source of excitation. Our attempt here is not to provide an exhaustive catalogue, but to demonstrate the wide range of what can be sexual.

We each have a unique appearance to our sexual anatomy and an even more unique experience of sexual feelings and interactions. As we have repeatedly stressed, the variations — even anatomically — from one person to another are considerable. Unfortunately, some people are preoccupied with the notion that "biggest is best" and others believe that sexual satisfaction is mainly a matter of "pushing the right buttons." Instead, we believe that a mechanical view of sex often leads to a mechanical experience, whereas a view of sex as a matter of comfort, mood, and feelings combined with physical sensations and response is more likely to be fulfilling and fun.

CHAPTER THREE

Sexual Physiology

An orgasm is like a rocket ride. First the ascent, then the blackout, and after that the burst of light as the golden apple turns into the golden sun and azure skies with a slow parachute until the earth appears below, streams and meadows or a city street. You slowly touch and bounce up again. . . . Then you slowly touch once more, and then the afterglow and the deep refreshing sleep. (Berne, 1971, p. 233)

Orgasm can be a very mild experience, like a ripple or peaceful sigh; it can be a very sensuous experience where our body glows with warmth; it can be an intense experience with crying out and thrashing movements; it can be an ecstatic experience with momentary loss of awareness. (Boston Women's Health Book Collective, 1976, p. 45)

Sometimes I think orgasms are overrated. Getting there is more than half the fun. (Comment by a twenty-three-year-old woman)

PEOPLE INTERPRET their sexual responses in various ways, as these quotations show. But the basic details of how the body responds to sexual arousal are identical whether the stimulation comes from touching, kissing, intercourse, masturbation, fantasy, watching a movie, or reading a book. This statement does not imply that sex is just a mechanical process, any more than dancing or playing the violin is "only" mechanical because certain parts of the body are involved in these activities. Human sexual response is multidimensional, with input from feelings and thoughts, learning and language, personal and cultural values, and many other sources combining with our biological reflexes to create a total experience.

To understand the complexities of human sexuality, it is helpful to become familiar with the details of sexual physiology (the functions of our sexual anatomy). Learning about the various responses of the body during sexual arousal and about the forces that regulate them will increase your awareness of your own and your partner's responses and may clarify many misconceptions, myths, and questions about sex. It is also important to understand sexual physiology to comprehend many sexual disorders discussed later in the book.

SOURCES OF SEXUAL AROUSAL

When people talk about sexual arousal, they frequently say they are "turned on," "revved up," or "hot." Each phrase likens sexual arousal to an energy system, and as a starting point, this comparison is useful. From a scientific perspective, sexual arousal can be defined as a state of activation of a complex system of reflexes involving the sex organs and the nervous system. The brain itself, the controlling part of the nervous system, operates with electrical and chemical impulses "wired" to the rest of the body through the spinal cord and peripheral nerves. Signals from other parts of the body (like the skin, genitals, breasts) are integrated and focused in the brain, for without sexual thoughts, feelings, or images, sexual response is fragmentary and incomplete. At times, sexual arousal may be largely a cerebral event — that is, a person may be aroused while no visible physical changes are occurring elsewhere in the body. On other occasions, genital sensations can be so intense that they block out awareness of almost everything else.

Sexual arousal can occur under a wide variety of circumstances. It may be the result of voluntary actions such as kissing, hugging, reading a sexy book, or going to an erotic movie. Sexual arousal can also be unexpected, unwanted, or even alarming. Consider, for instance, the following situations: (1) a twelve-year-old boy gets an erection while taking a shower in a crowded all-male locker room at school; (2) a female college student who is an ardent feminist becomes sexually aroused while watching a rape scene in a movie; (3) a female medical student is sexually excited when she examines an elderly male patient; (4) a male lawyer is sexually aroused by discussions with a female client who hires him to help her obtain a divorce. These people may be embarrassed or uncomfortable temporarily, but unexpected sexual arousal is normal and happens to most of us occasionally.

The sources of sexual arousal are also varied. The process of getting "turned on" may be triggered by direct physical contact such as a touch or a kiss, or may be activated by a verbal invitation ("let's make love"), a nonverbal message ("body language"), or a visual cue (such as nudity or a particular clothing style). It may also spring from fantasies or the most everyday occurrences — clothing rubbing against the genitals, the rhythm of a moving vehicle, or taking a bath or shower. Sexual arousal occurs in all age groups, from infants to the elderly, and it occurs when we are asleep as well as when we are awake. Men have about a half dozen erections during a night's sleep (the erections usually last five to ten minutes), and women have similar episodes of vaginal lubrication during sleep. These reflex responses occur automatically and are not controlled by the specific content of dreams.

THE SEXUAL RESPONSE CYCLE

Before the 1960s, relatively little was known about the way the body responds during sexual arousal. Scientists were not convinced by Kinsey's claims that some women had more than one orgasm at a time, and it was thought that vaginal lubrication was produced by glands in the cervix and Bartholin's glands. The mechanisms controlling erection and ejaculation in the male were incompletely understood. As a matter of propriety, sexual response was studied in animals, not people. In this climate, the results of an investigation of sexual physiology based on direct laboratory observation of more than 10,000 episodes of sexual activity in 382 women and 312 men first appeared as a book titled *Human Sexual Response*, by Masters and Johnson.

The findings of this study indicated that human sexual response could be described as a cycle with four stages: *excitement, plateau, orgasm,* and *resolution.* These stages correspond to varying levels of sexual arousal and describe the typical responses

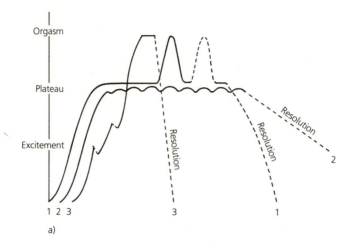

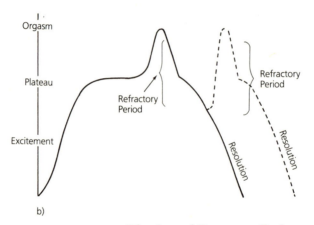

FIGURE 10.   The Sexual Response Cycle

(a) Three representational variations of female sexual response. Pattern 1 shows multiple orgasm; pattern 2 shows arousal that reaches the plateau level without going on to orgasm (note that resolution occurs very slowly); and pattern 3 shows several brief drops in the excitement phase followed by an even more rapid resolution phase.

(b) The most typical pattern of male sexual response. The dotted line shows one possible variation: a second orgasm and ejaculation occurring after the refractory period is over. Numerous other variations are possible, including patterns that would match 2 and 3 of the female response cycle.

Source: From Masters and Johnson, 1966. © 1966 by William H. Masters and Virginia E. Johnson.

people have during sexual function. Although it is convenient to use the cycle as a model for descriptive purposes, remember that the stages are arbitrarily defined. They are not always clearly separated from one another and may vary considerably both in one person at different times and between people. Bear in mind also that the physiological processes of sexual response are not simply mechanical movements detached from thoughts or feelings but are part of the sexual involvement and identity of the whole person.

Although the sexual response cycle usually follows a consistent pattern of progression, the simplified schematic patterns of sexual response may vary widely, as Figure 10 reveals. Sometimes excitation is rapid and leads quickly to orgasm. On other occasions, excitement mounts slowly over a period of hours — while having a romantic, intimate meal, for example — and the rest of the cycle may seem brief in comparison. The plateau stage may not always lead to orgasm, as the high levels of arousal that characterize this phase may dissipate; and a person may slip back to the excitement phase. If sexual stimulation stops, a person may also drift back into an unaroused state.

There are two basic physiologic reactions during human sexual response. The first is *vasocongestion,* an increased amount of blood concentrated in body tissues in the genitals and female breasts. The second is increased *neuromuscular tension* or *myotonia.* Here, tension does not refer to a negative physical state ("feeling tense") but to a buildup of energy in the nerves and muscles. Myotonia occurs throughout the body in response to sexual arousal, not simply in the genital region. Although there are some differences in male and female sexual response, many details are similar. The physiology of sexual response is also the same for heterosexuals and homosexuals.

Before we discuss the specific details of sexual response, a note of caution is in order. It is often tempting to equate the speed, size, and strength of sexual responses (such as erection, vaginal lubrication, or muscular contractions during orgasm) with the gratification a person experiences or with his/her proficiency as a lover. This is like saying that a bowl of chili is "better" than sirloin steak simply because chili causes a faster and larger secretion of digestive juices than steak. In both cases ("better" digestive response, "better" sexual response), the degree to which one experience is "better" than the other depends on your perspective *and* on your personal satisfaction.

Excitement

Excitation results from sexual stimulation, which may be physical, psychological, or a combination of the two. Sexual responses are like other physiologic processes that may be triggered not only by direct physical contact but by vision, smell, thought, or emotion. For example, thinking about food, smelling fresh baked goods, or watching a television commercial may prompt salivation and gastric acid production; and fear may activate a complex set of reflexes, including sweating, a faster pulse rate, and increased blood pressure.

The first sign of sexual excitation in the female is the appearance of vaginal lubrication, which starts ten to thirty seconds after the onset of sexual stimulation. Vaginal lubrication occurs because vasocongestion in the walls of the vagina leads to moisture seeping across the vaginal lining in a process called *transudation*. Beads of vaginal secretion first appear as isolated droplets, which flow together and eventually moisten the entire inner surface of the vagina. Early in the excitement stage the quantities of fluid may be so small that neither the woman nor her partner notices it. As vaginal lubrication increases, the fluid sometimes flows out of the vagina, moistening the labia and the vaginal opening, but this depends on the woman's position and the type of sexual play going on.

The consistency, quantity, and odor of vaginal lubrication vary considerably from one woman to another and also vary in the same woman from time to time. Contrary to commonly held beliefs, the amount of vaginal lubrication is *not* necessarily indicative of the woman's level of sexual arousal, and the presence of vaginal lubrication does not mean the woman is "ready" for intercourse. Vaginal lubrication makes insertion of the penis into the vagina easier and smoother and prevents discomfort during intravaginal thrusting.

Other changes also occur in women during the excitement phase. The inner two-thirds of the vagina expand, the cervix and uterus are pulled upward, and the outer lips of the vagina flatten and move apart (see Figure 11). In addition, the inner lips of the vagina enlarge in diameter, and the clitoris increases in size as a result of vasocongestion. A woman's nipples typically become erect during the excitement phase as a result of contractions of small muscle fibers. Late in the excitement phase (again as a result of vasocongestion), the veins on the breasts become more visible and there also may be a small increase in breast size.

The most prominent physical sign of sexual excitation in men is erection of the penis, which usually occurs within a few seconds after sexual stimulation starts (Figure 12). Although this response may seem very different from vaginal lubrication, they are parallel events that both occur because of vasocongestion. Erection results from the spongy tissues of the penis rapidly filling with blood. It is not certain at present whether engorgement occurs because the veins that drain the penis cannot keep up with this rapid filling or if special structures called "polsters" in the blood vessels of the penis limit outflow. Whatever the exact mechanism, the increased size and firmness of the erect penis are due to increased fluid pressure: erection can thus be viewed in simplest terms as a hydraulic event. Despite this seeming mechanical simplicity, a man may be physically and/or psychologically aroused and not have a firm erection, particularly under conditions of anxiety or fatigue. Contrary to some common misconceptions, there is neither a bone in the human penis (although erections are sometimes called "boners") nor a penis muscle that controls the process of erection.*

In addition to erection, the skin ridges of the scrotum begin to smooth out and the testes are partially drawn toward the body. Late in the excitement phase, the testes increase slightly in size. Nipple erection occurs during excitation for some men but not others.

Although many people think of male sexual response as nearly instantaneous and constant, in real life it does not always happen this way. Literary descriptions of a "pulsating," "throb-

---

* Many animals have a penis bone (*os penis*). The late Dr. Francis Ryan, a zoologist who taught a popular course on comparative anatomy at Columbia University, delighted in waving a large, baseball-bat-sized bone in the air and asking, "Does anyone know what this is?" After no response from his all-male class, he would declare, "This, gentlemen, is the *os penis* of the Arctic Whale." He would pause then for effect before saying, "Life in the Arctic is a stiff proposition."

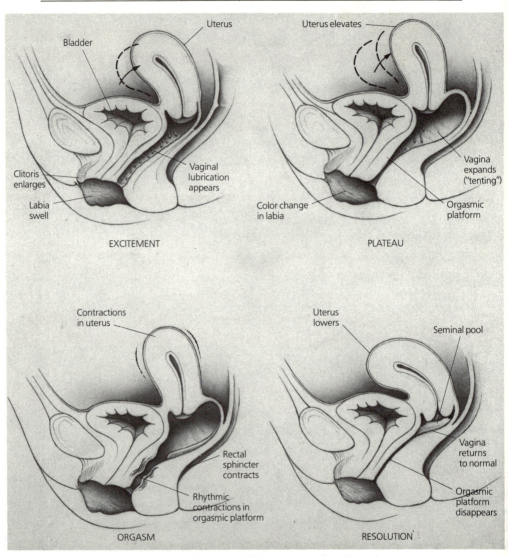

Bladder

Uterus

Uterus elevates

Clitoris
enlarges

Vaginal
lubrication
appears

Labia
swell

EXCITEMENT

Color change
in labia

Vagina
expands
("tenting")

Orgasmic
platform

PLATEAU

Contractions
in uterus

Uterus
lowers

Seminal pool

Rectal
sphincter
contracts

Rhythmic
contractions in
orgasmic platform

ORGASM

Vagina
returns
to normal

Orgasmic
platform
disappears

RESOLUTION

FIGURE 11.   Internal Changes in the Female
Sexual Response Cycle

bing," or "steel-hard" penis are common but often fictional. As
Zilbergeld (1978, p. 24) observes, in our unrealistic expectations,
"The mere sight or touch of a woman is sufficient to set the
penis jumping, and whenever a man's fly is unzipped, his penis
leaps out. . . . Nowhere does a penis merely mosey out for a look
at what's happening." In other words, a man is expected to be

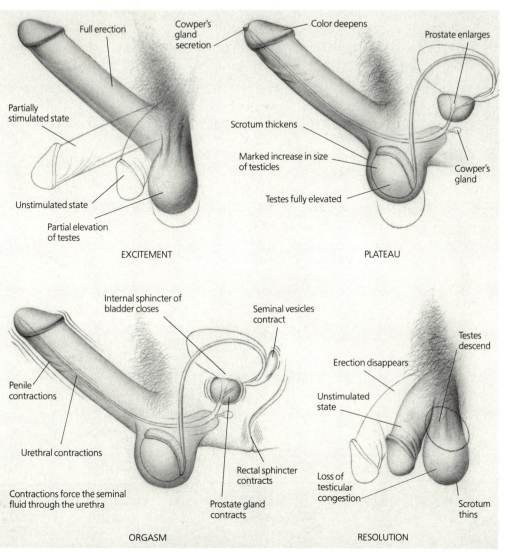

FIGURE 12. External and Internal Changes in the Male
Sexual Response Cycle

instantly erect at the drop of a bra, which of course creates a
dilemma for anyone who finds that his arousal is not so dramatic
or visible.

As we have seen, the physical changes of the excitement phase
are neither constant nor always increasing for men and women.
Mental or physical distractions can and often do decrease the

buildup of sexual tension that is the hallmark of excitation. A honking horn, a knock on the door, an inopportune telephone ring, a shift in position, a muscle cramp, and a growling stomach are among the innumerable possible distractions. In addition, changes of tempo or manner of direct sexual stimulation can also temporarily disrupt sexual arousal, just as too much of a particular caress may temporarily cause a dulling of sensations.

Some people become upset or worried if their initial sexual arousal does not build steadily to a shattering peak. If a man's erection recedes even briefly, he may think "I'm losing it" or his partner may wonder "What am I doing wrong?" If a woman's vagina seems to get dry or her nipples lose their erection, she (or her partner) may have the same concerns. As a result, sexual spontaneity is likely to be lessened and awareness of body sensations is reduced. In such situations, the initial worry often becomes a self-fulfilling prophecy.

Vasocongestive mechanisms of sexual arousal wax and wane in everyone, just as most biological processes fluctuate a bit. An erection may be diminishing in firmness or size, or vaginal lubrication may seem to cease, although physical sensations and neuromuscular tension indicate that the man and woman are clearly nearing the plateau phase of the sexual response cycle. In this example, if the partners become alarmed or give up because they "see" that their physical response is less than what they want or expect it to be, they are not really giving themselves a chance.

Plateau

In the excitement phase, there is a marked increase in sexual tension above baseline (unaroused) levels. As Figure 10 shows, in the plateau phase high levels of sexual arousal are maintained and intensified, potentially setting the stage for orgasm. The duration of the plateau phase varies widely. For men who have difficulty controlling ejaculation, this phase may be exceptionally brief. In some women, a short plateau phase may precede a particularly intense orgasm. For other people, a long, leisurely time at the plateau level is an intimate and erotic "high" that may be a satisfying end in its own right.

During the plateau phase in women prominent vasocongestion in the outer third of the vagina causes the tissues to swell.

This reaction, called the *orgasmic platform,* narrows the opening of the vagina by 30 percent or more (Figure 11). One reason penis size is not so important to a woman's physical stimulation during intercourse is that her outer vagina or orgasmic platform "grips" the penis if plateau levels of arousal are reached.* During the plateau phase, the inner two-thirds of the vagina expand slightly more in size as the uterus becomes more elevated in a process known as "tenting." The production of vaginal lubrication often slows during this phase as compared to excitation, especially if the plateau phase is prolonged.

The clitoris pulls back against the pubic bone during the plateau phase. This change, coupled with the vasocongestion occurring in the vaginal lips, hides the clitoris (Figure 13) and partially protects its head from direct touch. No loss of clitoral sensation occurs during these changes, however, and stimulation of the mons or the labia will result in clitoral sensations. (Marriage manuals in the 1950s all seemed to instruct the male that finding and stimulating the clitoris was the key to sexual responsiveness. We wonder how many men panicked when this phase of clitoral retraction made the clitoris seem to disappear.)

The inner lips enlarge dramatically as a result of engorgement with blood, doubling or even tripling in thickness. As this happens, the inner lips push the outer lips apart, providing more immediate access to the opening of the vagina. Once this reaction has occurred, vivid color changes develop in the inner lips. The inner lips of women who have never been pregnant range from pink to bright red, while those in women who have been pregnant range from bright red to a deep wine color because of greater vascular supply delivering more blood flow to this area. Masters and Johnson noted in 1966 that if effective sexual stimulation continues once this "sex skin" color change appears, orgasm invariably follows. In more than 7,500 cycles of female sexual response, an orgasm never occurred without the preceding color change of the inner lips.

Late in the excitement phase, the areola of the female breast begins to swell. During the plateau phase, swelling continues to the point that the earlier nipple erection usually becomes ob-

---

* As we mentioned in chapter 2, some women may find a large penis to be important for *psychological* stimulation, and some women claim to receive more physical stimulation from a larger penis.

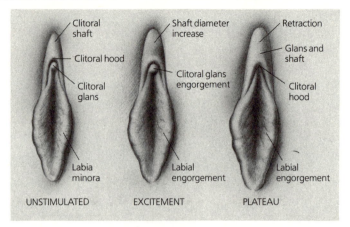

FIGURE 13.   The Clitoris and Labia in the Female
Sexual Response Cycle

*In the plateau phase, the clitoris seems to disappear beneath its hood, but it
is actually quite engorged. The orgasmic phase is omitted because of lack of
information.*

scured (Figure 14). Increases in breast size during the plateau
phase are striking in women who have not breast-fed a child,
averaging 20 to 25 percent above baseline levels. For women
who have previously nursed a child, this increase is much less
pronounced or nonexistent because of their more developed
venous drainage. However, this does not reduce erotic sensations in the breasts.

Late in the excitement phase or early in the plateau phase, a
reddish, spotty skin color change resembling measles develops
in 50 to 75 percent of women and about one-fourth of men.
This "*sex flush*" generally begins just below the breastbone in the
upper region of the abdomen and then spreads rapidly over the
breasts and front of the chest. It may also appear on other parts
of the body, including the neck, buttocks, back, arms, legs, and
face. The sex flush results from changes in the pattern of blood
flow just below the surface of the skin.

During the male's plateau phase (Figure 12), the diameter of
the head of the penis near the coronal ridge increases slightly.
This area often deepens in color because of pooling of blood.
Vasocongestion also causes the testes to swell, becoming 50 to
100 percent larger than in the unstimulated state.

As sexual tension mounts toward orgasm, the testes not only

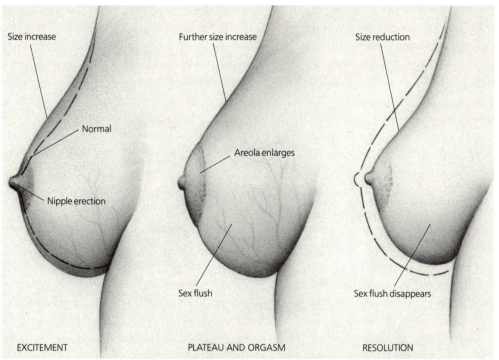

Size increase

Normal

Nipple erection

EXCITEMENT

Further size increase

Areola enlarges

Sex flush

PLATEAU AND ORGASM

Size reduction

Sex flush disappears

RESOLUTION

FIGURE 14. Breast Changes during the Female
Sexual Response Cycle

*After orgasm, the rapid reduction in swelling of the areola often makes it
appear as though the nipple has again become erect.*

continue to elevate but also begin to rotate forward so that their
back surfaces rest in firm contact with the perineum (the area
between the scrotum and the anus). Full elevation of the testes
indicates that orgasm is imminent. In some men, particularly
over age fifty, the testes elevate only partially. This seems to
cause a reduction in ejaculatory pressure.

Small amounts of clear fluid may sometimes appear from the
male urethra during the plateau phase. This fluid is thought to
come from Cowper's glands and, as mentioned in the previous
chapter, occasionally carries live sperm. Many men experience a
sensation of internal pressure or warmth during the plateau
phase that corresponds to vasocongestion in the region of the
prostate gland and seminal vesicles.

In addition to the sensations and changes just described, men
and women experience several changes throughout their bodies

during the plateau phase. A generalized increase in neuromuscular tension is particularly apparent in the buttocks and thighs.
The heart rate increases, sometimes leading to a prominent
awareness of heart thumping inside the chest. Breathing also
becomes faster and the blood pressure increases, too.

Orgasm

If effective sexual stimulation continues late in the plateau
phase, a point may be reached where the body suddenly discharges its accumulated sexual tension in a peak of sexual
arousal called *orgasm*. Orgasm is sometimes called climax or coming. Eric Berne (1971, p. 3) observed, "Climax started off as a
decent enough word, but it has been so overworked on the news-
stands that it now sounds like the moment when two toasted
marshmallows finally get stuck to each other." We also prefer
the word "orgasm." Biologically, orgasm is the shortest phase of
the sexual response cycle, usually lasting for only a few seconds
during which rhythmic muscular contractions produce intense
physical sensations followed by rapid relaxation. Psychologically,
orgasm is usually a time of pleasure and suspended thought —
the mind turns inward to enjoy the personal experience.

Orgasms vary not only for one person at different times but
also for each individual. Sometimes orgasm is an explosive, ecstatic avalanche of sensations, while others are milder, less intense, and less dramatic. While "an orgasm is an orgasm is an
orgasm," one orgasm may differ from another just as a glass of
ice water tastes better and is more satisfying if you are hot and
thirsty than if you are cool and not thirsty at all. Different in-
tensities of orgasms arise from physical factors such as fatigue
and the time since the last orgasm as well as from a wide range
of psychosocial factors, including mood, relation to partner, activity, expectations, and feelings about the experience.

For all these reasons, trying to define or describe orgasm is a
difficult task because each individual's subjective experience includes a psychosocial as well as physiological dimension. Measuring intense muscular contractions during one orgasm does not
mean that it is necessarily perceived as "better than" another
orgasm with less intense bodily changes. A milder physiological

orgasm may be *experienced* as bigger, better, or more satisfying than a physiologically more intense one.

Until the mid-twentieth century, many people (including some medical authorities) believed that women were not capable of orgasm. This belief undoubtedly reflected a cultural bias: sex was seen as something the man did *to* the woman for his own gratification. Women were told for centuries to "do their wifely duties" by making themselves available to their husbands for sex, yet were also cautioned that "proper" women did not enjoy sex. Since a sign of physical pleasure or orgasm was thought to be "unladylike," it followed that women were not able to have orgasms. In other words, females were told "You can't have any physical sexual release, and even if you can, you shouldn't." It is now clear, however, that orgasm occurs in both sexes.

Orgasm in the female is marked by simultaneous rhythmic muscular contractions of the uterus, the outer third of the vagina (the orgasmic platform), and the anal sphincter (Figure 11). The first few contractions are intense and close together (at 0.8 second intervals). As orgasm continues, the contractions diminish in force and duration and occur at less regular intervals. A mild orgasm may have only three to five contractions, while an intense orgasm may have ten to fifteen.

Orgasm is a total body response, not just a pelvic event. Brain wave patterns show distinctive changes during orgasm and muscles in many different body regions contract during this phase of sexual response. In addition, the sex flush achieves its greatest intensity and its widest distribution at the time of orgasm.

Women often describe the sensations of an orgasm as beginning with a momentary sense of suspension, quickly followed by an intensely pleasurable feeling that usually begins at the clitoris and rapidly spreads throughout the pelvis. The physical sensations of the genitals are often described as warm, electric, or tingly, and these usually spread through the body. Finally, most women feel muscle contractions in their vagina or lower pelvis, often described as "pelvic throbbing."

Despite a popular misconception, most women do not ejaculate during orgasm. The erroneous belief that women ejaculate probably stems from descriptions in erotic novels of fluid gushing from the vagina as a woman writhes and moans at the peak moment of sexual passion. Such descriptions are not particularly accurate.

Recently, however, it has been suggested that a somewhat different form of female ejaculation occurs. Various workers have claimed that some women expel semen-like fluid from the urethra at the time of orgasm. It has been theorized that this fluid may come from a "female prostate," rudimentary glands (Skene's glands) around the urethra near the neck of the bladder that derive embryologically from the same tissues that develop into the prostate gland in males. In fact, some suggest that this "female prostate" is the anatomical site of the "G spot," but this idea — although stirring considerable controversy and conjecture — has not yet been proven scientifically. And, while a report on one woman with this ejaculation-like phenomenon indicated that the fluid was not urine, another more detailed study of six other women who "ejaculated" showed that the fluid they expelled was indistinguishable from urine.

Further confusion in this area is caused by uncertainty over the number of women who have this ejaculation-like response. Perry and Whipple, who have been among the most vociferous self-proclaimed "experts" on this subject, initially claimed that "perhaps 10 percent of females" had this response, but later reported that they were finding "closer to 40 percent" of women "had ever experienced" female ejaculation. In our own studies, a survey of approximately 300 women aged eighteen to forty revealed only fourteen who claimed to note any gushing or expulsion of fluid at orgasm. This observation is certainly more in keeping with our experience with well over a thousand women in our sex therapy program, where there have been only a handful of reports of fluid "ejaculated" by women with orgasm. However, we *have* observed several cases of women who expelled a type of fluid that was not urine.

Although it is clear that at least *some* women experience this ejaculation-like reponse, it should be realized that a number of these cases represent a condition called *urinary stress incontinence* in which urine is expelled from the urethra due to physical straining such as occurs with coughing, sneezing, or sexual arousal. Since this condition is usually correctable either by the use of Kegel exercises or minor surgery, medical evaluation is warranted if a woman is bothered by such a response.

Orgasms in men, unlike those in women, occur in two distinct stages. In the first stage of orgasm, the vas deferens (each of the two tubes that carry sperm) and the prostate and seminal vesicles begin a series of contractions that forces semen into the bulb of the urethra (Figure 12). The man experiences a sensation of *ejaculatory inevitability* — that is, the feeling of having reached the brink of control — as these contractions begin. This sense of inevitability is quite accurate because at this point ejaculation cannot be stopped.* In the second stage of the male orgasm, contractions of the urethra and penis combine with contractions in the prostate gland to cause ejaculation or the spurting of semen out of the tip of the penis. The external appearance of semen does not occur until several seconds after the point of ejaculatory inevitability because of the distance semen must travel through the urethra.

During ejaculation, the neck of the urinary bladder is tightly shut to ensure that semen moves forward and to avoid any mixing of urine and semen. The rhythmic contractions of the prostate, perineal muscles, and shaft of the penis (creating the physical force that propels semen on its journey) occur initially at 0.8 second intervals, just as in women, and account for the spurting of the semen during ejaculation. After the first three or four contractions of the penis, the intervals between contractions become longer and the intensity of the contractions tapers off.

Male orgasm and ejaculation are not one and the same process, although in most men and under most circumstances the two occur simultaneously. Orgasm refers specifically to the sudden rhythmic muscular contractions in the pelvic region and elsewhere in the body that effectively release accumulated sexual tension and the mental sensations accompanying this experience. Ejaculation refers to the release of semen, which sometimes can occur without the presence of orgasm. Orgasm without ejaculation is common in boys before puberty and can also occur if the prostate is diseased or with the use of some drugs. Ejaculation without orgasm is less common but can occur in certain cases of neurological illness.

* Women do not have a consistently identifiable point of orgasmic inevitability that corresponds to the stage of ejaculatory inevitability in the male response cycle. Distractions can interrupt women's orgasms, whereas if the male has reached "inevitability," orgasm occurs no matter what.

In *retrograde ejaculation,* the bladder neck does not close off properly during orgasm so that semen spurts backward into the bladder. This condition occurs in some men with multiple sclerosis or diabetes, or after certain types of prostate surgery. There are no harmful physical effects, but infertility results, and the man may have a different sensation during ejaculation.

The subjective experience of orgasm in men starts quite consistently with the sensation of deep warmth or pressure (sometimes accompanied by throbbing) that corresponds to ejaculatory inevitability. Orgasm is then felt as sharp, intensely pleasurable contractions involving the anal sphincter, rectum, perineum, and genitals, which some men describe as a sensation of pumping. A different feeling, sometimes called a warm rush of fluid or a shooting sensation, describes the actual process of semen traveling through the urethra. In general, men's orgasms tend to be more uniform than women's although all male orgasms are certainly not identical.

During the orgasmic phase in both sexes, there are high levels of myotonia evident throughout the body. Late in the plateau phase or during orgasm, the myotonia is often visible in facial muscles, where a grimace or frown may be seen. While this expression is sometimes viewed by a partner as an indication of displeasure or discomfort, it is actually an involuntary response that indicates high levels of sexual arousal. Muscle spasms or cramps in the hands or feet may also occur late in the plateau phase or during orgasm, and at the peak of orgasm, the whole body may seem to become rigid for a moment.

While many controversies about the nature of orgasms exist, there are several that deserve special mention. The first controversy originated with Freud, who believed that there were two types of female orgasm, a clitoral and a vaginal orgasm. Freud stated that clitoral orgasms (those originating from masturbation or other noncoital acts) were evidence of psychological immaturity, since the clitoris was the center of infantile sexuality in the female. Vaginal orgasms (those deriving from coitus) were "authentic" and "mature" since they demonstrated that normal psychosexual development was complete. In his essay "Some Psychological Consequences of the Anatomical Distinction Between the Sexes," Freud wrote that "the elimination of clitoral sexuality is a necessary precondition for the development of femininity." Many women were considered neurotic or were pushed into psychoanalysis because of this view.

Physiologically, all female orgasms follow the same reflex response patterns, no matter what the source of sexual stimulation. An orgasm that comes from rubbing the clitoris cannot be distinguished physiologically from one that comes from intercourse or breast stimulation alone. This does not mean that all female orgasms feel the same, have the same intensity, or are identically satisfying. As discussed earlier, feeling and intensity are matters of perceptions, and satisfaction is influenced by many factors.

Some women prefer orgasms that occur as a result of intercourse, while others prefer masturbatory orgasms. Those who prefer coital orgasms often say that the overall experience is more satisfying, but the actual orgasm is less direct and intense. *The Hite Report* notes that many women find masturbatory orgasms to be more satisfying than coital ones, perhaps because the woman is not affected by her partner's style, needs, or tempo. In other reports, attempts have been made to differentiate between "vulval orgasm," "uterine orgasm," and "blended orgasm" or other classifications of orgasmic types. Recently, Ladas, Whipple, and Perry have claimed that stimulation of the "G spot" produces a completely different type of orgasm from stimulation of the clitoris: one in which no orgasmic platform forms, and in which the uterus, instead of elevating and expanding the inner portion of the vagina, "seems to be pushed down and the upper portion of the vagina compresses." However, data to support these claims have not yet been published. Despite the continued controversy about "types" of female orgasms, the idea that one type is immature or less good than another has been generally discarded.

A second controversy about female orgasm is the question of whether or not all women in good health are able to experience a coital orgasm without any other type of simultaneous stimulation. While Masters and Johnson and others such as the psychiatrist Mary Jane Sherfey and the psychologist Lonnie Barbach believe all women have this ability, some sexologists believe that there may be a group of women who do not. Helen Kaplan seems to favor the latter view when she says that "this pattern may represent a normal variant of female sexuality, at least for some women." And various studies from Kinsey's day until the present show that the number of women who experience orgasm regularly during intercourse is about 40 to 50 percent. Many authorities believe that lack of coital orgasm is usually caused by factors such as anxiety, poor communication between partners, hostility, distrust, or low self-esteem. However, if certain females are incapable of experiencing a sexual reflex because of physiological factors, it would have implications in diagnosing and treating some women's sexual problems.

Another controversial area has to do with the role of the muscles surrounding the vagina in orgasm. Both Arnold Kegel (a surgeon who was the inventor of the "Kegel exercises") and other workers, including Perry and Whipple, claim that the condition of the pubococcygeus muscle (PC muscle) is an important determinant of the occurrence of orgasms in women. However, other studies fail to document any correlation between PC muscle strength and female orgasmic responsiveness and have also found that using the Kegel exercises did not improve orgasmic responsivity in nonorgasmic women.

Finally, although many observers believe that most women don't feel that orgasm is a necessity for sexual satisfaction, a recent study by Waterman and Chiauzzi found that "orgasm consistency was significantly related to sexual satisfaction in females but not in males." While this doesn't mean that women who have the most frequent orgasms are happiest sexually, it does imply that not having orgasms (or not having them very often) may correlate with sexual dissatisfaction.

Resolution Phase

There is a major difference between male and female sexual response immediately following orgasm. Generally, females have the physical capability of being *multiorgasmic* — that is, they can have one or more additional orgasms within a short time without dropping below the plateau level of sexual arousal (Figure 10a, pattern 1). Being multiorgasmic depends on both continued effective sexual stimulation and sexual interest, neither of which is consistently present for most women. For this reason, some women never experience multiple orgasms, and others are multiorgasmic in only a small fraction of their sexual experiences. It is unusual for a woman to have multiple orgasms during most of her sexual activity.

Interestingly, multiple orgasm in females seems to occur more frequently during masturbation than intercourse. This may reflect several factors: (1) the relative ease of continuing sexual stimulation, (2) the lack of distraction by concerns about one's partner, and (3) the more frequent use of sexual fantasy by women during masturbation as compared to intercourse.

Men, on the other hand, are not able to have multiple orgasm if it is defined in the same way. Immediately after ejaculation, the male enters a *refractory period* (Figure 10b), a recovery time during which further orgasm or ejaculation is physiologically impossible. A partial or full erection may be maintained during this refractory period, but usually the erection subsides quickly. There is great variability in the length of the refractory period both within and between individual males, and it may last anywhere from a few minutes to many hours. For most males, this interval usually gets longer with each repeated ejaculation within a time span of several hours. In addition, as a man gets older, the refractory period gets longer. In 1978, Robbins and Jensen reported on 13 men (only one of whom they studied in the laboratory) who said they had multiple orgasms by withholding ejaculation, but their claims have not yet been fully substantiated. However, it does appear that at least a few men have the capacity to have multiple orgasms before a true refractory period sets in, although it should be stressed that this does not happen once ejaculation has occurred.

The period of return to the unaroused state is called the resolution phase. In this phase, which includes the refractory period in men, the anatomic and physiologic changes that occurred during the excitement and plateau phase reverse. In females, the orgasmic platform disappears as the muscular contractions of orgasm pump blood away from these tissues. The uterus moves back into its resting position, the color changes of the labia disappear, the vagina begins to shorten in both width and length, and the clitoris returns to its usual size and position (Figure 11). If the breasts enlarged earlier in the response cycle, they decrease in size at this time, and their areolar tissue flattens out faster than the nipples themselves, giving the impression that the nipples are again erect (Figure 14). Stimulation of the clitoris, the nipples, or the vagina may be unpleasant or irritating during the post-orgasmic phase.

In males, erection diminishes in two stages. First, as a result of orgasmic contractions that pump blood out of the penis, there is a partial loss of erection. In the slower second stage of this process, genital blood flow returns to baseline (unaroused) patterns. The testes decrease in size and descend into the scrotum, moving away from the body, unless sexual stimulation is continued (Figure 12).

As both men and women return to their unaroused state, the "sex flush" disappears and prominent sweating is sometimes noticeable. A fast, heavy breathing pattern may be present just after orgasm, accompanied by a fast heartbeat, but both recede gradually as the entire body relaxes.

If there has been considerable excitement but orgasm has not occurred, resolution takes a longer time. Although certain changes occur quickly (such as disappearance of the orgasmic platform in women and the erection in men), there is sometimes a lingering sensation of pelvic heaviness or aching that is due to continued vasocongestion. This may create a condition of some discomfort, particularly if high levels of arousal were prolonged. Testicular aching ("blue balls") in men and pelvic congestion in women may be relieved by orgasms that occur during sleep or by masturbation. Although *nocturnal emissions* ("wet dreams") in young males are well known, females also can experience orgasm during sleep.

COMMON MYTHS ABOUT SEXUAL RESPONSE

In the preceding chapter, we discussed a number of misconceptions related to sexual anatomy and sexual satisfaction (e.g., a bigger penis provides more stimulation to the female during intercourse). In light of the physiologic responses just considered, we can now debunk some other common myths about sex.

One commonly held belief is that males have a greater sexual capacity than females. The reverse is actually true. From the viewpoint of physical capability, females have an almost unlimited orgasmic potential, while men, because of the refractory period, are unable to have a rapid series of ejaculations. (While women do not have a true refractory period, orgasmic potential is undoubtedly restricted by fatigue. There may be other physiologic limitations not known at present.) Many males also find it difficult to obtain another erection shortly after ejaculation. From a mechanical perspective then, their capacity to participate in repeated intercourse usually does not match that of the opposite sex.

Another misconception about sexual response is that the male can *always* tell if his female partner had an orgasm. At times, the male may be unaware of his partner's orgasm because he is caught up in his own feelings of arousal or because he doesn't recognize the physical signs of female orgasm. This may occur either because his expectations are inaccurate, because he doesn't know what to expect, or because the accuracy of his sense of vision is lessened during high levels of sexual excitation. Some males are fooled by a partner who "fakes" orgasm by means of loud moans and groans, intense pelvic thrusting, heavy breathing, and voluntary contractions of the outer portion of the vagina.

The notion that all orgasms are intense, earth-shattering, explosive events is another widespread sexual misconception that can probably be traced to the literary imagination. Although the reflex mechanisms of orgasmic response are fairly uniform, some orgasms are mild, fluttery, or warm, while others are blockbusters. These differences arise from variations in the person's physical state such as being tired, tense, having a sore throat or headache, or from variations in the emotions that accompany the sexual experience. The sensations of any physiologic process — drinking a glass of water, eating a meal, breathing, urinating, or sex — vary in different times and circumstances.

In the 1950s, the idea that "mutual orgasm" (both partners experiencing orgasm at the same time) was the ultimate peak in sexual pleasure became popular and was advocated enthusiastically in numerous marriage manuals. Many people tried to "fine tune" the timing of their responses, but working at sex usually resulted in a loss of spontaneity and fun. While mutual orgasm can be exhilarating, each person can be so wrapped up in his or her own response that the experience of the partner's orgasm is missed.

HORMONAL REGULATION OF SEXUAL FUNCTION AND BEHAVIOR

The physiologic processes of sex are not only vascular and neuromuscular. An important part of sexual physiology is under the control of the endocrine system, which consists of ductless glands that produce chemical substances called *hormones*. Hormones are secreted directly into the bloodstream, where they are carried to tissues on which they act. Some hormones, such as cortisol (made in the adrenal glands, which lie just above the kidneys), are necessary for life itself and influence a wide range of body functions. Other hormones are required for reproduction or sexual development. Here, we will consider the hormones that influence our sexual function.

The most important hormone in sexual function is *testosterone*. This hormone, sometimes called the male sex hormone, is actually present in both sexes. In a normal man, 6 to 8 mg of testosterone are produced per day, with more than 95 percent manufactured in the testes and the remainder in the adrenal glands. In a woman, approximately 0.5 mg of testosterone is made daily in the ovaries and the adrenals.

Testosterone is the principal biologic determinant of the sex drive in both men and women. Deficiencies of testosterone may cause a drop in sexual desire, and excessive testosterone may heighten sexual interest. In men, too little testosterone may cause difficulty obtaining or maintaining erections, but it is not certain whether testosterone deficiencies interfere with female sexual functioning apart from reducing sexual desire. However, there is no evidence whatsoever to suggest that because women have less testosterone than men, they have lower sexual interest. Instead, it seems that men and women have different levels of behavioral sensitivity to the effects of this hormone, with women actually being more sensitive to small quantities in their circulation.

Estrogens, sometimes called female hormones, are also present in both sexes and are made primarily in the ovaries in women and in the testes in men. In women, they are important from a sexual viewpoint in maintaining the condition of the vaginal lining and in producing vaginal lubrication. Estrogens also help to preserve the texture and function of the female breasts and the elasticity of the vagina. In men, estrogens have no known function. It does not seem that estrogens are important determinants of female sexual interest or capacity, since surgical removal of the ovaries does not reduce the sex drive in women nor lessen sexual responsivity. Too much estrogen in males, however, dramatically reduces the sexual appetite and can cause difficulties with erection and enlargement of the breasts.

Progesterone, a hormone structurally related to both the estrogens and testoterone, is also present in both sexes. The effects of progesterone on sexual behavior and function have been studied primarily in animals, where it appears that large amounts suppress sexual interest. Some authorities speculate that it may also act as a sexual inhibitor in humans.

Two decades ago, it was thought that the "master gland" of endocrine function was the *pituitary,* an acorn-sized structure lying beneath the brain. It is now clear that the regulatory role of the pituitary is more like a relay station and that a portion of the brain itself — the *hypothalamus* — has primary control over most endocrine pathways.

The hypothalamus produces a substance called *gonadotropin releasing hormone* (GnRH) that controls the secretion of two hormones made in the pituitary gland that act on the gonads (ovaries and testes). *Luteinizing hormone* (LH) stimulates the Leydig cells in the testes to manufacture testosterone; in the female, LH serves as the trigger for ovulation (release of an egg from the ovary). *Follicle stimulating hormone* (FSH) stimulates the production of sperm cells in the testes; in the female, FSH prepares the ovary for ovulation.

The hypothalamus acts much like a thermostat in regulating hormonal function (Figure 15). Instead of reacting to temperature, as a thermostat does, the hypothalamus reacts to the concentrations of hormones in its own blood supply. For example, in adult males the amount of testosterone "registers" in the hypothalamus. If the amount is high, production of GnRH is turned off, leading to a drop in LH secretion by the pituitary. The decrease in LH in the bloodstream quickly results in re-

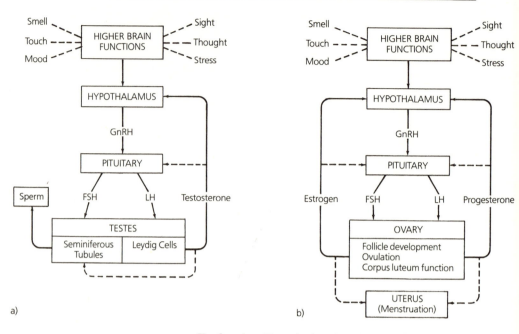

FIGURE 15.   Endocrine Regulation in the Adult
Male (a) and Female (b)

duced production of testosterone in the testes, and therefore
lower amounts of testosterone are secreted into the blood. When
the amount of testosterone reaching the hypothalamus drops
below a certain level, it triggers the secretion of GnRH into the
pituitary. The pituitary responds to this signal by sending more
LH into the circulation, where it will soon reach the testes and
cause an increased rate of testosterone production.

It is tempting to try to understand sexual behavior in terms of
hormones. In many animal species, patterns of sexual interaction are tightly regulated by hormonal events, which control
both the sexual receptivity of the female (her willingness to
mate) and the sexual interest (courtship behavior) of the male,
as well as male mounting and penile thrusting. Testosterone
and estrogen have been found in mammals, birds, reptiles, am-
phibians, and fish; in all of these groups, actions of sex hormones on the brain appear to be important determinants of sexual
behavior.

In humans, however, there is a more complicated relationship between hormones and sexual behavior. Although a marked testosterone deficiency usually reduces sexual interest in men or women, there are cases where this effect is not seen. Similarly, although many men with subnormal testosterone levels have difficulty with erection, others continue to have completely normal sexual function. Women who have low amounts of estrogen in their bodies do not usually lose their ability to be sexually aroused or to have orgasms. People's sex hormone levels do not "predict" their sexual behavior or interest.

MENSTRUATION

Menstruation is a flow of blood that occurs about once a month in most women from approximately ages twelve to forty-eight. Although menstruation is a normal part of the female reproductive cycle, it is a subject of considerable misunderstanding and taboo. In ancient times, a menstruating woman was regarded as unclean and liable to pollute foods she handled, or as contagious and liable to cause illness or even death in others. In the modern era, menstruation is sometimes seen as a physical and emotional handicap that makes women "inferior" to men. Whether menstruation is called "the curse," "the monthlies," or "being on the rag," it is often referred to in negative terms.

The menstrual cycle is traditionally described as starting with the first day of menstrual flow (cycle day one) and ending the day before the next menstruation begins. The length of the cycle varies, normally ranging from twenty-one to forty days and averaging about twenty-eight days. Very few women are so regular that they can accurately and consistently predict the length of each cycle on the basis of their past pattern of cycle lengths.

The menstrual cycle consists of three phases, which we will describe in terms of an "average" twenty-eight-day cycle. The *follicular phase* is the first portion of the menstrual cycle. Ovarian follicles, oval arrangements of cells around a young egg, begin to mature as FSH stimulates them. At the start of this phase, estrogen and progesterone levels are quite low and the uterus sheds its lining, resulting in three to six days of menstrual flow.

Menstrual flow consists of a small amount of blood combined with tiny bits of tissue from the lining of the uterus, and the entire amount is usually only 2 or 3 ounces (4 to 6 tablespoons).

Midway in the follicular phase (around cycle days seven to ten), estrogen output from the ovaries increases, which acts with FSH to prepare the developing follicle for ovulation. Estrogen also causes the lining of the uterus to thicken, or proliferate, because of growth of glands, connective tissue, and blood vessels. Just before ovulation, estrogen levels reach a broad peak, which acts on the hypothalamus to trigger the surge of LH and FSH released from the pituitary a day or two afterward.

Ovulation, release of the egg from the ovary, typically occurs at about the fourteenth cycle day in most twenty-eight-day cycles. However, we have studied cycles in which ovulation ranged from the ninth day to the nineteenth day of a cycle of this length, and in some menstrual cycles, ovulation does not occur at all. Ovulation usually follows the LH peak by twelve to twenty-four hours. The ovulatory phase is the shortest phase of the menstrual cycle.

The third portion of the menstrual cycle is the *luteal phase,* which encompasses the time from immediately after ovulation until the start of the next cycle. This phase is named for the corpus luteum, the mass of cells left in the ovary after the follicle ruptures during ovulation. The corpus luteum produces large amounts of progesterone and estrogen, leading to increased levels of these hormones in this portion of the menstrual cycle. The progesterone causes the small blood vessels in the thickened endometrium to develop and produces coiling in the endometrial glands, changes that prepare the uterus to receive a fertilized egg if pregnancy occurs. The progesterone also registers in the hypothalamus, where it shuts off output of GnRH, resulting in a rapid decline in LH and FSH from the pituitary. Unless the egg is fertilized, the corpus luteum degenerates ten to twelve days after ovulation and its hormone production drops drastically. The next menstrual flow occurs as the lining of the uterus sheds in preparation for regrowth in the next cycle. Menstruation is thus the result of abrupt withdrawal of hormone stimulation.

As part of our evolutionary heritage, there is some basis for assuming that women should be most interested in sex around the time of ovulation, when pregnancy can occur. In most animal species, the female follows this pattern of behavior rather exclusively, avoiding sexual contact at all other times. Some studies document a pattern of increased sexual activity around the presumed time of ovulation and do not find such a "peak" in women using birth control pills, which block ovulation, but these findings appear to have been a result of imprecise determination of when ovulation occurred. Other reports find no evidence of heightened female sexual interest during ovulation.

A recent well-designed study by Schreiner-Engel and co-workers (1981) found objective laboratory evidence of higher levels of vaginal vasocongestion during the follicular and luteal phases of the menstrual cycle. Subjective reports of sexual arousal were also higher at these times than during the ovulatory phase, providing further confirmation that an ovulatory peak in female sexual responsivity is unlikely. Another recent study confirmed that female sexual activity and sexual interest peak in the follicular phase well before ovulation. Nevertheless, individuals differ in unique ways, so that some women may well find themselves feeling sexier at mid-cycle while others are most interested in sex earlier or later in their cycles.

Taboos about sexual intercourse during menstruation are still a part of everyday life for many people. In some cases, avoiding sexual contact is a matter of religious practice. For example, Orthodox Jews are supposed to abstain from sex for seven days after the end of menstrual flow, and sex is resumed only after the woman has immersed herself in a ritual bath, called the *mikvah*. In other cases, abstinence seems to stem from cultural and psychological sources, as Delaney, Lupton, and Toth note: "A man is as likely to be sexually aroused by a woman when she is menstruating as he is at any other time. But the blood of the menstruating woman is somehow dangerous, magical, and apparently not something he wants to get on his penis." The following comments illustrate the broad range of feelings people have about sexual activity during menstruation:

> I often find that I feel sexiest when I'm having my period, so making love is particularly enjoyable then.

> I somehow feel like it's not right to have sex when my girlfriend is menstruating. I don't know why, but I just feel funny about it.

Jill and I love to have sex during her periods. It's a particularly passionate time because we don't need to use any birth control then.

In a word, I'm embarrassed about it. I feel like I'm not completely clean when I'm flowing, and tampons make me dry inside, so I really prefer waiting until my period is done.

The notion that sexual activity, including intercourse, is "dangerous" to either partner during menstruation has no basis in fact. Yet some people feel that intercourse is messy during menstruation and restrict their sexual experiences to noncoital options. It appears as though attitudes toward sex during menstruation are beginning to change, however, since younger people seem to be less affected by the negative attitudes concerning such activity than their parents' generation was.

CHAPTER FOUR

Birth Control

Young American women are now delaying childbirth until later ages than their mothers did, but in many cases this delay is achieved by aborting an unintended pregnancy — almost two-thirds of the million and a half abortions in the United States in 1980 were obtained by women under age 25. (U.S. Bureau of the Census, 1983, Henshaw and O'Reilly, 1983).

Currently, about 4.9 million couples of reproductive age in the United States (about 9 percent of couples in this age group) rely on vasectomy for contraception. Nevertheless, there are few reversible methods of contraception available for use by males, so that females are generally left to shoulder the burden of contraceptive responsibility.

If current rates of world population growth continue for the next five years, the world population will increase by about 375 million people, which is about the same as the current population of all the cities of Latin America and Africa combined.

AS THESE facts show, methods for preventing unwanted or unintended pregnancy are either not used, misused, or used sporadically by many people. Those who rely on "faith" and "hope" as their primary contraceptives are playing a risky form of reproductive roulette, as many discover to their surprise. Others depend on unreliable contraceptive techniques like douching, using astrological charts, or using tampons. Contrary to common misbelief, female orgasm is not a requirement for

TABLE 1
Contraceptive Methods Used by Sexually Active
Women Aged 15–44 in the United States

Method	Percent
Pill	27
Sterilization, female	19
Sterilization, male	13
Condom	12
IUD	6
Diaphragm	5
Spermicide	4
Withdrawal	3
Rhythm	2
Other	less than 0.5
None	8

Source: Modified from Forrest & Henshaw, in *Family Planning Perspectives*, vol. 15, 1983, Table 5.

conception, and making love while standing up or crossing your fingers doesn't prevent sperm from reaching the egg.

Not too long ago, contraceptive products could not be displayed or sold in certain states, abortions were illegal, and birth control methods were not very reliable. Although times have changed remarkably, people are now faced with a vast number of birth control products or methods, each with its own advantages and disadvantages (Table 1). In trying to make a personal choice, people may have a number of questions. Why use contraception? How does each method work? How well does it work? What are the medical risks? How might each method affect my sexuality? What possibilities do I have if I'm already pregnant? In the following pages, we will examine the currently available methods of birth control in an effort to answer some of these questions.

WHY USE CONTRACEPTION?

An individual's or couple's decisions about birth control can depend on many factors, including age, future plans, marital or relationship status (including trust and cooperation), finances, religious beliefs, sexual attitudes, health, and prior experiences.

Not using birth control if you are sexually active is a specific kind of personal decision, just as choosing to use birth control, whatever the reason, is a personal decision.

The main reason for using birth control is to prevent an unwanted pregnancy. Not only is an unwanted pregnancy likely to cause emotional turmoil and health risks, it also may present financial burdens. Often, unwanted pregnancies occur in young teenagers or women over thirty-five, times when health risks during pregnancy are highest. The social and economic costs may also be high at these same ages, as these two quotations show:

> *A twenty-two-year-old woman:* I got pregnant when I was fifteen and had my baby the day before my sixteenth birthday. My parents wanted me to finish school, and they took care of my son for a while when he was little, but then my father died and I had to drop out to go to work. Now, it doesn't look like I'll ever marry — who'd want to have me?

> *A thirty-seven-year-old woman:* I'd been married for fifteen years when I got pregnant again at thirty-five. My two other kids were fourteen and twelve, and I was finishing a two-year course to be a court stenographer. I had to change my plans completely and become a mother again, and it was no cup of tea, believe me. Now my husband has filed for divorce, and I'm sure the baby was part of the cause.

Of the 1.2 million teenagers who become pregnant in the United States each year, more than 400,000 obtain abortions. Many others drop out of school or enter into hasty marriages where the odds of divorce are high, the chances of getting a good job are low, and ending up on welfare is common. Others try to raise a child alone or with the assistance of relatives, but this plan often proves more difficult than it might at first seem. An unplanned pregnancy at any age may also disrupt career plans, and economic costs are created by terminating the pregnancy or raising the child. There is also an emotional cost to an unwanted pregnancy. Feelings of foolishness, guilt, anger, or helplessness may strain or break a relationship ("It was *all your fault!*"), or may create later sexual problems.

Of course, there are other reasons for using contraception,

including the wish to space pregnancies, limit family size, avoid potential genetic disorders or birth defects, protect the mother's health, and allow women more control over planning their lives. Contraception also permits people to enjoy a sexual relationship without making commitments to marriage or parenthood. Limiting reproduction also has major social and philosophical consequences in a world of limited natural resources where overpopulation exerts political and psychological effects and environmental issues are of prominent concern.

EVALUATING CONTRACEPTIVE EFFECTIVENESS AND SAFETY

The decision to use contraception and the choice of one method over another depend primarily on two practical matters: how well it works (its effectiveness) and its health risks (its safety).* Evaluating these two issues is complicated. *No one contraceptive method is always best or safest.*

When evaluating effectiveness and safety, remember that information from various sources may be biased. The popular media, for instance, are eager to report the latest "news" about the real or suspected hazards of a contraceptive method. Yet the story is usually condensed to a few paragraphs in the newspaper or is crammed into less than sixty seconds of TV or radio broadcast time. Scientific accuracy or caution is often lost in a process of oversimplification, misinterpretation, and unwarranted conclusions. In addition, much of the research on the effectiveness and safety of birth control methods is paid for by the drug companies that manufacture them. These companies have an obvious interest in presenting their merchandise in a way that will boost sales. Finally, all scientific studies are not equivalent in their applicability to *you*. In general, studies about people close to your age, cultural background, and socioeconomic status are more meaningful than studies about other groups. For example, if you are a twenty-two-year-old single American woman, you

* The choice of method may also depend, in individual cases, on factors such as cost, availability without a prescription, aesthetic preferences, and sometimes even the ease of concealment.

cannot put much faith in the findings of a study about thirty-five-year-old married women in Lapland.

Understanding some other aspects of evaluating effectiveness can also be helpful. First, it is important to distinguish between two factors: theoretical versus actual effectiveness. The *theoretical effectiveness* of a particular method is how it *should* work if used correctly and consistently, without human error or negligence. The *actual effectiveness* is what occurs in real life, when inconsistent use or improper technique ("user failure") combines with failures of the method alone ("method failures"). For example, if a couple runs out of condoms on a week-long camping trip yet continues to have intercourse, the woman's subsequent pregnancy is not counted as a method failure. But if she conceives even though each time she had intercourse, she has used a contraceptive foam exactly according to instructions, her pregnancy qualifies as a method failure.

Second, for most types of contraception, the longer a person uses a particular method, the more effective it becomes. The reason is that people improve their technique and become more accustomed to using the method regularly.

Third, effectiveness rates for almost every nonsurgical contraceptive method vary depending on whether a couple uses the method to *prevent* pregnancy or to *delay* (space) pregnancy. Failure rates are generally 50 to 100 percent higher for delay compared to prevention, since there seems to be less consistency in method use.

There are other difficulties in assessing the safety of contraceptive methods. First, there are often wide differences in the frequency of side effects reported by different investigators. Their results reflect differences in research design, choice of control groups, different characteristics in the populations studied (such as age, health, socioeconomic status), and the methods investigators use to identify a problem (self-administered questionnaire, personal interview, laboratory testing). Second, there are some relative aspects to the safety question. How important is avoiding pregnancy? Are the side effects of a contraceptive method more or less serious than the risks of pregnancy and childbirth? How do the risks of a contraceptive method compare to other health risks (such as the risk of getting cancer or having high blood pressure) or to risks of everyday life? These questions will be addressed in more detail as we review the safety and side effects of each method of contraception.

METHODS OF CONTRACEPTION

Birth Control Pills
(Oral Contraceptives)

The introduction of birth control pills in 1960 revolutionized contraceptive practices around the world. Millions of women turned enthusiastically to this convenient and effective method of preventing pregnancy, but within a decade reports of serious side effects of the pill began to appear and the popularity of this method declined substantially. Now, after twenty-five years of observation, what are the facts about the pill?

There are two types of oral contraceptives currently in use: a *combination pill,* which contains a synthetic estrogen and a progesterone-like synthetic substance called "progestogen," and a *minipill* with progestogen only in low dosage. This discussion will focus on combination pills (unless otherwise specified) because they are most commonly used.

Birth control pills prevent pregnancy primarily by blocking the normal cyclic output of FSH and LH by the pituitary, thus preventing ovulation. In addition, the progestogen makes implantation difficult by inhibiting the development of the lining of the uterus, and it also thickens the cervical mucus, decreasing the possibility that sperm can get through.

Birth control pills are taken one per day for twenty-one days beginning on the fifth day of the menstrual cycle (that is, four days after a period begins). Some brands of birth control pills are packaged with seven inactive pills (usually of another color) which the woman takes on a daily basis to complete the cycle, whereas with other brands the woman must remember to resume her pills one week later.

If a pill is missed, *two* pills should be taken the next day. If two pills are missed, it is likely that the pill will not work properly and an *alternate* form of contraception must be used to prevent pregnancy.

Birth control pills are the most effective nonsurgical method of contraception. Among women who use the pill consistently, only one pregnancy will occur among 200 women during a year. Among all women using the pill — including those who sometimes forget to use it — two or three pregnancies will occur in 100 women annually.

The minipill, which is taken every day, even during menstruation, is less effective than the combination pill. If it is used perfectly, one or two pregnancies will occur in 100 women during a year. In actual use conditions — including the occasional forgotten pill — five to ten pregnancies occur in 100 women using the minipill for a year.

Birth control pills have now been used by more than 150 million women around the world and have probably been studied more intensively than any other medication in history. Despite scare stories that appear regularly in the press, the evidence shows that the pill has more health benefits than risks for many users. For example, there is no reliable evidence that birth control pills cause cancer, and they actually protect against cancer of the ovaries and of the uterine lining.* Furthermore, women who use birth control pills are only one-fourth as likely to develop benign (non-cancerous) breast tumors as non-users, one-fourteenth as likely to develop ovarian cysts, one-half as likely to develop rheumatoid arthritis, and two-thirds as likely to develop iron deficiency anemia.

Other beneficial side effects of the pill have been noted. Many women find that the pill reduces the amount of menstrual flow and produces more regular cycles with less menstrual cramping. Acne can be improved by the pill (although sometimes it is worsened), premenstrual tension can be reduced, and pelvic inflammatory disease — a serious cause of infertility — is only half as common in pill users as non-users.

The most common bothersome side effects of the pill mimic those encountered in pregnancy: nausea, constipation, breast tenderness, minor elevations in blood pressure, edema (swelling), and skin rashes (including brown spots on the face called

---

* Two points should be made: (1) an earlier type of birth control pill, the *sequential pill*, was banned because it *was* found to cause cancer of the uterus; (2) as it may take decades for certain types of cancer to develop, it cannot be concluded that the combination pill *does not* cause cancer. However, it has recently been estimated that more than 2,000 cases of endometrial cancer are prevented each year by use of the pill.

chloasma). Other relatively minor side effects include weight gain or loss, an increased amount of vaginal secretions, and an increased susceptibility to vaginal infections.

Less common but more troublesome side effects caused by the pill include high blood pressure (4 percent of users), diabetes (1 percent of users), migraine headaches and/or eye problems (0.5 percent of users), and, rarely, jaundice or liver tumors. Because the pill causes alterations in liver function, women with hepatitis or other forms of liver disease should avoid it. If taken during pregnancy, the pill may cause birth defects, but if used before pregnancy, the pill does not have this effect. Some reports have linked the pill to depression, while others have not.

The most serious risks to women using birth control pills are disorders of the circulatory system: the overall death rate from circulatory diseases is approximately four times higher in users than non-users. However, to put this finding in perspective, only one in 27,000 women using the pill annually dies from this cause.

There are three different types of circulatory problems involved. The most common, affecting about one in 1,000 users each year, is the formation of a blood clot in a vein (usually in the legs). Most often this clot results in only minor discomfort caused by inflammation and swelling, but there is a risk that a piece of the clot may break off and cause serious damage in the lungs or brain. These problems, called *thromboembolic diseases,* are two to four times more common among pill users than non-users but account for only two or three extra deaths per 100,000 users annually. Birth control pills also increase the risk of heart attacks, but the excess risk is almost entirely found in users who smoke and/or users older than thirty-five. Women in their thirties who smoke and use the pill are about seven times more likely to have a heart attack than users who do not smoke, but pill users who do not smoke are less at risk than smokers who do not use the pill. Finally, an association between pill use and strokes has also been found, but even with an increased risk, stroke is a rare disorder in women under forty-five.

A few final words about the safety of oral contraceptives are in order. Safety is a relative matter, not just a statistical equation. For example, a woman who has intercourse frequently receives more contraceptive protection from the pill than one who does not. While other contraceptive methods may seem "safer," she may choose to risk certain health problems to be more certain of not becoming pregnant. Another example applies to women in developing countries, where maternal death rates during pregnancy and childbirth are much higher than in countries with advanced medical care systems. In this situation, oral contraceptives are far safer than no birth control even over age thirty-five. The safety of birth control pills should also be looked at in comparison to risks of everyday living such as auto accidents or sports injuries. In this case the figures may seem somewhat different: only one pill-related death occurred in 46,000 women under thirty-five who were non-smokers in 200,000 *woman-years* (one woman-year is equivalent to a woman using a particular contraceptive method for 12 calendar months) of use of the pill. By contrast, nine deaths from auto accidents would be expected in a group of this size over this period of time. Confirming this view, the largest prospective American study of pill users found that overall, the risks of oral contraceptive use appear to be minimal.

Evaluating the effects of birth control pills on sexuality is tricky for a number of reasons. First of all, the pills used during the 1960s had much higher amounts of estrogen than today's pills have, and different effects seem to relate to the level of estrogen. Older research generally noted that long-term use of the pill had negative sexual effects, but more recent studies do not support this view. Second, there are many individual reasons why the pill helps or hinders a woman's sexual feelings and function. Some reasons are biological: a woman who develops a vaginal infection may have sexual problems as a result, or a woman who has less premenstrual and menstrual cramping may feel better physically and thus may be more sexually receptive and responsive. Some reasons are psychological. If a woman expects to have sexual side effects from the pill, her prophecy often becomes self-fulfilling. Some women may feel guilty about using the pill because of religious beliefs or conflicts between wanting children and using contraception. Alternatively, the psychological security of using a highly effective method of contraception reduces the fear of pregnancy and may therefore improve sexual interest and enjoyment (after all, it is harder to enjoy something when you are afraid). One woman told us, "After switching to the pill, I found that sex became more fun. I had my first orgasm during intercourse less than two months later." Since the pill also requires no interruptions at the time of sexual activity, it permits a degree of spontaneity and intimacy that other methods may not.

Most women using oral contraceptives today do not report significant changes in sexual interest, behavior, or enjoyment. Approximately 10 percent experience improved sexuality, which is about the same as the percent who note decreased sexual interest or responsiveness.

IUDs

The IUD, or *intrauterine device,* is a small plastic object that is inserted into the uterus through the vagina and cervix and then continuously kept in place. There are many different models of IUDs (Figure 16), which vary in shape, size, and composition. For example, some contain copper filaments (the Copper 7 and the Copper T) and some contain a hormone (the Progestasert T slowly releases a synthetic form of progesterone into the uterus).

While IUDs became popular in this country only in the last twenty years, they are a modern application of an ancient concept. For centuries, Arab and Turkish camel drivers put a large pebble into the camel's uterus for contraceptive purposes, since a pregnant camel on a long desert trek would not be very helpful. The first IUD designed for humans, a ring made of silkworm gut, did not gain much attention when introduced in 1909. In the late 1920s a ring of gut and silver wire was developed by Grafenberg, a German physician, and enjoyed some popularity. Early models of the IUD, however, were condemned because of the risk of pelvic infection and fell into disrepute.

By 1978 about 6 percent of married women of reproductive age in America used the IUD, while about 20 percent did in Scandinavia and approximately half of all women using contraceptives adopted it in China. On a worldwide basis, approximately 60 million women use the IUD.

The exact way the IUD works is not known. The most plausible explanation is that it interferes with implantation of the fertilized egg in the lining of the uterus. This outcome may result from a local inflammatory reaction or from interference with chemical reactions inside the uterus that affect implantation. IUDs containing progesterone also alter the development of the lining of the uterus so that implantation is unlikely.

The IUD must be inserted into the uterus (Figure 16) by a trained health care professional after determining that the woman is not pregnant and does not have gonorrhea or other pelvic infections. Inserting an IUD can cause abortion in a pregnant woman and can push bacteria from an infection into the uterus or Fallopian tubes. Insertion is usually done during a menstrual period, since it is a fairly reliable sign that the woman is not pregnant, but it can be done at other times.

Inserting an IUD usually causes only brief discomfort, but some women prefer to be given a short-acting painkiller. Taking two or three aspirin tablets about an hour before insertion provides some pain relief and may theoretically decrease cramping because it blocks release of prostaglandins. The woman must be shown how to check the plastic thread that comes through the mouth of the cervix to be sure the IUD is in place. If this thread cannot be located, or if it seems longer than it was before, the woman must return for a checkup.

IUDs are sometimes expelled from the uterus so that they no longer provide contraceptive protection. Expulsion rates are lower in women who use copper- or progesterone-containing devices (about six or seven per 100 women during a year) compared to a rate of approximately fifteen per 100 women during a year for other IUDs. Expulsion rates are higher in younger women, in women who have had no children, and during menstruation. One-fifth of expulsions go unnoticed, and this accounts for one-third of the pregnancies among IUD users.

Even though they may be expelled, IUDs are highly effective, with only one to six pregnancies occurring in 100 women using IUDs for one year. One study showed that women using IUDs to *prevent* pregnancy had a failure rate of 2.9 percent, whereas women using IUDs to delay or space pregnancy had a failure rate of 5.6 percent.

Perforation of the uterus (puncturing the wall of the uterus) is the most serious risk of using an IUD. If it occurs, it is almost

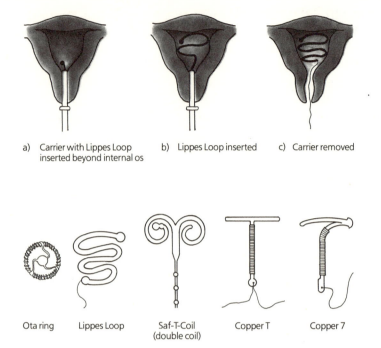

a) Carrier with Lippes Loop   b) Lippes Loop inserted   c) Carrier removed
    inserted beyond internal os

Ota ring   Lippes Loop   Saf-T-Coil   Copper T   Copper 7
                                       (double coil)

FIGURE 16.   The IUD
Insertion of an IUD (a–c) and several models of IUDs currently in use.

always at the time of insertion. This problem, which occurs in about one in 1,000 insertions, can cause sudden pain and bleeding or, more rarely, may be without immediate symptoms. Perforation usually requires surgery to prevent damage to the intestines, since the IUD usually perforates through the uterus into the abdomen. Recently, it has been found that inserting an IUD while a woman is breast-feeding causes a tenfold increase in the risk of perforation of the uterus, so it seems advisable to choose a different method of birth control during this time.

The most common side effects of the IUD are increased bleeding from the uterus and cramping pain. Menstrual periods are typically heavier and longer (except in women using IUDs with progesterone, which decreases menstrual flow), and there is more likely to be spotting between periods. This increased blood loss can cause anemia. Cramps and bleeding can be serious enough to have the IUD removed (about 10 percent of women find this necessary) but usually lessen after the first three months.

IUD users have about a four times higher risk of pelvic infection than other women. This risk is present not only at the time of insertion but also with continued use of the device. Bacteria, viruses, or fungus infections may enter the uterus by "climbing" the tail of the IUD, and there is a chance that the infection will spread to the Fallopian tubes and ovaries, which can cause permanent scarring and infertility.

If a pregnancy occurs in a woman with an IUD, there is about a 5 percent chance that the pregnancy is ectopic or misplaced. This poses a major threat to the woman's health if not properly detected. Uterine pregnancies in IUD users have a 30 percent risk of spontaneous abortion if the IUD is removed, compared to a 50 percent abortion risk if the device is left in place. IUDs, however, do not cause any birth defects.

The IUD is well suited to women who want a highly effective contraceptive method that requires no active participation on their part and is easily and promptly reversible. It may be particularly appropriate for women who cannot use birth control pills for medical reasons and for the mentally retarded, who may not be able to remember to take a pill or use a mechanical method of birth control. IUDs should not be used by women who are pregnant or who have active pelvic infection, bleeding disorders, anatomical abnormalities of the uterus or cervix, or abnormal pelvic bleeding. Women who have had an ectopic pregnancy, severe menstrual disorders, or any disease that suppresses normal immunity to infection should not use IUDs.

Like birth control pills, the IUD is effective in preventing pregnancy and does not interfere with sexual spontaneity or mood. Yet the IUD can cause pain during intercourse. The woman may experience pain if the IUD is not in the right position or if there is pelvic infection or inflammation; the man may have pain at the tip of the penis or along the shaft because of irritation from the tail of the IUD in the vagina. Both abdominal cramping and persistent bleeding can lessen a woman's interest in sexual activity, but these side effects occur in only about 10 percent of users. Infrequently, women who use IUDs find that orgasm may cause intense, unpleasant cramping because of uterine contractions around the IUD.

Diaphragms

The diaphragm is a round, shallow dome of thin rubber stretched over a flexible ring. After a spermicidal (sperm-killing) jelly or cream is applied inside the dome and around the inner part of the rim, the diaphragm must be inserted inside the vagina and positioned so that it completely covers the cervix. Before the introduction of birth control pills, the diaphragm was the most widely used method of contraception, and it is still popular today.

The diaphragm is a mechanical barrier that blocks the mouth of the cervix so that sperm cannot enter. Because this blocking effect is not very reliable by itself, the use of a spermicide is required to kill sperm that manage to swim inside the rim of the diaphragm.

Diaphragms come in different sizes and must be properly fitted by a trained health care professional to match the anatomy of the user. The size and position of the cervix and the size and shape of the vagina must be taken into account to achieve a proper fit; the actual fitting is done using a set of graduated flexible rings to select the right size. After the fitting, the woman is shown how to insert the diaphragm either manually or with a plastic inserter. Some clinicians recommend that a woman be refitted if she gains or loses ten pounds. However, a recent study found that more women whose weight was stable needed refitting (Fiscella, 1982), so the matter is not a clear one. We recommend that women be refitted for a diaphragm after a pregnancy or on an annual basis. If there is a weight change of 15 pounds or more, it also may be advisable to be refitted.

The diaphragm can be inserted up to two hours before intercourse and should remain in place at least six hours afterward. If it is worn for more than two hours before intercourse, the effectiveness of the spermicide may drop. For this reason, a full application of spermicidal jelly, cream, or foam should be placed in the vagina before intercourse in these circumstances. If the diaphragm is removed less than six hours after intercourse, it is possible that live sperm in the vagina may reach the cervix and swim up into the uterus.

The effectiveness of this method is not as high as birth control pills or IUDs, with failure rates in most reports ranging from six to twenty pregnancies in 100 women using this device during a year. Yet since the effectiveness of this method depends on the regularity of its use, the woman's motivation and memory are important factors. A woman who leaves her diaphragm at home and finds herself unexpectedly in a sexual situation might have a "user" failure rather than a "method" failure. In one study with a highly motivated, experienced population of users, the failure rate was only 1.9 per 100 woman-years. Another problem that influences effectiveness is that the diaphragm may slip out of position during sexual play because of improper insertion, a poor fit, or expansion of the inside of the vagina and movement of the uterus during sexual excitation. Even a well-fitted and properly inserted diaphragm can become dislodged during intercourse when the woman is on top of the man or if the penis is removed and reinserted into the vagina during plateau levels of arousal. Even a tiny hole in the dome of the diaphragm will permit sperm to enter, so it is important to inspect it carefully and to avoid use of Vaseline on it, since this can cause deterioration of the latex. Despite these problems, consistent use of the diaphragm with a condom provides a rate of effectiveness comparable to birth control pills. In addition, the diaphragm seems to offer protection against cervical cancer and against some sexually transmitted diseases, although recent reports suggest that women who use diaphragms have an increased rate of urinary tract infections.

The only potential side effects are: (1) possible allergic reactions to the rubber in the diaphragm or to the spermicide itself, and (2) the chance of introducing infection into the vagina if the diaphragm is not clean. Both problems are infrequent. In addition, it is not advisable to wear a diaphragm for more than twelve hours at a time because this seems to cause an overgrowth of bacteria in the vagina and cervix that is linked to the toxic shock syndrome (see chapter 19). The diaphragm has no effects on hormones or physical processes of the body and poses no danger to later fertility.

Women who have pelvic disorders affecting the vagina or cervix or who do not like to touch their genitals should not use the diaphragm.

The primary sexual difficulty with use of the diaphragm is inconvenience. For example, either partner may experience loss of sexual arousal while the woman takes time to insert and check the diaphragm if she did not insert it before sex play began. Some couples solve this problem by making the insertion of the diaphragm a part of their sexual play preliminary to intercourse. Either partner may find the diaphragm or the process of its insertion unaesthetic. In addition, a diaphragm that is too large may cause pain during intercourse and some men complain that intercourse feels "different" with the diaphragm in place. There is a sexual advantage, however: during menstrual periods, the diaphragm can be used to provide a "reverse barrier" that contains flow during sexual activity.

The Cervical Cap

A device related to the diaphragm is the *cervical cap,* which fits snugly over the cervix and stays in place by suction. This method of birth control is now enjoying an upsurge of interest in the United States, although it has had greater popularity in England in the last twenty years. The cervical cap can be worn either for short periods like the diaphragm or for several weeks at a time. Currently, a custom-molded cervical cap that contains a small valve that permits the release of menstrual flow and cervical secretions is being studied; if found to be safe and effective, this type of cap would have the advantage of being able to be worn continuously for months at a time without having to be removed.

The effectiveness of the cervical cap has not been extensively studied but it seems to be somewhat lower than the diaphragm. In one recent study of women at a university clinic, the pregnancy rate was 19.6 per 100 woman-years. In some cases, the cap has reportedly caused discomfort to the male during intercourse and has dislodged from the cervix during intercourse.

Condoms

The condom (also called a rubber, safe, or prophylactic) is a thin sheath of latex rubber or tissue from a lamb's intestine that fits snugly over the penis. Currently, it is the only effective non-surgical birth control method for males and it does not require a prescription. Condoms can be purchased in drugstores, by mail order, at family planning clinics, and, in some locations, from coin-operated vending machines in men's rooms of bars and gas stations.

Condoms are usually rolled up into plastic or foil packets and come in many different styles. Some are lubricated, some come in a variety of colors, and some have tiny ribs or bumps on their surface that supposedly provide more stimulation to the female as intravaginal thrusting occurs. One of the more recent innovations in condom design is the introduction of a condom coated with a spermicide combined with a lubricant. Condoms are designed with either a round end or a reservoir end, a small receptacle to catch the semen.

The condom has been relatively ignored as a method of contraception for a number of reasons:

> Physicians rarely recommend the condom because its use requires no medical expertise; researchers rarely undertake to study or improve it because they see no challenge in such a simple device; and the media rarely accept advertisements or even provide news coverage because they are still reluctant to associate contraception with sexual intercourse. (Dumm, Piotrow, and Dalsimer, 1974)

The condom, which works by preventing sperm from entering the vagina, must be unrolled onto the erect penis shortly before intercourse. If it is put on too early, it may tear if rubbed against the sheets (or dirt, sand, or car seat) or it may be accidentally punctured by a fingernail. If it is put on shortly before ejaculation, some drops of pre-ejaculatory fluid containing live sperm might have already entered the vagina. If a round-end condom is used, a little extra space should be left at the tip to catch the ejaculate. After ejaculation, the condom should be held at the base of the penis so it does not slip or spill while being withdrawn. If leakage occurs for any reason it is wise to put a spermicidal foam or cream in the vagina immediately. A condom should not be tested before use by inflating, stretching, or being filled with water. It is impossible to detect microscopic holes that would be large enough to allow sperm to pass through, and you *may* inadvertently damage the condom while you are trying to check it.

In theory, the condom is a very reliable method of birth control — when used properly and consistently. Note that condoms are *not* reusable, and homemade condoms of plastic wrap *do not work!* Fewer than five pregnancies occur in 100 couples using condoms for a year. Furthermore, if condoms are consistently used in combination with a vaginal spermicide, the effectiveness is even better, becoming almost as good as the pill. In actual practice, however, couples using the condom as their only method of contraception often have unwanted pregnancies — overall failure rates of ten to twenty per 100 couple-years are commonly reported. The main problem is not defective condoms (they must meet rigorous testing standards of the Food and Drug Administration) or improper use, but *inconsistent* use. Many people using this method sometimes "take chances" instead of using condoms during every act of intercourse.

One other aspect of condom use that is increasingly important today is that it provides considerable protection against many sexually transmitted diseases. Very infrequently, the material the condom is made from can cause burning or irritation to the genitals. Otherwise, condoms have no health risks at all.

There are several sexual disadvantages to using condoms. Putting the condom on may interrupt sexual spontaneity, although some couples make this a shared moment of sexual play. Many men complain of lessened sensations in the penis while wearing a condom (one student told his class "It's like playing the piano wearing mittens"), and some men have difficulty maintaining an erection while trying to put on the device. The condom is a poor contraceptive choice for a man having problems with erection, since it calls attention to the degree of erection and may seriously increase "performance anxiety" (see chapter 18). Also, unless the penis is removed from the vagina soon after ejaculation, loss of the erection makes spillage of semen more likely. This can certainly interfere with the intimacy and mood of the moment.

There are also some sexual advantages to the condom. The

female may be pleased because her partner is not leaving the responsibility for birth control to her. Lubricated condoms can make intercourse more comfortable if vaginal lubrication is a problem. There is little or no postcoital drippiness for the woman whose partner uses condoms. And use of a condom may help many men who have difficulty controlling ejaculation.

Spermicides

Spermicides, or vaginal chemical contraceptives, come in many varieties, including foams, jellies, creams, and tablets or suppositories. They do not require a prescription and are available at drugstores and family planning clinics. Spermicides should not be confused with feminine hygiene products, often displayed beside them, which have no spermicidal effectiveness.

Spermicides work in two ways: their active chemical ingredient kills sperm, while the material containing this ingredient provides a mechanical barrier that blocks the entrance to the cervix. Because of the individual differences in the many products on the market, the manufacturer's instructions regarding use and effectiveness must be consulted. This information is given in the package insert that comes with each product. *These products are not identical in how they are used or in their reliability in preventing pregnancy.* In general, foams and suppositories are much more effective than creams or jellies, which should be used only with another method of birth control such as a diaphragm or condom. All spermicides require proper placement in the vagina. If used properly and regularly, some spermicidal products can be very effective. In several studies, less than five pregnancies occurred for every 100 couples using these methods for one year. Failure rates tend to be about three times higher than this in actual use as a result of inconsistent use of the method and improper following of the manufacturer's instructions. Failures occur when a couple inserts the spermicide in the vagina incorrectly, has intercourse a second time without using more of the product, or overlooks the time limits of product effectiveness. In addition, failures can occur if the spermicide has become outdated; for this reason, users should check the expiration date stamped on the spermicide package prior to use.

Burning or irritation of the vagina or penis occurs in about one in twenty people using spermicides, but this problem can often be alleviated by changing to a different product. Although one study found a higher rate of birth defects in the infants of women who had used a spermicide in the ten months before conception, the researchers stressed that these results were not conclusive, and a more recent investigation found no evidence supporting the existence of such a problem. However, it now appears that women who use spermicides after becoming pregnant are more likely than other women to have miscarriages.

On the positive side, there is mounting evidence that spermicides provide some protection against sexually transmitted diseases such as gonorrhea and also protect against pelvic inflammatory disease.

The use of spermicides may require interrupting the spontaneous flow of sexual activity, but insertion can be made a part of sexual play. With spermicidal suppositories, unlike aerosol foams, ten to fifteen minutes must elapse before vaginal distribution of the product has occurred (depending on the brand) and intercourse is "safe" — so these should not be used if you are in a hurry! Vaginal chemical contraceptives may discourage cunnilingus (oral stimulation of the vulva and vagina) because most products do not have a very pleasant taste. Finally, some people find that these products are messy and provide too much vaginal lubrication (one woman described the sensation as "sloshy").

The Contraceptive Sponge

The newest type of birth control device to become widely available (approved by the Food and Drug Administration in April 1983) is a soft, disposable contraceptive sponge, which is inserted into the vagina. Now sold exclusively under the "Today" brand name, the two-inch-by-one-inch round product is made of polyurethane permeated with the commonly used spermicide, nonoxynol-9. The contraceptive sponge is sold over the counter in drugstores and doesn't require a prescription or need to be fitted by a physician. Its early reception by consumers has been enthusiastic.

There are three different ways in which the sponge functions. First and most important, it carries the spermicidal effect of nonoxynol-9 (which is the active chemical ingredient in many contraceptive creams, foams, and gels). Second, the sponge also functions as a mechanical barrier, partially preventing sperm from entering the mouth of the cervix. Finally, the sponge is also thought to trap and absorb sperm, although the importance of this action is uncertain.

The contraceptive sponge is inserted into the vagina before intercourse. The sponge is first dampened with about two tablespoons of water and then squeezed gently until foam appears (this activates the spermicide in the sponge). Many users report that it is considerably easier to insert than a diaphragm and, since it can be inserted up to eighteen hours before intercourse, it is convenient as well. (Insertion can be done either manually or with an applicator.) As another major advantage, the sponge retains its contraceptive effectiveness for twenty-four hours without any need for the reapplication of spermicide, so it provides contraceptive protection regardless of how many times the user has intercourse. The sponge is removed by pulling on a small ribbon attached to one of its sides.

The sponge has been found to have an actual failure rate of about 15 percent, according to the F.D.A., which makes it approximately equivalent to a diaphragm in this regard. As with the diaphragm or spermicides, many of the failures are due to improper use — such as removing the sponge too soon after intercourse — and the user's motivation also seems to play a role. The effectiveness rate can be materially improved by combining use of the sponge with use of a condom.

Women who have used the sponge report a number of distinct advantages compared to the diaphragm. In addition to mentioning how easy it is to insert (which is nearly a universal reaction), many women are pleased to have an inexpensive, disposable, contraceptive method available that is unobtrusive, tasteless, and odorless, and lasts for twenty-four hours. In addition, the sponge reportedly isn't as messy or drippy as some spermicidal products.

The commonest side effect of the sponge is mild irritation of the vagina or penis, which occurs in about 3 to 5 percent of users. Although the sponge hasn't been in widespread use for a long enough time to assess its safety fully, the only serious health risk to come to attention at the time of this writing is the rare occurrence of toxic shock syndrome in a few women who used the sponge. Whether this was a coincidental finding or a direct relationship isn't clear at present; thus, users should read the package insert carefully to determine if any new information arises on this (or other) health effects.

Because the sponge can be inserted either hours before sexual activity or at the last moment before intercourse, convenience makes it especially advantageous from a sexual viewpoint. In addition, men and women who have tried the sponge almost invariably say they don't feel it during intercourse, and many couples also report that it doesn't interfere with cunnilingus. But the sponge's most noticeable sexual benefit may be that once it is inserted, it provides protection for multiple acts of intercourse. As one woman told us, "That means we can make love at night and then again in the morning without my having to get up and get ready all over again."

Rhythm

Rhythm methods of contraception depend on periodic abstinence from intercourse during times of the menstrual cycle when fertility is most likely. These methods are the only methods of birth control approved by the Roman Catholic Church, which considers them "natural" rather than "artificial."

The *calendar method* involves identifying "safe days" in the menstrual cycle (days where intercourse will not lead to pregnancy) based on the lengths of previous cycles. The underlying assumption is that ovulation will occur approximately fourteen days before the start of the next menstrual period. Careful hormone studies of the menstrual cycle, however, indicate that this assumption is not always correct.

To calculate the "unsafe" (fertile) period, a record must be kept of the length of each menstrual cycle for a minimum of six months. The first day of the unsafe period is determined by subtracting 18 from the length of the shortest cycle, and the last day of the unsafe period is found by subtracting 11 from the number of days in the longest cycle. For example, if a woman's shortest cycle is 26 days and her longest is 32 days during her "record keeping" time, she must abstain from intercourse beginning on cycle day 8 ($26 - 18 = 8$) and continue abstention until day 21 ($32 - 11 = 21$). Thus, the unsafe days would be days 8 to 21, inclusive, a time of 14 days when intercourse would not be permitted.

The *temperature method* involves daily recording of the basal body temperature (BBT) to pinpoint the time of ovulation. Intercourse is not allowed from the day menstrual flow stops until two to four days after the temperature rise. (If no temperature rise is detected in a full menstrual cycle, which sometimes happens, users of this method must observe total abstinence from sexual intercourse.)

The *ovulation method,* also known as the Billings method, depends on changes in cervical mucus to indicate the probable days of fertility during the menstrual cycle. The start of the fertile time is signaled by the appearance of a mucus discharge within the vagina which is whitish or cloudy and of tacky consistency. A day or two before ovulation, greater amounts of mucus are produced in a clear and runnier form with a stretchy consistency very similar to egg white. Intercourse is thought to be "safe" four days after the clear mucus begins, when the mucus has returned to a cloudy color.

The overall effectiveness of the rhythm methods leaves a great deal to be desired. The calendar method is unquestionably the least reliable among this group (failure rates are approximately 15 to 45 per 100 woman-years), and unless the woman's cycle is very regular, long periods of abstinence from intercourse will be required. Temperature methods are inaccurate because they are difficult to interpret, and in about 20 percent of ovulatory cycles, the BBT chart does not indicate ovulation. The World Health Organization found in 1978 that the ovulation method was "relatively ineffective for preventing pregnancy" based on carefully designed studies done in five different countries, with an overall failure rate of 19.4 per 100 woman-years. This is probably because many women have difficulty noting the the cyclic changes in their cervical mucus, and women who have a vaginal infection (which may itself create a discharge) usually cannot use this method. In a more recent study of 725 women, an overall failure rate of 22.3 per 100 woman-years was noted.

Most couples using the rhythm methods do not experience

major sexual difficulties. However, some couples may develop sexual problems because the need for abstinence creates unusual pressures to have intercourse on "safe" days regardless of whether they feel like it or not. Fear of pregnancy may also lead to sexual difficulties.

Sterilization

The highest degree of contraceptive protection currently available short of absolute abstinence is found in the use of surgical procedures to prevent pregnancy (*sterilization*). The popularity of these operations for both men and women has increased considerably in America in the last fifteen years. One recent estimate indicates that among all married American couples, about one-quarter will use sterilization within two years after the birth of their last wanted child and by ten years after their last child, more than half will undergo sterilization.

Sterilization procedures are appealing because they are safe, effective, and permanent. Their permanence can be a drawback, however, if there is a change in feelings or circumstances (e.g., death of a child or spouse, divorce) that leads a person to want more children. Although there is some possibility of reversing the sterilization, these procedures are far from guaranteed. Anyone thinking about sterilization as a method of birth control should thoroughly consider its probable irreversibility. One way of retaining the option to reproduce after male sterilization is to use a sperm bank to store several samples of frozen semen produced before sterilization. If reproduction becomes desirable at a later time, the semen samples are thawed and used in artificial insemination.

Female Sterilization

More than one hundred different types of operations can be used to achieve female sterilization. Almost all block the Fallopian tubes to prevent the union of sperm and egg. *Tubal ligation* (tying the tubes) is rarely done by itself today, since it is not as effective as other methods that also cut, clip, or otherwise block the tubes. Many of these operations are still called "tubal ligation" even though more is done.

Tubal ligation is frequently achieved by using a *laparoscope*, a tubelike instrument with lights and a viewer that is inserted through the abdominal wall. The tubes are cut and cauterized (burned) through this instrument. If laparoscopy can be done through the naval, no scar is visible; as only a one-inch incision is required for this operation, it is often called "BAND AID sterilization." A similar instrument can be inserted through the back end of the vagina or through the vagina and uterus to perform a tubal ligation.

Laparotomy, an operation involving a four- or five-inch incision through the abdomen, can also be used to perform a tubal ligation. It is rarely used for sterilization unless there is another reason for abdominal surgery, a technical difficulty, or a danger to the woman because of a medical problem. Female sterilization also results from *hysterectomy* (removal of the uterus) or *ovariectomy* (removal of the ovaries), but these operations are generally done for other reasons with sterility occurring as a by-product.

Female sterilization by most techniques offers almost foolproof protection against pregnancy. In rare instances, the cut ends of the tubes may rejoin, leading to a pregnancy, but the most common cause for "method failure" is when the woman is already pregnant (but no one knows it) when the operation is done. Side effects are infrequent (less than 5 percent) and are usually limited to the first few days after surgery, when infection or bleeding may be a problem.

Most women have no sexual difficulties after sterilization, which affects neither their hormones (since their ovaries are intact) nor their sexual anatomy. A few women may run into problems. If a woman undergoes sterilization involuntarily (e.g., if it was a decision she was pushed into by her husband or by health or economic circumstances), she may develop a reduced interest in sex for psychological reasons. Some women find that after sterilization they no longer feel the same way about sex (or have lowered sexual responsiveness) because they feel "incomplete" or "less than a woman." This reaction is particularly possible in a woman whose religious background considers sex unnatural or sinful if it is separated from its reproductive potential. On the other hand, some women show increased sexual interest after sterilization due to no longer fearing pregnancy.

Male Sterilization

Vasectomy is a simple surgical procedure that consists of cutting and tying each vas deferens (the tube that carries sperm). The operation is usually done using local anesthesia in a doctor's office or clinic and requires only about fifteen or twenty minutes. The man can then go home, although it is usually advisable to avoid strenuous physical activity for a day or two afterward.

Vasectomy does not stop sperm production; it blocks the passage of sperm from the testes to the upper part of the vas deferens. Sperm then accumulate in the epididymis (a mass of tubes at the back of each testis), where they are engulfed and destroyed by cells called phagocytes. Some sperm appear to leak out into the scrotum, where they disintegrate. Because some sperm are already above the point where the vasectomy was done, it usually takes six to eight weeks (about a dozen ejaculations) after this operation before the ejaculate is sterile (contains no sperm). For this reason, the man and his sex partner should use another method of birth control until at least two consecutive semen exams show that no sperm are present.

Vasectomy does not affect hormone production and does not interfere in any way with erection or ejaculation. The amount of the ejaculate is almost unchanged, since secretions from the testis and epididymis account for less than 5 percent of the volume of seminal fluid. There are no long-range health risks created by vasectomy.

Vasectomy is the simplest and safest form of surgical contraception: the failure rate is a very low 0.15 percent. Failures occur either because of unprotected intercourse before all sperm have disappeared from the ejaculate, mistakes in performing the operation, or, very rarely, because the cut ends of the vas have grown back together. The rate of medical complications such as bleeding or infection is less than 5 percent, although minor swelling, pain, and temporary skin discoloration due to bruising occur in about half of men after this procedure. Postoperative ice packs and a well-fitted athletic supporter can reduce discomfort.

Less than one man in twenty reports decreased sexual pleasure after a vasectomy, while close to half claim increased pleasure and about a quarter report an increased frequency of

intercourse. In fact, many couples find that sex is more spontaneous and fun after a vasectomy since they don't need to worry
about becoming pregnant or using other contraceptive methods.
Men who have a history of sexual problems are not always good
candidates for vasectomy, since they may be more prone to developing psychological impotence or other difficulties after such
surgery. For some couples, there may be a drop in marital har-
mony after a vasectomy, because the man expected special recognition or gratitude from his wife in return for having the
surgery or because the wife resents not being able to have additional children. Occasionally, a man who has been well adjusted
to a vasectomy may become impotent (usually temporarily) if he
divorces and then remarries, particularly if he and his new wife
would like to have children. More typical, however, is this comment from a thirty-year-old man six months after his vasectomy:
"It was simple and quick and did the trick. Now sex is *really*
making love, without worrying about 'accidents.' "

Other Methods of
Contraception

Withdrawal (coitus interruptus) is the removal of the penis from
the vagina before ejaculation occurs. This is a difficult form of
birth control in which even perfect timing (which is not always
possible) does not give perfect results. If ejaculation occurs before the penis is withdrawn completely, or if drops of semen
spurt into the opening of the vagina, pregnancy can occur. As
live sperm can also be in the pre-ejaculatory fluid long before
ejaculation (or even if the male doesn't ejaculate), this method is
chancy at best, with a failure rate of 20 to 25 percent. This
method can also be a frustrating form of birth control for both
parties, who may find that it seriously interrupts the spontaneity
of the sexual interaction. However, it is clearly better than not
using any contraceptive when nothing else is available.

Douching, using a liquid to flush the vagina, is a poor method
of contraception, since sperm can quickly penetrate the cervical
mucus, where they are unaffected by douching. The failure rate
is more than 42 percent.

Breast-feeding inhibits ovulation in some women (possibly via
the effect of high prolactin levels on the ovaries) but this method

of birth control is very unreliable. Women who do not want to become pregnant soon after childbirth should begin using a more effective means of contraception right away. Some studies of nursing mothers not using any other method of birth control indicate pregnancy rates of over 50 percent.

Abstinence from intercourse is, needless to say, a highly effective way to prevent pregnancy if practiced consistently. Some couples voluntarily limit their sex lives to noncoital acts, but this does not seem to be a very popular choice for most heterosexual couples.

POSTCONCEPTION METHODS OF BIRTH CONTROL

The "Morning After" Pill

Various types of estrogen used in high doses shortly after conception can prevent implantation and provide a form of "after the fact" birth control. This approach can be useful in special situations (such as rape or leakage of a condom) but it is not suitable for routine contraceptive use. Most commonly, diethylstilbestrol (DES) is given to the woman for five days, beginning as soon after unprotected intercourse as possible and within seventy-two hours. The effectiveness of using this timetable is extremely high: the pregnancy rate is less than one percent. The primary side effects are nausea (sometimes accompanied by vomiting), breast tenderness, and alterations in the menstrual cycle. Used on a short-term basis, as just outlined, DES or other forms of estrogen are unlikely to have any lasting effect on the woman or on a fetus if a pregnancy occurs. Although DES used during later stages of pregnancy has been linked to abnormalities in the reproductive systems of children of mothers receiving this drug, there is no indication that this effect applies to its use in preventing pregnancy. Nevertheless, the risks of "morning after" pills are not entirely known at present, and it should not be assumed that they are completely safe.

Menstrual Extraction

This technique has been pioneered by the women's self-help movement. Just when a period is expected, a thin, flexible plastic tube is inserted through the cervix into the uterus, and suction (using either a syringe or a pump) is applied to draw out the endometrial lining. If the woman is pregnant, the tiny embryo is easily withdrawn. There may be mild cramping, but no anesthetic is usually used. Although further research is required, there do not seem to be any major side effects when this procedure is done under proper conditions by a health care professional.

Abortion

Abortion is the termination of pregnancy before the fetus is able to survive outside the uterus. Abortion can be either *spontaneous* — that is, when a medical problem ends a pregnancy — or *induced,* when the contents of the pregnant uterus are removed intentionally. On a worldwide basis, it has been estimated that 30 to 55 million induced abortions are performed annually.

Abortion has been practiced since early times. In ancient Greece, where it was thought the fetus had no soul, Plato suggested in *The Republic* that abortion be used in cases of incest or older parents, and Aristotle recommended abortion as a way to limit family size. In 1800, there were no laws against abortion in America, and by the mid-1800s, advertisements for abortion services appeared in respectable newspapers, journals, and religious magazines. In 1869 Pope Pius IX issued a decree declaring abortion sinful and banning it entirely, thus providing support for an antiabortion campaign that had been already started by the newly formed American Medical Association. By 1900 abortion was illegal all over America. Although an illegal abortion trade flourished, with incredibly high rates of bleeding, infection, and death among women getting such "help," it was not until the late 1960s that a few states changed their abortion laws. In January 1973 a decision by the U.S. Supreme Court established the legal right of women to choose whether or not to have an abortion.

Today, the abortion issue is argued heatedly by various groups. The "Right-to-Life" movement opposes abortion on the grounds of protecting the rights of the unborn fetus, while "Pro-Choice" advocates want women to have freedom of choice in controlling their own bodies and futures. Public opinion on these issues is divided. The complex aspects of rights and responsibilities as they affect individuals and society are not simple or easily summarized and are beyond the scope of this discussion.

The abortion method chosen depends mainly on the length of the pregnancy. During the first trimester, the most common method is called *vacuum aspiration* or *suction*. After widening the mouth of the cervix, a small plastic tube is inserted into the uterus. The tube is connected to an electric pump that removes the fetal tissue, placenta, and membranes from the uterus by suction. The procedure usually requires only ten to fifteen minutes.

During the fourth and fifth months of pregnancy, the safest method for abortion is *dilatation and evacuation* (D&E), a method that is similar to vacuum aspiration. After the mouth of the cervix is widened (dilated), vacuum suction is applied to remove part of the contents of the uterus. However, because at this later stage of pregnancy not all of the tissue can be removed by suction alone, a forceps is used to remove additional material until the uterus is completely empty. In addition, a metal instrument called a curette may be used to gently scrape the walls of the uterus to ensure that all remnants of extra tissue have been removed.

Second-trimester abortions are also sometimes done by using chemicals to stimulate contractions of the uterus, causing expulsion of the fetus and membranes. *Prostaglandin-induced* abortions are done by injecting prostaglandins into the amniotic sac, dripping them slowly into a vein, or placing them in the vagina. In *saline-induced* abortions, 7 ounces of a saline (salt) solution are injected into the amniotic fluid. Both methods require hours before completion, have the disadvantages of physical and emotional discomfort during contractions of the uterus, and require vaginal delivery of the dead fetus.

Less commonly, surgical procedures are used for abortions.

Dilatation and curettage (D&C) involves dilating the cervix and then gently scraping the lining of the uterus with a metal instrument (the curette) to extract the fetal tissue, placenta, and membranes. A D&C can be done until about fifteen weeks after the last menstrual period and, unlike the previously mentioned methods, requires the use of general anesthesia. A *hysterotomy* is an operation like a cesarean section which can be used throughout the second trimester. It also requires general anesthesia and is rarely used.

Although some people believe that eating certain foods, vigorous exercise, or other "do it yourself" techniques wll lead to abortion, this is not true. Attempts at inducing abortion by inserting objects such as a wire coathanger or knitting needle into the uterus are *extremely dangerous* and can lead to fatal infections or hemorrhage.

Modern abortion techniques are less risky to the mother than a full-term pregnancy: fewer than four deaths occur per 100,000 induced abortions compared to approximately twenty deaths per 100,000 pregnancies. In fact, among healthy women (those with no serious pre-existing medical problems) the risk of dying from a legal abortion in the United States is less than one per 100,000 procedures. Abortions done during the first trimester are simplest and safest, as shown in Table 2; after this, complications such as bleeding, infection, and perforation of the uterus are more common. There is no solid evidence that having a properly done abortion causes later infertility. Recent reports do indicate that having two or more induced abortions leads to a higher rate of miscarriages in subsequent pregnancies and may increase the rate of prematurity and low-birth-weight infants.

The emotional benefits of abortion outweigh the psychological risks for most women. Serious problems requiring psychiatric referral occur in fewer than 3 women in every 100,000 who have had an abortion. Nevertheless, short-lived feelings of guilt, sadness, and loss are common in women who have had abortions; preabortion and postabortion counseling is often effective in helping women deal with these reactions.

One relatively neglected aspect of the abortion experience is the *male's* reaction. A 1984 survey by Shostak, McLouth, and Seng of the impact of abortion on men shows that many men try

TABLE 2
Safety of Legal Abortions, by Weeks of
Pregnancy, United States 1972–1978

Weeks of Pregnancy	Number of Deaths per 100,000 Abortions
Up to 8	0.5
9–10	1.4
11–12	2.3
13–15	6.7
16–20	13.9
21 or more	17.5

Source: Centers for Disease Control, *Abortion Surveillance 1978*, issued November 1980, p. 48.
Note: Eighty-eight percent of all legal abortions in the United States during these years were done before the thirteenth week of pregnancy.

to approach the abortion decision in an abstract, intellectual way but later find themselves having to deal with feelings of hurt, guilt, or anger. Few abortion centers offer counseling services for men, perhaps reflecting the view of one abortion counselor who told us: "The imperial male attitude is just too hard to deal with in many cases. It's the woman who's pregnant, and it's her feelings I worry about. Let the men go find a psychiatrist on their own if they need to." While not all abortion counselors share this view, many point out that men don't seem interested in their services.

THE PSYCHOLOGY OF CONTRACEPTIVE USE

Why do some people who do not want a pregnancy avoid the use of contraception or misuse the form of birth control they have selected? One common reason is inadequate knowledge. Illustrating this point, researchers from Johns Hopkins University found that about half of sexually active teenage women aged fifteen to nineteen thought they could not become pregnant. Similarly, a survey of urban mothers showed that only one-third knew when during the menstrual cycle a woman is most likely to become pregnant, and for mothers under age twenty-one, only 10 percent were correct in their knowledge.

Inadequate knowledge does not explain why many people who understand the mechanics and risks of conception still do not use birth control or use it haphazardly. This phenomenon raises a fundamental question about personal motivation and contraceptive practice. Clearly, some people avidly want to avoid pregnancy, while others only want to delay the timing of a pregnancy or do not mind if they happen to conceive (in this discussion, we will not deal with couples wanting a pregnancy, since contraceptive use does not really apply). In general, the more motivated a person is to avoid pregnancy, the more likely he or she will choose and consistently use a highly effective birth control method. But this motivation may also be affected by other factors, such as forgetfulness, denial ("it can't happen to me"), dissatisfaction with a particular method, wanting to please a partner, or concern about medical risks. This comment from a twenty-three-year-old woman illustrates a commonly encountered situation:

> I'd been using the pill for about two years, but I read about its risks so I decided to switch. I got pregnant the third time I used a diaphragm. I'm back on the pill now, older but wiser.

Not using contraception is sometimes a sign of personal embarrassment about sexuality in general or about a particular method. One twenty-four-year-old woman told us that she had tried to insert a diaphragm for a half hour and finally just gave up because she felt foolish about it. Other people worry that someone may discover their contraceptive devices, giving away that they are, or plan to be, sexually active. While this concern obviously affects teenagers who may be afraid of a parent finding out that they use contraception, it may also apply to a woman whose date accidentally discovers a container of foam in her purse. She may have no intention of having intercourse or *any* sexual activity with him, but the "discovery" may be taken to mean that she is willing or even expecting to.

For other people, it is okay to have sex in the heat of passion, but *planning* for it makes it immoral, mechanical, or unromantic (these people are often called parents). Many people also use contraceptives on a fairly regular basis with only occasional lapses into sexual brinksmanship. In these instances they decide to indulge their soaring sexual feelings (and not call Time Out) or to forgo contraception to "have more fun" or "make it more natural." Couples using rhythm methods sometimes have a particular problem with sexual brinksmanship, since sex may seem more enticing on the days when abstinence is expected.

In some cases hostility, power struggles, or differences in the reproductive goals of two sexual partners may lead to disuse or misuse of contraception. The man who insists on using withdrawal may make sex unpleasant and even threatening to his partner. The woman can attempt to manipulate her partner by limiting her sexual availability ("My IUD is cramping" or "I have an infection"), or by seeking a pregnancy to entrap her partner in marriage or to prevent separation or divorce. Who assumes responsibility for contraception may be a heated issue, particularly if the woman has done so for a long while and then wishes to turn this job over to her partner.

One last aspect of the psychology of contraceptive use will be mentioned. When sexual problems exist in a relationship, couples frequently become sloppy in their attention to birth control or stop using contraception altogether. Sometimes they believe their sexual difficulty results from or is worsened by their contraceptive method. In other cases, they think that when sex is a problem conception is unlikely (of course, they *may* be mildly shocked to discover how wrong they are). In yet other instances, discontinuing the use of birth control may be used as a psychological ploy to symbolize that "sex really is not so important" or, conversely, to attempt to make sexual contact more intimate and exciting.

FUTURE TRENDS IN CONTRACEPTION

There is general agreement that currently available contraceptive methods are less than perfect from the viewpoints of

safety, reversibility, effectiveness, and ease of use. To improve the range of choices, considerable research is under
way.

Not surprisingly, the most intense interest focuses on the possibility of developing a male birth control pill. In theory, this
could be accomplished by using drugs to: (1) block sperm production in the testes, (2) interfere with sperm maturation in the
epididymis, (3) impair sperm transport at any point along their
journey from testis to epididymis to vas deferens to urethra, or
(4) reduce sperm motility or ability to undergo capacitation. Various drugs have already been found to accomplish each task, but
so far they have either proved to have incomplete effects or to
have unacceptable side effects. For example, combinations of
estrogen and progestogen (as used in the "female" pill) are effective in blocking sperm production but cause a profound drop
in sexual interest in most males and frequently lead to impotence. Pills combining testosterone and a progestogen are currently under investigation, since they do not seem to cause the
circulatory side effects estrogen does and do not usually interfere with sexual function. However, these pills are not highly
effective in inhibiting sperm production for long periods of
time.

A recent report from China indicates that an extract from
cottonseed oil called gossypol suppresses sperm production.
While the incidence of side effects with gossypol seems to be
acceptable, with only about 10 percent of men developing minor
problems such as dizziness, fatigue, dry mouth, and digestive
tract difficulties, and only 5 percent reporting decreased libido
or potency problems, it is not clear if gossypol's effects on sperm
production are completely reversible.

An even more promising recent development has been the
creation of a synthetic form of inhibin, a protein substance made
in the testes that provides feedback to the hypothalamus and
pituitary gland controlling the production of FSH. Since it is
thought that inhibin suppresses sperm production without affecting sexual function, clinical trials with this substance may
provide the ultimate breakthrough in the search for an effective,
reversible male contraceptive.

The search for additional methods of male contraception also
includes several other possibilities:

1. Developing a vaccine or drug to block production of FSH in the pituitary, since this hormone controls sperm production but has no known effects on sexual function;
2. Developing a vaccine to impair the enzymes released by the acrosome (chemical reservoir) of the sperm, since these enzymes are required for the sperm to penetrate the egg;
3. Using ultrasound to produce temporary sterility by blocking sperm production.

Improvements in methods of female fertility control are also being studied currently by many groups of scientists. One of the most promising approaches is the attempt to develop an anti-pregnancy vaccine that is activated only in the event of conception and that works in a reliable and unobtrusive fashion. Vaccines directed against HCG, a hormone secreted by the placenta (and the hormone detected by pregnancy tests), are being tested at the present time and seem to prevent pregnancy without interfering with ovarian or menstrual cycles.

An attempt is also under way to identify a compound that could block progesterone production by the corpus luteum (the part of the ovarian follicle that produces hormones in early pregnancy). This compound would prevent implantation and induce menstruation whether or not pregnancy occurs. Research is also being conducted to develop a pill, liquid, or vaginal tampon that could be self-administered at the expected time of menstruation to induce menstrual flow and evacuate the contents of the uterus. Various synthetic analogues of prostaglandins that cause uterine contractions are being investigated for this purpose, but while they appear to be at least 90 percent effective, they are typically accompanied by an unacceptably high rate of negative side effects. Another idea is to block signals from the hypothalamus to the pituitary gland that trigger LH release. An important step in the initiation of ovulation would thereby be disrupted. Theoretically, blocking tubal transport of the fertilized egg could also reduce the chances of implantation, but this possibility has not been thoroughly studied.

Attention has also been focused on improving hormonal methods of female birth control. These include developing pills without estrogen (to avoid unwanted side effects) and using long-acting forms of contraceptive hormones. These hormones would be either injected or placed in small implants under the skin or in plastic vaginal rings that slowly release hormones. In fact, it will soon be possible for a woman wishing long-term contraceptive protection to carry a five-year supply of hormonal birth control with her all the time in an implant placed under the skin in the upper arm. This implant, called the NORPLANT system, actually consists of six small Silastic capsules packed with the hormone levonorgestrel — the same progestogen used in many minipills — which are thought to provide about the same degree of contraceptive effectiveness as the pill. The contraceptive, which is slowly released from the capsules, doesn't always block ovulation but instead prevents pregnancy in two ways: by thickening the cervical mucus (making it difficult for sperm to swim up the Fallopian tubes) and by blocking implantation. The major drawback at this time is that many users experience abnormal menstrual bleeding (but this problem generally decreases after about one year of use).

A synthetic form of progesterone called *medroxyprogesterone acetate,* or MPA, is currently used as a long-acting contraceptive in other countries. (In the United States, its trade name is Depo-Provera.) It is given by injection once every three months (so the woman does not need to remember to use it) and is about as effective as combination birth control pills. Its drawbacks include possible infertility after long-term use, reported toxicity in animal studies, and the newly reported finding that MPA may cause cancer of the uterus. However, two recent studies found no evidence to support a definite link between MPA and any form of cancer in humans.

Various techniques for developing a non-surgical method of female sterilization are also being explored. These methods, which are irreversible, involve plugging up the place at which the Fallopian tubes open into the uterus. Material that causes thick scar tissue to form at the tubal openings is inserted through the vagina and cervix with a special instrument that doesn't require making an incision or using anesthesia. Studies of the effectiveness of this approach are now under way.

Finally, a pocket-size electronic device that flashes a red light when a woman is fertile may be useful to those interested in practicing natural family planning. The device consists of a thermometer and a microcomputer that keeps track of a woman's temperature changes during her menstrual cycle, displaying a green light on "safe" days and a red light on days near the time of ovulation. While there are no published studies as yet on the accuracy of this device (or similar devices that claim to detect ovulation by a litmus paper test, measuring the thickness of cervical mucus, and other similar means), this type of approach bears further investigation, particularly since its use would be acceptable to the Roman Catholic Church.

CHAPTER FIVE

Childhood Sexuality

CHILDHOOD has been called "the last frontier in sex research" because there is little reliable data about sexual behavior during this formative time. Studies based on interviewing adults about what they did or how they felt during childhood are distorted by faulty recall, exaggeration, and omissions due to embarrassment or the wish to seem "normal." Attempts to interview children or administer questionnaires to them about their sexual attitudes and behavior have often been thwarted by community outrage about "putting nasty ideas in children's minds" and accusations of undermining the moral fabric of our society. Except for some limited cross-cultural data from primitive societies in which childhood sex play is permitted and data from a few instances of direct observation, we are forced to rely on guesswork and inference in this important area.

Prior to the work of Freud and some early sexologists around the turn of the century, childhood sexuality was seen either as nonexistent or as something to be repressed because of its sinful and dangerous nature. These contradictory views still exist, but at least some parents today have come to regard the developing sexuality of their children in a more matter-of-fact, accepting way. Other parents are uncomfortable with any form of sexual interest or behavior in their children for several reasons. They worry that it is abnormal; they are uncertain about how to deal with it; or they are dealing with sexual conflicts within themselves.

By learning about the typical patterns of sexual development during childhood, parents (or prospective parents) can become more effective in helping their children learn about sexuality in a comfortable, unthreatening way.

SEX IN INFANCY

Ultrasound studies have provided some evidence that reflex erections occur in developing baby boys for several months before birth, while they are still within the uterus. Many newborn baby boys have erections in the first few minutes after birth — often, even before the umbilical cord is cut. Similarly, newborn baby girls have vaginal lubrication and clitoral erection in their first twenty-four hours, so it is clear that the sexual reflexes are already operating at the very start of infancy and probably even before birth.

An important phase of infantile sexuality comes from the sensuous closeness of parent and child through holding, clinging, and cuddling. This parent-child bonding begins at birth and extends to include nursing, bathing, dressing, and other physical interactions between parents and their newborn child. A child who is deprived of warm, close bonding during infancy may experience later difficulties forming intimate relationships or, more speculatively, in being comfortable with his or her sexuality.

Very young infants respond quite naturally to a variety of sources of physical sensation with signs of sexual arousal. For example, it is common for baby boys to have firm erections while they are nursing. While this is alarming to some parents, who see it as somehow abnormal or perverse, the fact is that the sensation of cuddling close to the warmth and softness of the mother's body and having the intense neurological stimulation of suckling (the lips are well endowed with sensory nerve endings) combine to send messages to the brain that are interpreted as pleasurable and that activate sexual reflexes. Clitoral erection and vaginal lubrication in baby girls also occur commonly during nursing, indicating that this pattern is not restricted to one gender (although penile erection is more visible and thus more likely to be noticed). Similar signs of reflex sexual activation may occur when babies are bathed, powdered, diapered, or playfully bounced around. It is important to recognize, however, as Martinson (1981, p. 26) points out, that "the infant is too young to be consciously aware of the encounter, and therefore no socio-sexual erotic awakening can be said to occur." How parents respond to observing these sexual reflexes during infancy may be part of the child's earliest sexual learning: the parent who is shocked or disapproving is apt to react in a manner that conveys discomfort, while parents who react calmly give children a message of acceptance regarding sex.

As any observant parent knows, baby boys and girls begin to touch or rub their genitals as soon as they develop the necessary motor coordination. Kinsey and others have reported that this sometimes leads to orgasm in infants less than one year old. The question is, what meaning does this behavior have? Is the infant simply exploring his or her body, with an equal likelihood that equally accessible parts (elbow, tummy, genitals) will be touched? Or is there a sexual component to such behavior, with a genuine sense of pleasure leading to repeated self-stimulation?

Although infants cannot answer these questions for us, the evidence seems to support the latter view. Helen Kaplan notes that babies "express joy when their genitals are stimulated" (Kaplan, 1974, p. 147). Bakwin points out that "infants show extreme annoyance if efforts are made to interrupt them" during masturbation and adds that self-stimulation is done "many times during the day" (Bakwin, 1974, p. 204). By the third or fourth month of life, genital stimulation is accompanied by smiling and cooing. By one year of age, genital play is commonly observed when the infant is naked or bathing. Genital play is more common in infants reared in families than in infants reared in nurseries, suggesting that parent-child bonding plays a major role in the development of subsequent sexuality.

The parents of very young children react to these displays of sexual behavior in a variety of ways. Some are amused, some are surprised, and some are alarmed — particularly if they do not realize that this is a completely normal developmental pattern.

SEX IN EARLY CHILDHOOD (AGES 2 TO 5)

By age two, most children have begun to walk and talk and have established a sense of being a boy or girl. There is unquestionable curiosity about body parts, and most children discover (if they have not already) that genital stimulation is a source of pleasurable sensations. Genital play first occurs as a solitary activity and later in games like "show me yours and I'll show you mine" and "doctor." In addition to rubbing the penis or clitoris manually, some children stimulate themselves by rubbing a doll, a pillow, a blanket, or some other object against their genitals.

Conversations with three-year-old boys and girls indicate that they are well aware of the sensual feelings of genital stimulation, although these feelings are not labeled by them as erotic or sexual (concepts the child does not yet understand). The following comments from our files illustrate this point:

> *A three-year-old girl:* When I rub my 'gina it's nice and warm. Sometimes it tickles. Sometimes it gets real hot. [Note: this child referred to her entire genital area as her " 'gina" and was specifically describing manual rubbing of the mons and clitoris which she practiced at least a half-dozen times a day. From age 2½ to age 3½, she preferred to go bottomless so she could have easy access to her genitals and frequently took off her underpants to achieve this goal.]

> *A three-year-old boy:* Look at my wiener! I can make it stand up. I rub it and it stands up and it feels good. Sometimes I rub it a lot and it feels *very, very* good. Sometimes I just rub it a little. And then it feels a little good. [This boy was very proud of his "wiener," which he liked to show to visitors. His parents told us that he stimulated his penis "several times a day" that they knew of and were pretty certain that he also pursued this activity in private.]

At about the same time, children also become aware of parental attitudes of disapproval of genital play and may be confused by parents who encourage them to be aware of their bodies but exclude the genitals from such awareness. While it is important for parents to educate their children about socially appropriate behavior (e.g., it is not acceptable to show or fondle your genitals in public places), some parents try to stop all forms of their child's sexual experimentation by saying "That's not nice" or "Don't touch yourself down there," or by nonverbal communications such as pushing the child's hand away. The negative message that the child gets in such situations may be among the earliest causes of later sexual difficulties. This attitude is compounded by many children's assumption that their genitals are "dirty" from messages received during toilet training. The emphasis on cleanliness in the bathroom ("wipe yourself carefully," "wash your hands after you go") conditions the child to see genital function in negative terms, even though it actually represents a legitimate health concern of parents.

By age four, most children in our society begin asking questions about how babies are made and how birth occurs. Some parents respond with matter-of-fact answers, while others are obviously uncomfortable and reluctant to discuss this information at any length. Children have a pretty good idea of what bothers mommy or daddy, so they may react either by not asking such questions at all *or* by bombarding one or both parents with questions to see them squirm.

Four-year-olds generally have vague and somewhat magical notions about sex. They often believe the "stork-brings-the-baby" explanation without any further questioning, or, if given a more accurate explanation of reproductive facts, interpret them in unique ways. For example, four-year-olds are quite literal in thinking that the mommy's egg from which a baby grows is just like the ones bought by the dozen in the grocery store. Similarly, some four-year-olds presented with a "daddy-plants-a-seed-in-mommy's-body" explanation of conception and pregnancy are convinced that there is a patch of dirt inside the mother's body that must be periodically watered and weeded for the baby to grow. This way of viewing sexual matters reflects the four-year-old's concrete, literal view of the world in general.

Children who attend nursery school or day care centers before reaching school age are apt to confront many situations with sexual overtones. For instance, Billy and Peter, each four years old, have to be told repeatedly that it's not appropriate to kiss each other while they're playing. In the same nursery school class, Gerry amuses himself by sneaking up behind a girl and pulling up her skirt ("So I can see her underpants," he explains with a lot of giggling). Both girls and boys express considerable interest in bathroom functions and bathroom etiquette, and both sexes are very willing to try out new "dirty" words, a common practice that tends to alarm parents more than teachers.

At age five, when most children enter kindergarten, the opportunity to relate to age-mates in a structured environment leads to modesty, and sex games decrease in frequency. Children of this age become fascinated with learning words about sexual parts that they have not heard before, and jokes about sex and genital function begin to make their rounds, often heard first from a slightly older child and then repeated. The five-year-old may not understand the joke but laughs heartily (sometimes at the wrong line) to cover this up. As Money observes, when frank, direct information about sex is not available to a child, sexual jokes become the most important source of sex education for both girls and boys. Since even young children quickly learn the difference between a "clean" and "dirty" joke, this leads to the attitude that sex is dirty.

At this age children also begin to form ideas about sex based on their observations of physical interactions between parents — seeing Mommy and Daddy hugging and kissing, and obviously enjoying it, is a pretty good advertisement for the pleasures of physical and emotional intimacy. On the other hand, seeing parents constantly fighting or hearing one tell the other "don't touch me" can have just the opposite effect on the child's view of intimacy.

SEX AND THE SCHOOL-AGE CHILD

Six- and seven-year-old children have usually acquired a clear understanding of basic anatomic differences between the sexes and typically show a strong sense of modesty about body exposure. Parental attitudes and practices regarding nudity in the home undoubtedly influence the child's self-consciousness, but at the same time the natural curiosity of childhood is likely to emerge in games like "hospital" or "playing house" that permit sexual exploration. These games may involve simply inspecting each other's genitals or may include touching, kissing, rubbing, or inserting objects into the rectum or vagina.

Sexual experimentation includes activities with children of the same sex and the opposite sex. One purpose of this behavior is seeking knowledge: "How different am I from others who are like me?" and "How different are members of the opposite sex from me?" Another purpose is testing the forbidden to see what happens: who finds out, how they react, what can I get away with, and so on. These two components are interrelated, since forbidden knowledge is usually more alluring than easily available knowledge.

Childhood participation in such games is probably nearly universal, although available studies (mainly based on recall data) give much lower estimates. For example, Kinsey found that about 45 percent of adult women recalled participating in some form of sex play by age twelve, and 57 percent of adult males recalled similar experiences. In a survey of children aged four to fourteen, 35 percent of the girls and 52 percent of the boys reported some homosexual play. In a more recent survey, parents of six- and seven-year-old children noted that 76 percent of their daughters and 83 percent of their sons had participated in some sex play, with more than half of their known experiences involving play with siblings. One explanation for this relatively low rate of recalling childhood sex play is suggested by a theory of the Austrian sex researcher Ernest Borneman. According to Borneman, puberty characteristically is accompanied by a form of amnesia — blocked memories — about prior sexual experiences. Borneman suggests that this selective amnesia might explain why so many parents are shocked when they find their children playing sexual games: The adults have no recollection of such experiences in their own lives, even though they probably occurred. While there is as yet no substantiation of Borneman's thesis by others, it is an intriguing suggestion.

Childhood sex play is not psychologically harmful under ordinary circumstances and is probably a valuable psychosocial experience in developmental terms. However, psychological harm *can* come from harsh parental reaction. When children are discovered in sex play, either solitary or with others, negative parental reaction may be difficult to understand but easy to perceive. From the child's viewpoint, play is play, but for the parent who discovers a child masturbating or engaging in sex play with others, SEX in capital letters flashes across the scene. The parent who reacts with ominous predictions or threats that continuing such "bad" behavior will lead to dire consequences is frightening the child. The parent who says "that's dirty" may be interpreted very literally by the child, sowing the seeds of an attitude that may persist into adulthood.

Parental reactions to the discovery of sex play in school-age children frequently operate on a double standard. Girls are often cautioned strongly against sexual play, especially with boys. Boys, on the other hand, tend to get mixed messages from their parents; they may be warned or even punished for such activity, but there is a hint of resignation or even pride in the attitude that "boys will be boys." One father described the sexual escapades of his seven-year-old son and a female classmate by saying "Good for him, he's getting an early start." The unspoken permission for boys to follow their sexual curiosity (except in homosexual situations, where parents consistently react in a negative way) is only rarely found directed to school-age girls in American society. With the arrival of puberty, parents seem to react with even more of a double standard toward the sexual behavior of their sons and daughters.

Freud's concept of a period of sexual latency during late childhood — a time when sexual interests and impulses are diverted into nonsexual behaviors and interests — is no longer accepted by many sexologists. Money (1980) says that this is a time of sexual prudery when participation in sex play simply goes underground. Cross-cultural studies clearly show that if a society is not repressive toward childhood sex rehearsals, such play continues and may even be more frequent during the preadolescent years. Kinsey's data also shows that sexual experimentation does not stop or even slow down during this period. A detailed study of childhood sexuality conducted by Goldman and Goldman in 1982 involving interviews with over 800 children ages five and over in Australia, North America, Britain, and Sweden also provides no indication of a phase of childhood development where sexual development is suspended. In fact, these researchers note: "Discrediting Freud's latency period theory, overwhelming evidence was produced which reveals children from age five to fifteen to be increasingly interested in exploring sexual topics in linear progression with age." Perhaps the available evidence is best summed up in the following passage:

> Children pursue the course of their psychosexual development in blithe disregard of an expected sexual latency. Their only nod in the direction of the theoretical expectations is that they have learned to play according to adult rules. They learn to fulfill the letter of the law, even as they proceed secretly in their own ways. (Gadpaille, 1974, p. 189)

The sexual experiences of older children may be infrequent and less important than other events in their lives but may include the entire range of possible sexual acts, including attempts at intercourse that are sometimes successful. Masturbation occurs in private as well as in heterosexual or homosexual pairs or groups; sexual play with animals and objects has been noted; and oral or anal sex has been reported. By ages eight or nine, there is little question that children have awareness of the erotic element of such activities, and it is no longer accurate to think of these as "play" only. Sexual arousal is more than a by-product of these deliberate activities and is willfully sought, not just an accidental happening. Erotic arousal may be accompanied by sexual fantasies, and in some instances, falling in love occurs. These encounters can help children learn how to relate to others, with important consequences for their adult psychosexual adjustment.

Many parents are unaware that homosexual play among children, as well as heterosexual play, is a normal part of growing up. Homosexual play does *not* lead to adult homosexuality, although many parents worry unnecessarily on this point.

Another common form of childhood sexual behavior is sexual contact between siblings. While technically such behavior may be called *incest* — sexual activity between relatives — it seems unnecessarily pejorative to label the "look-see" games of a five-year-old boy and his six-year-old sister in so heavy-handed a fashion. Nevertheless, it can be difficult to decide when factors such as age differences between siblings, aggressive components of the sexual behavior, or exploitive or coercive elements should lead sexual contacts between siblings to be viewed as innocent play, a form of childhood learning, or a matter for parental action.

Finkelhor's data on sex between siblings is the most detailed non-clinical sample available. In brief, he found that 13 percent of college students surveyed admitted to childhood sexual activity with a brother or sister (a figure he considered an underestimate). Approximately three-quarters of these relationships were heterosexual (brother-sister), while one-quarter were homosexual (brother-brother or sister-sister). Additional findings from this survey included:

1. Sexual contact between siblings was not restricted to young children only; 73 percent of the experiences reported occurred when at least one sibling was over age eight.
2. The most common type of sexual activity between siblings was genital touching; only 4 percent included intercourse. Among younger children, looking at one another's genitals was the primary form of sex play.
3. There was considerable variability in the duration of these activities. One-third were single occurrences, while 27 percent continued off and on for at least one year.
4. In a quarter of the experiences, some type of force was involved (with girls being victimized most often).
5. Almost one-quarter of the experiences involved siblings who were at least five years apart in age.

These findings suggest that it may be necessary for sexologists to rethink previous notions about sexual contact between siblings as an innocent form of play. A situation where one sibling is much older than the other (by four years or more) or where force is used (which may be much more common than previously realized) is almost invariably exploitive and thus is cause for concern. However, parents must use their judgment in dealing with such situations. One episode of genital touching between an eleven-year-old girl and her consenting seven-year-old brother is not the same as a pattern of forced sex between siblings. Furthermore, it is important to realize that parental displays of alarm and horror on learning of their children's incestuous activities are not only inappropriate but may sometimes be harmful to the children. Finally, if an exploitive incest situation is discovered, obtaining psychological counseling for the victim may be advisable.

Since there is now considerable evidence that incest victimization often has negative long-term psychological consequences, including sexual problems in adulthood, parents may want to consider ways to minimize the risk of such an occurrence. For example, it may be helpful to discourage siblings who are more than two years apart from bathing together, and it also is advisable to avoid having an older sibling share a bedroom with a much younger one. A more detailed discussion of incest, including information about parent-child incest, appears in chapter 16.

As long as aggressive or coercive behavior is not involved, it is unlikely that isolated instances of childhood sexual activity are abnormal. It is not very helpful for parents to react to the discovery of childhood sex play with alarm or punishment. A matter-of-fact approach that includes understanding and age-appropriate sex education (while maintaining the parents' rights to set limits) is likely to be more effective than threats and theatrics in helping the child undergo healthy psychosexual growth.

SEX EDUCATION

In the last decade, there has been an unusual amount of controversy on the topic of sex education. While almost everyone seems to agree that teaching children about sex is necessary, there is much disagreement about what should be taught, *where* it should be taught, and who should do the teaching.

The background can be summarized as follows. A number of studies indicate that only a minority of parents provide meaningful quantities of sex education for their children. American teenagers, for example, report that they learned most of what they know about sex from their friends, not their parents. Until relatively recently, this problem seemed to polarize communities into two groups: those who favored sex education in schools to prevent lack of knowledge and those who insisted that sex education in the schools was unnecessary and unwise. Opponents of sex education in the schools argued that: (1) exposing children to information about sex would liven their sexual curiosity and draw them prematurely into sexual behavior; (2) teaching about sex is so closely linked to moral and religious values that it should be done at home or in a religious setting; and (3) the quality of materials and teaching in public school sex education was uneven at best, and quite poor in many cases.

Today, although opposition to sex education in the schools continues, its tone is somewhat muted. Seventy-seven percent of American adults believe sex education should be taught in schools, and when such courses are given, less than 5 percent of

parents ban their children from attending. An increasing number of school systems have some form of sex education (often
called "Family Life Education") offered in the curriculum, and
three states — New Jersey, Maryland, and Kentucky — as well
as the District of Columbia now require it. Perhaps even more
encouraging is a broad coalition of community-oriented programs, including the Y.M.C.A. and Y.W.C.A., the Girls Clubs of
America, the Salvation Army, Four-H, Campfire, Inc., and a
number of other youth-serving groups who have now begun to
implement sex education programs geared to both children and
parents.

Despite these signs of progress, there are still a number of
problems with sex education today. Certainly one of the most
pressing dilemmas is that relatively few American fathers play
an active role in providing their children with age-appropriate
sex information. Another aspect that requires attention is the
fact that sex education, beyond the most rudimentary "birds and
bees" facts of anatomy and reproduction, is often ignored by
parents and schools alike until a child reaches adolescence. Since
children are exposed to a great deal of information about sex at
an earlier age — through television shows, movies, books, and a
host of other sources — parents run the risk of allowing them to
interpret what they see as accurate depictions of what sex is all
about, which may have unfortunate consequences. Put another
way, this is education by default.

Despite the fact that some parents are vigorously opposed
to sex education for children, parents don't really have a
choice about whether their children get sex information: they
can only choose whether or not to participate in the sex education that is already taking place. Realizing this, and wanting
to do a good job in providing sex education at home, many
parents approach this task with great trepidation, being at
once unsure of how to begin, uncertain of what to say, and worried that they'll overload or frighten a child with inappropriate
detail.

The truth of the matter is that teaching children about sex
need not be different from teaching them about lots of other
things; you don't need to have a Ph.D. in agriculture to teach
children about gardening, for example. And just as you
wouldn't wait for a child to ask you about the alphabet before

exploring the A-B-C's, don't wait to talk about sex, either — take the initiative in talking about this topic.

Here are some straightforward suggestions for parents to keep in mind when it comes to sex education:

1. When you discuss sex with your child, try to do it in a matter-of-fact manner, the way you'd talk about anything else.

2. Avoid lecturing about sex. While it may relieve your anxiety to cover the whole topic in a 15-minute talk, young children don't usually have a long enough attention span for this approach and also need to ask questions about what they're learning.

3. Be sure that your discussions include more than just biological facts. Children need to learn about values, emotions, and decision-making, too.

4. Don't worry about telling a child "too much" about sex. Children will almost always tune out what they don't understand; in most cases, it will just go over their heads.

5. When your child uses four-letter words, calmly explain their meaning, and then explain why you don't want him or her to use those words. For example, you might say, "Other people get upset if they hear those words," or "I don't think that's a very good way of explaining how you feel." Remember that laughing or joking about your child's four-letter words will usually encourage repeat performances.

6. Try to use correct terminology for sexual body parts instead of using terms like "pee pee" for penis or "bottom" for vagina.

7. Even preschool-age children should know how to protect themselves from sexual abuse. This means that you need to let them know that it's okay to say "No" to an adult. Here's a good example of how this might be discussed with a four or five year old:

> You know, there are big people out there who have a hard time making friends with other big people. So sometimes they make friends with kids. And *that's* OK, but sometimes they ask kids to do things big people shouldn't ask kids to do. Like, they ask them

to put their hands down their pants, or to touch each other sexually. I love you a lot, and if anyone ever asks you to do that, or asks you to do something you think is funny and asks you to keep it a secret, I want you to say "No" and come tell me right away. (Sanford, 1982, p. 13).

8. Don't wait until your child hits the teenage years before discussing puberty. Physical changes like breast development, menstruation, and wet dreams commonly occur before age ten.

9. Be sure to discuss menstruation with boys, as well as girls, and be sure that girls understand what an erection is. Also, don't leave topics like homosexuality and prostitution out of your discussions. Most children see and hear these subjects mentioned on television or read about them and have a natural curiosity about what they are.

10. Help your child feel comfortable in coming to ask you questions about sex. Don't embarrass a child or tell him or her "you're too young to understand that now." If a child is old enough to ask questions, he or she *needs* to understand it at some level.

11. If you don't know the answer to a question your child has asked, don't be afraid to say so. Then either look it up or call on someone, such as your family doctor, who can help you with the necessary facts.

12. After you've tried to answer your child's question, check to see if your answer is understood. Also see if you've told him or her what he or she really wanted to know and give a chance to ask more questions that may arise from the answer you've provided.

We believe that waiting until a child's teenage years to provide him or her with sex education is waiting too long. Educating *all* children in an age-appropriate fashion about sexuality will ultimately help them make informed, responsible sexual choices in their lives and play an important role in the long-term prevention of sexual disorders.

CHAPTER SIX

Adolescent Sexuality

ADOLESCENCE, the period from ages twelve to nineteen, is a time of rapid change and difficult challenge. Physical maturation is only one part of this process because adolescents face a wide variety of psychosocial demands: becoming independent from parents, developing skills in interacting well with their peers, devising a workable set of ethical principles, becoming intellectually competent, and acquiring a sense of social and personal responsibility, to name just a few. At the same time this complex set of developmental challenges is being met, the adolescent must also cope with his or her sexuality by learning how to deal with changing sexual feelings, deciding whether to participate in various types of sexual activity, discovering how to recognize love, and learning how to prevent unwanted pregnancy. It is no wonder that the adolescent sometimes feels conflict, pain, and confusion.

On the other hand, adolescence is also a time of discovery and awakening, a time when intellectual and emotional maturation combine with physical development to create increasing freedom and excitement. Adolescence is not simply a period of turmoil, as older theory states, but is just as likely to be a time of pleasure and happiness as a turbulent, troubled passage to adulthood. The paradoxical nature of adolescence is particularly visible in the sexual sphere. For a comprehensive overview of adolescent sexuality, we will consider both the biological processes of puberty and psychosexual aspects of the teenage years.

PUBERTY

Puberty can be simply defined as a period of change from biological immaturity to maturity. In this transition, dramatic physical changes occur such as the adolescent "growth spurt," the development of secondary sex characteristics, the onset of menstruation (*menarche*), and the ability of the male to ejaculate. In addition, this is a time when fertility for both sexes is achieved and important psychological changes occur.

Some people think that puberty happens overnight — as a student put it, "One morning, you wake up with pimples" — but actually, the maturing process lasts anywhere from one and one-half to six years. The blueprint for the physical changes that occur during puberty was actually established before birth, when the hypothalamus and pituitary gland were programmed by hormones for a later "awakening."

Before puberty begins, there is a change in output of LH and FSH levels during sleep, due to increased production of the releasing factor from the hypothalamus that controls these hormones. Gradually, the gonads begin to enlarge as they slowly respond to this stimulation (imagine the fine-tuning you might have to do to start up a complicated piece of machinery that had not been used for eight or ten years). Shortly thereafter, sex hormone output begins to rise. During puberty, testosterone levels in boys increase ten to twenty times, while in girls they stay fairly constant. In girls, estrogen levels gradually increase eight to ten times above those found in childhood. Rising levels of adrenal hormone output before and during puberty trigger not only growth of pubic hair and axillary (underarm) hair but the increase in skin oil production that often leads to acne. These hormone changes account for many of the physical changes of puberty.

Everyone probably remembers a teenage friend who went away over a summer and came back much taller. This was not a result of sunshine, exercise, or clean living — it was simply the *adolescent growth spurt*. The explanation of this explosive growth rate is simple — it occurs because rising sex hormone levels in puberty temporarily cause bone growth.

The adolescent growth spurt typically occurs two years earlier in girls than in boys (on average, age twelve versus fourteen). As a result, girls are usually taller than boys of the same age from about age eleven to fourteen. (Remember those seventh-grade dances where girls seemed to tower over their partners, especially if they wore high heels?) The self-consciousness this may cause fortunately lessens by mid-adolescence as males "catch up" and usually go on to become somewhat taller than females.

The adolescent growth spurt does not always start at the same time in all parts of the body. For example, growth of the foot typically begins about four months before growth in the lower leg, so an adolescent's feet may seem disproportionately large. Not realizing that more harmonious relative proportions will be restored, some teenagers are upset at this pattern.

There is no correlation between an "early" growth spurt and final adult height (which is controlled partly by genetics). This finding should comfort a young teenager who is the class "shrimp"; by age sixteen or seventeen the "shrimp" may be as tall as or taller than his or her classmates.

Sexual Maturation in Girls

The first physical sign of puberty in girls is usually the beginning of breast development, which occurs as early as age eight or as late as age thirteen. In the earliest stage of breast growth, there is only a small mound of tissue (the "breast bud"), but gradually the nipple and areola enlarge and the contour of the breast becomes more prominent. Some girls have fully mature breasts before their twelfth birthdays, but others do not reach this stage until age nineteen or even later. Breast growth is controlled by estrogen levels and heredity.

The appearance of pubic hair (first lightly pigmented and sparse, gradually becoming darker, coarser, curlier, and more abundant) usually starts shortly after breast growth begins. By this time, the vagina has already begun to lengthen and the uterus is slowly enlarging. Menarche usually occurs as breast growth nears completion, and almost invariably comes after the peak growth spurt.

In the United States, the average age at menarche is 12.8 years for whites and 12.5 for blacks. It is interesting to note that a century ago, menstruation first occurred at an average age of sixteen or older; since then, the age at menarche has been consistently decreasing on a decade-by-decade basis. Socioeconomic factors, climate, heredity, family size, and nutrition all influence the age of menarche. Most recently, this decrease has leveled off, perhaps reflecting more uniform nutritional conditions.

Researchers studying the onset of menstruation believe that menarche occurs only when a minimum percentage of body fat is present. In support of this theory, female long distance runners and ballet dancers have been noted to have delayed menarche or actually to stop having periods when they are in rigorous training, just as girls who have anorexia nervosa, the self-starvation disease, typically stop having periods too.

The age of menarche varies widely from one girl to another, occurring as early as age eight and as late as sixteen or later. In the first year after menarche, menstrual cycles are frequently irregular and ovulation usually does not occur. It is possible, however, to begin ovulating with the very first menstrual cycle, and young adolescents who engage in intercourse need contraception.

One other aspect of female puberty not mentioned often is that vaginal secretions are likely to increase because of the changing hormone status. Some vaginal lubrication occurs because of sexual excitation whether this results from daydreams, reading, or sexual activity. But vaginal lubrication can also appear spontaneously, without a direct connection to sexual thoughts or acts. The sensations of vaginal wetness may be curious, pleasing, shameful, or alarming to the young teenager.

Sexual Maturation in Boys

The physical signs of puberty in boys are also controlled by hormone changes, but puberty usually starts one to two years later than in girls. The earliest change during puberty is growth of the testes, resulting from LH stimulation and subsequent testosterone production. The increasing levels of testosterone also stimulate growth of the penis and the accessory male sex organs (prostate, seminal vesicles, and epididymis). Ejaculation is not possible before puberty because the prostate and seminal vesicles do not begin to function until they receive appropriate hormone signals.

Boys begin to undergo genital development at an average age of 11.6 years and the genitals reach adult size and shape at an average age of 14.9. In some boys, genital development occurs rapidly (in about a year), while in others it may take up to five and one-half years. Sperm production (which begins in childhood) becomes fully established during puberty, so fertility is present.

There is no exact counterpart in male puberty to menarche, but wet dreams seem to have a parallel degree of psychological importance. Kinsey and his colleagues (1948) reported that one-quarter of fourteen-year-olds and nearly two-thirds of seventeen-year-olds had this experience, yet many pubertal boys are not told about the possibility of nocturnal emissions and are surprised, puzzled, or frightened upon discovery of the event. The ejaculatory experience itself or the resulting sensation of wetness or stickiness may awaken the boy having a nocturnal emission, and — just as uninformed girls may view their initial menstrual flow as a sign of illness — he may become anxious about disease or injury. Whether informed or not about this experience, the pubertal boy may attempt to "hide the evidence" of a stained sheet or pajamas to avoid embarrassment or questioning by his parents.

Growth of pubic hair begins around the time of genital development and is usually followed a year or two later by the appearance of facial and axillary hair. Facial hair growth is an important event because the earlier changes of male puberty are usually less visible than breast development in the female and beard growth is a visible sign of "becoming a man." Facial hair growth begins at the corners of the upper lip with a fine, fuzzy appearance, and then spreads to form a mustache with coarser texture. Hair next appears on the upper cheeks and just below the lower lip, and last of all develops on the chin. Body hair also appears during puberty, and chest hair continues to grow for a decade or more after this time.*

Deepening of the voice is another change of puberty and is

---

* The degree of facial and body hairiness in both sexes is controlled by genetic factors in addition to hormones. As a result, some people are hairier than others. This is often a source of embarrassment to women who compare themselves to the apparently "hairless" women in *Playboy* or *Vogue*, or to men who are exposed to macho figures of heavily mustached men with an abundant crop of chest hair in various advertisements.

caused by testosterone stimulation of the voice box, or *larynx*. As the larynx grows, the boy's voice may go through an awkward period of breaks and squawks, which may be a source of embarrassment. Like age at menarche, the average age of this voice change in boys has decreased, from eighteen years in 1749 to about 13.5 years today. Breast enlargement, or gynecomastia, is commonly seen in male puberty but is characteristically transient.

Hormone differences between adolescent boys and girls also cause differences in body shapes. For instance, the average seventeen- or eighteen-year-old boy has a leaner body and more muscle mass than his female counterpart. This is because estrogens cause accumulation of fat under the skin, while testosterone stimulates muscle growth. The structure of pelvic bones is also different in males and females, with the wider female pelvis creating a properly sized birth canal.

Pubertal Hormones and Sexuality

During puberty, rising hormone levels contribute to an activation of sexual sensations and erotic thoughts and dreams for boys and girls. John Money in 1980 described the role of hormones as follows: "the correct conception of hormonal puberty is that it puts gas in the metaphorical tank and upgrades the model of the vehicle, but it does not build the engine nor program the itinerary of the journey."

The relationship between pubertal hormones and sexual behavior is shown in the finding that boys who undergo "late" puberty (around ages fifteen or sixteen) generally have less and later teenage sexual activity — including masturbation and intercourse — than boys who have "early" puberty (around ages twelve or thirteen). Kinsey and his colleagues pointed out this pattern and we have some preliminary data showing that it is probably true. If testosterone levels of the pubertal boy increase the frequency or intensity of erections, for example, he may possibly have a heightened awareness of sexual sensations. Increased testosterone in the blood may also influence the brain itself to activate sexual feelings or thoughts or to lower the threshold for external triggers that activate such feelings or thoughts. Boys with higher testosterone levels, then, are more likely to be more physically developed and sexually active. Shorter, less muscular, later-maturing boys may experience a social handicap. While having sexual feelings, they may feel less confident about their abilities and therefore "lag" in sexual behavior.

In parallel fashion, girls who undergo "late" puberty seem to have a lower rate of early adolescent sexual activity than girls who complete puberty at ages twelve or thirteen. Although a lower frequency or later age of participation in sexual activity might be explained by purely psychological or social factors (e.g., less physically developed girls may be more shy or self-conscious about sex with a partner), it appears that masturbation is less frequent and occurs later in late-maturing compared to early-maturing adolescent girls. Recently, a large cross-cultural study found similar evidence. According to Udry and Cliquet (1982), data from five different countries show that girls who are younger at menarche tend to have intercourse and to give birth at earlier ages than girls with later menarche.

In contrast to the findings linking sexual activity to "early" puberty, when pubertal changes occur before age nine — a condition called *precocious puberty* — there is usually no accompanying change in sexual behavior. This is probably because the hormonal stimulation alone is not enough to initiate new behavior patterns without a state of psychosexual readiness that the younger child simply hasn't attained.

While precocious puberty is usually a rare condition, occurring in about one in 10,000 children, there has recently been an epidemic of early sexual development among children in Puerto Rico. Premature breast development in infants and pre-schoolers, sometimes accompanied by the onset of periods, is the most common problem, with an estimated 3,000 Puerto Rican children (about one in fifty) having been affected between 1972 and the end of 1983. Many physicians suspect (but have been unable to prove) that the problem is caused by eating chicken that contains estrogen. Despite the fact that the Food and Drug Administration banned the use of estrogen to stimulate growth of food animals over a decade ago, tests have shown that some chickens raised in Puerto Rico contain high levels of this hor-

mone. In addition, a majority of the children who stop eating
chicken (a relative staple of the Puerto Rican diet) have had a
major reduction of their abnormal anatomical development.

PSYCHOSEXUAL ASPECTS OF ADOLESCENCE

Sexual fantasies and dreams become more common and explicit
in adolescence than at earlier ages, often as an accompaniment
to masturbation. One study found that only 7 percent of adolescent girls and 11 percent of adolescent boys who masturbated
never fantasized, and about half reported using fantasy most of
the time during masturbation. Fantasy seems to serve several
different purposes in adolescence: it can add to the pleasure of
a sexual activity, be a substitute for a real (but unavailable) experience, induce arousal or orgasm, provide a form of mental
rehearsal for later sexual experiences (thus increasing comfort
and anticipating possible problems just as rehearsing any other
kind of activity can), and provide a safe, controlled, unembar-
rassing means of sexual experimentation. Each of these functions of fantasy is a forerunner of ways in which sexual imagery
will continue to be used in adulthood by most people. For this
reason, the adolescent's experience in and exploration of the
range and uses of fantasy is important to her or his later sexual
existence and confidence.

As adolescents struggle to establish a sense of personal identity
and independence from parents and other authority figures,
interactions with their peer group (other people of about their
same age) become increasingly important. Teenagers look to
each other for support and guidance, vowing to correct the mistakes of the older generations, but quickly discover that their
peer group has its own set of expectations, social controls, and
rules of conduct. Thus the adolescents' need for freedom is
usually accompanied by a need to be like their friends, even
though these two needs sometimes conflict.

Peer group pressures vary from one community to another
and also reflect the ethnic and economic subcultures within each
community. In one group, the code of sexual conduct may be
very traditional, with a high premium on female virginity and
almost all sexual activity limited to "meaningful" relationships.

If this code is not followed by females, they will get a "reputation," which may tarnish their futures and make them prey for
boys looking for an "easy lay." In another group, sex may be
viewed as a status symbol — the "uninitiated" versus "those in
the know." This view often motivates members of the group to
participate in sexual activity to be accepted. The suggestion has
been made that a new tyranny of sexual values is emerging;
teenagers are expected by their peers to become sexually experienced at an early age and those who are not comfortable with
this pressure are viewed as old-fashioned, immature, or "uptight."

The teenager's sexual decision-making reflects individual psychological readiness, personal values, moral reasoning, fear of
negative consequences, and involvement in romantic attachments. These personal factors are often not compatible with
peer pressures and seem to be felt as limits more strongly by
adolescent females than by males in our society. It appears that
teenagers who engage in sexual intercourse and those who are
close to doing so place a high premium on personal independence, have loosened family ties in favor of more reliance on
friends, and are more apt to experiment with drugs or alcohol
and to engage in political activism than their contemporaries.

In seeking to become free from parental or adult control,
some adolescents see sex as a way of proving their ability to make
independent decisions and of challenging the values of the older
generation. Their freedom is not achieved so easily: adolescents
manage to acquire a sizable sexual legacy from the older generation complete with a persistent double standard and a strong
sense of sexual guilt. Teenage *attitudes* have changed more rapidly than behavior, since an attitude of equality between the
sexes is now fairly widespread; yet the old double standard persists in certain ways. The male is still expected to be the sexual
initiator; if the female assumes this role, she is likely to be viewed
as "aggressive" or "oversexed." Adolescents have not rid themselves of all sexual conflict, misinformation, and embarrassment;
instead, it seems they have sometimes traded one set of problems
for another.

While adults generally encourage adolescents to develop independence, our society puts teenagers in a double-bind. No
longer children, not yet adults, adolescents are expected to act

grown up in many ways, but this attitude usually does not extend to their sexual behavior. Many adults seem threatened by adolescent sexuality and try to regulate it in illogical ways: ban sex education in schools (it would "put ideas in their heads"), limit information about contraceptive methods ("keep them afraid of getting pregnant"), censor what teenagers read or can see in movies ("pure minds think pure thoughts"), invent school dress codes ("modesty conquers lust"), or simply pretend that adolescent sexuality does not exist.

Fortunately, not all parents adopt such a negative view of teenage sexuality, and in some instances parents take a much more liberal stance. Not only are there some parents who discuss sex very openly and assist their teenage daughters and sons in obtaining contraception, a few parents actually pressure their adolescent children into becoming sexually experienced. This attitude sometimes reflects the parents' desire to relive their own teenage years through the experiences of their children.

It is also important to realize that teenagers may create pressure for their parents by their sexual behavior. Most parents are concerned about the possibility of an unwanted teenage pregnancy, realizing that even if their son or daughter has access to contraception, that does not mean it will be effectively used whenever needed. Parents are also realistically worried about venereal disease. In addition, many parents are caught in a double-bind of their own: they do not want to seem old-fashioned and unduly restrictive but they genuinely believe in traditional values about sexual behavior that the teenager may have a hard time understanding. Interestingly, some parents become worried if their teenage child *does not* show any interest in the opposite sex, since they interpret this as a possible sign of homosexuality.

Most parents, regardless of their own sexual life-styles, have a tendency to be less permissive about premarital sex for their own children. Perhaps as a result, when parents are the primary source of sex education, adolescents have more traditional sex values and have higher rates of virginity.

PATTERNS OF SEXUAL BEHAVIOR

Discussing patterns of sexual behavior during adolescence depends on interpreting the data available from various researchers who collected information in different times and places using widely different sampling methods. The findings of Kinsey and his colleagues will be mentioned in each section on sexual behavior as a reference point, but readers should remember that these statistics are now more than thirty years old.

Masturbation

Kinsey and his colleagues found a marked difference between adolescent females and males in the incidence of masturbation. While 82 percent of fifteen-year-old boys had masturbated to orgasm, only 20 percent of fifteen-year-old girls had done so, and this pronounced difference persisted through the rest of the teenage years. In a survey published in 1973, Sorenson found that 39 percent of adolescent girls and 58 percent of adolescent boys had masturbatory experience. It appears that by age twenty, this figure rises to about 85 percent of males and 60 percent of females. Most recently, our own data from interviews with 580 women aged eighteen to thirty indicated that more than three-quarters had masturbated during adolescence, confirming indications of a trend to a higher incidence of masturbation in female teenagers since Kinsey's time.

In spite of this behavioral trend, guilt or anxiety about masturbation continues to plague teenagers. Fifty-seven percent of adolescent girls and 45 percent of adolescent boys reported such negative feelings "sometimes" or "often" according to Sorenson, and other workers have confirmed this finding. On the other hand, masturbation fulfills some important needs for adolescents: relieving sexual tension, providing a safe means of sexual experimentation, improving sexual self-confidence, controlling sexual impulses, combating loneliness, and discharging general stress and tension. The interplay of guilt and pleasure is shown in the following comment from a nineteen-year-old female college student:

I began to masturbate in a very fumbling, uncertain way when I was about 14. At first, it didn't do much for me, and I started to think that maybe it was wrong, like they taught me in Sunday school. Then one night I was reading a really sexy book and I started rubbing myself as I read. Suddenly I had this gigantic orgasm — it really took me by surprise I guess — and from then on I found masturbation a lot more enjoyable. I was glad to see that I could come like that and it gave me a lot of self-confidence. Now I never feel guilty about masturbating, and I use it to relax or just feel good.

Petting

Kinsey and his colleagues defined petting as physical contacts between females and males in an attempt to produce erotic arousal without sexual intercourse. Most authorities would narrow this definition a bit and not include kissing as a form of petting, and some define petting as sexual touching "below the waist" while any other sexual touching is called "necking."

Kinsey's team reported that by age fifteen, 39 percent of girls and 57 percent of boys had engaged in petting and that by age eighteen, these figures had risen to over 80 percent for both sexes. Only 21 percent of boys and 15 percent of girls, however, had petted to orgasm before age nineteen. Sorenson found that 22 percent of his sample had no sexual experience other than kissing, and 17 percent had some petting experience without having intercourse. More recently, interviews with sixty first-year college students about their high school sexual experiences showed that 82 percent had experience in genital stimulation with a partner; 40 percent of the women and half of the men reported having been orgasmic during petting.

"Necking" and "petting" must be looked at in terms of the changing trends in teenage sexual behavior discussed more fully in the next section. Along with engaging in most forms of sexual behavior at an earlier age, many of today's teenagers have moved away from older rituals of dating and "going steady" in favor of less structured patterns of social interaction. Widespread use of illicit drugs such as marihuana may have contributed to this change, and evidence from several sources indicates that adolescents who use such drugs are more sexually experienced than adolescents who do not use marihuana or other drugs.

Sexual Intercourse

The first time I had intercourse I was 17 years old. I was a senior in high school and I'd been going with a guy I really thought I loved. We had done just about everything else and it seemed kind of silly to stay a virgin any longer so one night we just went ahead. No big planning or discussion or lines, it just happened. I was nervous at first but it turned out really nice. From then on, we had intercourse two or three times a week and it was great sex. I have no regrets at all.

I was 16 the first time I tried to have intercourse. My girlfriend was younger but had done it before. I was so nervous I couldn't get it in, and then when she tried to do it it went soft. We tried for hours but no luck. I was really down. A few days later, we tried again, and it was smooth as silk. I felt really good then, like everything was okay.

My first time was very unpleasant. The boy I was with rushed and fumbled around and then came so fast it was over before it started. I thought. "What's so great about this?" For weeks afterward, I was afraid I had V.D. and had bad dreams about it.

The first experience of sexual intercourse can be a time of happiness, pleasure, intimacy, and satisfaction, or it can be a source of worry, discomfort, disappointment, or guilt as these descriptions show.

According to the available research data, the age of first sexual intercourse has declined in the last few decades, particularly for teenage girls (Table 3). In 1953 Kinsey and his colleagues reported that only 1 percent of thirteen-year-old girls and 3 percent of fifteen-year-old girls were nonvirgins; by age twenty, this figure had only increased to 20 percent. In contrast, in 1973 Sorenson found that nearly one-third of thirteen- to fifteen-year-old girls and 57 percent of sixteen- to nineteen-year-old girls were nonvirgins. Reporting in 1975, Jessor and Jessor noted that 26, 40, and 55 percent of girls in tenth, eleventh, and twelfth grades were no longer virgins. Even more recently, Zelnik and Kantner found that the prevalence of sexual intercourse among never-married American teenage women increased by nearly two-thirds between 1971 and 1979.

Statistics concerning the age of adolescent males' first sexual intercourse show less change over time. Kinsey and his colleagues reported that 15 percent of thirteen-year-old boys and 39 percent of fifteen-year-old boys were nonvirgins; by age twenty, this figure had increased to 73 percent. In 1973, Sorenson found that 44 percent of thirteen- to fifteen-year-old boys and 72 percent of sixteen- to nineteen-year-old boys were coitally experienced. According to Zelnik and Kantner, 56 percent of never-married seventeen-year-old males and 78 percent of never-married nineteen-year-old males were nonvirgins.

The types of relationships generally found between teenagers and their first coital partners are shown in Table 4. It is a mistake, however, to regard the lower age of first sexual intercourse as a sign of teenage promiscuity because many teenagers restrict themselves to one sex partner at a time. In fact, many adolescents who are no longer virgins have intercourse infrequently. For some teenagers, particularly those who "tried" intercourse as a kind of experimentation, once the initial mystery is gone, the behavior itself is far less intriguing. As a result, they may

TABLE 3
Percent of Unmarried American Teenage Females with Coital Experience

AGE	STUDY			
	Kinsey et al. (1953)	Sorenson (1973)	Zelnik and Kantner (1980)	
			[1971]	[1979]
13	1	9	—	—
14	2	15	—	—
15	3	26	14	23
16	5	35	21	38
17	9	37	26	49
18	14	45	40	57
19	17		46	69

have little or no sexual intercourse for long periods of time — sometimes waiting to meet the "right person." Teenagers in long-term romantic relationships are more likely to participate in coitus fairly regularly.

In the last few years, it has become apparent that among sexually experienced teenagers, a group is emerging who are disappointed, dissatisfied, or troubled by their sex lives. Given the name "unhappy nonvirgins" by Kolodny, this group includes an estimated 30 percent of adolescents who have had coital experience. In some cases, these are teens who had such high expectations of what sex "should" be that they feel like either failures or dupes when their actual experience is less than earth-shattering ecstasy. In other instances, these teenagers have experienced sexual dysfunctions that have prevented them from enjoying sex. Still others in this group enjoy sexual activity initially but become disillusioned when sex dominates their relationship ("That's all he ever wants to do now") or when their relationship breaks up and they feel that they've been used or manipulated. Many of these "unhappy nonvirgins" revert to abstinence as a means of coping, hoping that when they're older — or when

TABLE 4
Percentage Distribution of Women Aged 15–19 and of Men Aged 17–21, by Relationship with Their First Sexual Partner, According to Race

Relationship with first partner	Women			Men		
	Total (N=936)	White (N=478)	Black (N=458)	Total (n=670)	White (N=396)	Black (N=274)
Engaged	9.3	9.6	8.2	0.6	0.5	1.0
Going steady	55.2	57.6	46.5	36.5	39.2	21.9
Dating	24.4	22.2	32.6	20.0	20.2	19.0
Friends	6.7	6.0	9.4	33.7	30.2	52.4
Recently met*	4.4	4.6	3.3	9.3	9.9	5.7
Total	100.0	100.0	100.0	100.0	100.0	100.0

* Among women, this category includes a small number who reported some other relationship.

Source: Zelnik and Shah, 1983, p. 65. Reprinted with permission from *Family Planning Perspectives*, Volume 15, Number 2, 1983.

they meet the right person — things will be different. Others continue to be sexually active while deriving little, if any, enjoyment of sex.

Homosexual Experience

The Kinsey studies showed that it was fairly common for males to have at least one homosexual experience during adolescence, while considerably fewer adolescent females engaged in sex with another female. More recently, there seems to have been a moderate decline in adolescent homosexual experience. Sorenson found that 5 percent of thirteen- to fifteen-year-old boys and 17 percent of sixteen- to nineteen-year-old boys had ever had a homosexual experience, and 6 percent of all adolescent females he surveyed had at least one episode of homosexual activity. Hass reported in 1979 that 11 percent of the teenage girls and 14 percent of the teenage boys he studied had at least one sexual encounter with a person of the same sex, but noted that this was probably an underestimate because many respondents did not regard preadolescent "games" as sexual acts.

It is important to realize that an isolated same-sex encounter or a transient pattern of homosexual activity does not translate into "being homosexual." Most adolescents who have had some experience with homosexual activity do not see themselves as homosexuals and do not go on to homosexual orientation in adulthood. Nevertheless, some adolescents develop guilt or ambivalence about their sexual orientation as a result of a single same-sex episode and may experience emotional turmoil.

The teenager who is worried about being homosexual may deal with it in a variety of ways. Some avoid homosexual contacts while trying to reaffirm their heterosexual identity through dating and heterosexual activity. Others withdraw from all sexual situations. Still others look on themselves as bisexual, consider homosexual arousal a passing phase which they will outgrow, or seek help from a professional.

Some adolescents intuitively "feel" that they are homosexual or work through their initial confusion about sexual identity to accept their homosexuality in a positive way. These teenagers may seek out readings on the subject, contacts with other homosexuals, and a social introduction to the homosexual subculture. As we will discuss in chapter 14, these persons face some difficulties because of current attitudes toward homosexuality, and they may not choose to announce their sexual preferences to family or friends (referred to as "coming out") until a later time, if at all.

UNINTENDED TEENAGE PREGNANCY

More than one million pregnancies occur each year among American teenage females, which is equivalent to one adolescent pregnancy beginning every 35 seconds. Since the majority of these pregnancies are unplanned and unwanted, it is no surprise that they frequently create considerable psychological anguish, serious economic consequences, and even health risks that are too often ignored or misunderstood.

A few background statistics can highlight the scope of this epidemic.

- 30,000 pregnancies occur annually among girls under fifteen years of age.
- 400,000 American teenagers have abortions each year, accounting for more than one-third of all abortions performed in this country.
- Six out of ten teenage females who have a child before age seventeen will be pregnant again before age nineteen.
- With one out of twenty adolescent females having a baby each year, America's teenage birth rate is the highest in the western hemisphere, double the rate of Sweden, and is an astonishing seventeen times higher than Japan's.
- Four out of ten girls now fourteen years old will get pregnant in their teens.

These statistics show how widespread the problem of unintended teenage pregnancy is, but to understand *why* it is a problem, we need to examine some additional aspects of the consequences of teenage pregnancy. To begin with, there are increased health risks associated with teenage pregnancy, particularly among younger teens (those in the thirteen- to sixteen-year age group). For example, babies of teenage mothers have an increased chance of being underweight and are nearly twice as likely to die in infancy as those born to women in their twenties. In addition, teenagers tend to have more medically complicated pregnancies — including miscarriages, toxemia, and hemorrhage — as well as a higher risk of maternal death than women in their twenties.

Possibly even more alarming than these medical risks are the socioeconomic consequences of unintended teenage pregnancy. Even though it is now illegal to expel students who are pregnant or who are mothers from public schools, most teenage mothers who keep their babies drop out of school and don't return. Largely as a result of this abrupt withdrawal from formal education, women in this group are far less likely than their peers to enter the job market or to gain regular employment. It is no surprise, then, that these teenage mothers are overrepresented in poverty statistics and are apt to become largely dependent on government services and support.

Unmarried teenage girls who find themselves pregnant are confronted by a series of psychologically complicated choices, as well. They often get little or no support — either emotionally or financially — from the child's father. They must decide whether to have an abortion (which sometimes produces intense feelings of guilt and anguish) or have the baby. If they have the baby, they then must decide to keep it or put it up for adoption; today, fewer than 5 percent of unwed teenage mothers choose adoption as a course of action. In other cases, their partners may pressure them to do something they don't want to, thus creating additional pressures and uncertainties. Here's how one seventeen-year-old described her dilemma:

> When I found out I was pregnant, my boyfriend insisted that we get married and have the baby. I had no interest in marrying him or in being saddled with an infant at age eighteen, so I refused. But his parents hired a lawyer to try to stop me from having an abortion, and the whole thing wound up being a nightmare for me and my parents. Fortunately, I got the abortion and dumped my so-called boyfriend, so I'll be going to college next year instead of playing mommy.

Some teenagers, unlike the one quoted above, find themselves rushed into unanticipated marriage as a result of a pregnancy. Unfortunately, these marriages are much likelier than most to end in divorce or desertion, and there is a suicide risk among these young women that is considerably higher than in the general population.

There is relatively little research describing the consequences of unintended teenage *fatherhood*. This may be partly because it is difficult to identify these individuals for study and partly because they are not socially or economically linked to the pregnancy outcome in the same ways mothers are. However, the available evidence (summarized concisely in reports from the Alan Guttmacher Institute and the Ford Foundation) shows that males who become fathers while in their teens tend to have lower income and less educational attainment than peers who postpone fatherhood until their twenties. Nevertheless, the impact of teenage pregnancy is considerably less on males than on females.

Clearly, many adolescent males continue to regard the ultimate responsibility for contraception as the female's, generally feeling that an unintended pregnancy could have been prevented and thus is the "fault" of the female — in other words, that it is "not their worry." Others feel a joint responsibility that extends only to offering to share (or perhaps pay entirely) for the cost of an abortion; to them, this gesture is an honest acknowledgment of their involvement and willingness to help, but it is involvement of the most limited sort. In fact, as the noted sex educator Sol Gordon points out, "almost 90 percent of all teenage boys who make a teenage girl pregnant abandon her."

While there are no easy solutions to the problem of unintended teenage pregnancy, it appears that misinformation or complete lack of information is a key factor. At present, only one-third of American junior and senior high schools offer sex education courses, and many of those offered are remarkably incomplete. Since many of the sex education courses are given only to older teenagers, their preventive function is lessened considerably. Those people who believe that sex education should be taught in the home — while voicing a fine idea — overlook the reality of the situation today. Research indicates that only about 10 percent of parents discuss sexuality with their teens beyond simply saying "don't." On the other hand, a 1982 study by Zelnik and Kim demonstrates that among unmarried sexually active teenage women, those who have had sex education courses have fewer pregnancies than those who haven't.

Almost all authorities agree that greater responsibility for contraceptive use by the adolescent male is a major element in the effort to reduce the rate of unintended teenage pregnancy. First, educating males about contraceptive options at an early age seems warranted since studies suggest that this information leads to better contraceptive use. Although teenage males are generally unwilling to admit to ignorance or misinformation about sex, it is not unusual to find fifteen- or sixteen-year-old boys who believe that a diaphragm should be removed right after intercourse, or who don't know the fertile days in a female's menstrual cycle. Such education need not be restricted to schools — it can be done at home, in church-affiliated programs, or as part of community projects. Education must be practical, too, explaining how and where to purchase contraceptives, why it's important to discuss birth control with a partner, and why consistent contraceptive use is necessary.

Another important step is to provide males (as well as females) with a better view of how birth control practices relate to their own lives. For instance, teens must recognize how rigid sex roles or the risk of parental disapproval can influence their contraceptive behavior. In addition, teens need to be aware that the risk of contracting a sexually transmitted disease is materially reduced by use of certain contraceptive methods. This is important not only because it encourages the teenage male to use contraception, but because the male's expression of interest and concern about contraception encourages his partner to find and use an appropriate method as well. In addition, teenagers need incentives to engage in responsible birth control practices.

Some authorities suggest public campaigns geared at urging teens to say "no" to having intercourse. This approach might be effective with some adolescents, but would probably not be realistically effective with the majority of teenagers, given today's patterns of sexual behavior in our culture: it is hard to put a genie back in a bottle. Also, such an approach runs the risk of being repressive — it is, after all, an attempt to frighten teens into abstinence — and this may produce a backlash. Indeed, anti-drug and anti-cigarette campaigns have often been discredited by teens on this basis. In any case, since it is unlikely that the majority of sexually experienced teenagers will become celibate en masse, it is necessary to provide teens with positive role models toward appropriate contraceptive use and a more effective view of the ways in which responsibility in sexual behavior is important to their welfare.

CHAPTER SEVEN

Adult Sexuality

TRADITIONALLY, developmentalists have studied childhood and adolescence as a means of understanding adulthood. This viewpoint is useful in many ways, but it has inadvertently created the impression that development stops at a precise moment, leaving the adult a relatively static creature. Another view that can be traced back to the writings of psychologists Carl Jung and Erik Erikson stresses the developmental aspects of adulthood. According to this perspective, adulthood is a continuing pattern of learning, crisis, and choice. Daniel Levinson has suggested that the life cycle is a kind of journey:

> Many influences along the way shape the nature of the journey. They may produce alternate routes or detours along the way; they may speed up or slow down the timetable within certain limits; in extreme cases they may stop the developmental process altogether. But as long as the journey continues, it follows the basic sequence. (Levinson, 1978, p. 6)

Gail Sheehy has added to the view of adult developmental stages by conceptualizing them as a series of passages through fairly predictable crises that contribute to our growth. She observes:

> During each of these passages, how we feel about our way of living will undergo subtle changes in four areas of perception. One is the interior sense of self in relation to others. A second is the proportion of safeness to danger we feel in our lives. A third is our perception of time — do we have plenty of it, or are we

beginning to feel that time is running out? Last, there will be some shift at the gut level in our sense of aliveness or stagnation. These are the hazy sensations that compose the background tone of living and shape the decisions on which we take action. (Sheehy, 1976, p. 21)

We will only take an abbreviated look at adult sexuality here as the rest of this book is largely about the sexual experiences and sexual problems of adults. The notion of adulthood as a time of transition and development is a useful one to retain while reading subsequent chapters.

EARLY ADULTHOOD

The phase of early adulthood, from approximately ages twenty to forty, is a time when people make important life choices (marriage, occupation, life-style) and move from the relatively untested ambitions of adolescence to a personal maturity shaped by the realities of the world in which they live. For most people, it is a time of increasing responsibility in terms of interpersonal relations and family life.

In recent years, both in the United States and abroad, there has been a definite trend toward marriage at a later age than in past decades. As a result, many young men and women face an extended period of being single after adolescence that unquestionably changes patterns of sexual behavior from Kinsey's day. Today, most people in their twenties believe that becoming sexually experienced rather than preserving virginity is an important prelude for selecting a mate. Erikson remarked in 1968 that developing the capacity for intimacy is a central task for the young adult.

Young adults are generally less subject to "sexual peer pressure" than adolescents but more driven by an internal need to become sexually knowledgeable. Freedom from parental limits is accompanied by easier access to private surroundings (an apartment, a motel room, a vacation spot), which also creates more sexual opportunity. In this state of singlehood, several common patterns of sexual behavior can be seen. The *experimenter* seems to judge sexual experiences in terms of frequency, variety, and performance proficiency. She or he seems to view the world as a sexual smorgasbord and generally has the attitude that "now's the time to play, because later I'll settle down." The *seeker* strives to find the ideal relationship (and perfect marriage partner) by developing sexual relationships and hoping for the best. Living together can be a proving ground for a relationship begun on this basis. The *traditionalist* participates willingly and joyously in sex, but reserves intercourse for "serious relationships." The traditionalist may have several sexual partners before ultimately marrying, but does so one at a time. There are undoubtedly other patterns that can be identified, but these three seem to be most common.

The early years of adulthood are a time of sexual uncertainty for some and sexual satisfaction for others. Conflict can arise because of attitudes of sexual guilt or immorality carried over from earlier ages. The adolescent's concern with sexual normality has not fully disappeared, and the young adult continues to worry about his or her physique, sexual endowment, and personal skill in making love. Sexual identity conflicts may not yet have been resolved, and even for those who have come to accept themselves as homosexual or bisexual, social pressures and prejudices may cause some difficulty.

Despite the existence of such problems, young adults are more sexually active today than in the past. A major factor contributing to this change has been the relative disappearance of the old double standard that regarded premarital sexual experience as permissible for men but not for women. Thus, it is not surprising to see that the gender gap in premarital sexual experience has narrowed considerably from what it used to be.

Although the popular notion is that being young and single automatically leads to sexual happiness, the reality may be somewhat different. In one recent survey of 250 college students, for instance, 43 percent stated that they were concerned about being unable to find time for sexual relations, and 40 percent had problems with lack of privacy for sex. This survey also found relatively high rates of sexual dysfunctions: 37 percent of the students had difficulty becoming vaginally lubricated or getting erections at least half of the time; 30 percent of the females had trouble reaching orgasm; and 23 percent of the males ejaculated too quickly. College students are not the only ones who have

CATHY   **by Cathy Guisewite**

these sorts of problems, either. A 1983 *Psychology Today* survey found that 28 percent of men and 40 percent of women complained of lack of sexual desire, while almost 20 percent of both sexes admitted to fears about sexual adequacy.

Today's young adults are faced with some additional sexual conflicts that may represent a sort of backlash against the "anything goes" banner of the sexual revolution of the 1960s and 1970s. For example, while attitudes toward premarital sex have changed dramatically in the last three decades, having sex with a large number of partners is still somewhat frowned on. Furthermore, although most singles don't believe that love is necessary for good sex, there seems to be increasing disillusionment with casual sex or one-night stands.

This trend seems to be at least partly the result of increased awareness of the possibility of exposure to a sexually transmitted disease, such as the much-publicized genital herpes. Among young adult homosexual men, who have — as a group — typically participated in casual sex much more than their heterosexual age-mates, fear of contracting AIDS has also led recently to a reduction in the number of sexual partners and more interest in establishing "monogamous" relationships. Fear is not, however, the only element operating here. Many of the young adults we have interviewed are concerned with another aspect of casual sex: its relatively impersonal nature. The following remarks are typical of what we have been told:

> *A twenty-six-year-old man:* Having one-night stands was fun at first because there were no demands attached, no one's expectations to fulfill. But after a year or so I began to realize that something

was missing from these encounters — a sense of caring about the person I was making it with, or a feeling that she cared about me.

A thirty-year-old woman: You just can't compare the quality of sex with someone you hardly know and feel nothing for with the quality of sex in a caring relationship. Casual sex is just mechanical, one-dimensional release. Sex with someone I care about is warmer and psychologically far more satisfying.

Why are a number of young adults becoming disillusioned with having only casual sexual encounters? Peter Marin, in an article titled "A Revolution's Broken Promises," offers one interesting analysis of what may be occurring: he suggests that while loosening restraints on sexual behavior creates a climate of sexual freedom and choice, this freedom is not unequivocally positive. Sexual freedom can lead to disappointment, pressure, and conflict as well as satisfaction, so that "To the extent that it diversifies and expands experience, it also diversifies and multiplies the pain that accompanies experience, the kinds of errors that we can make, the kinds of harm we can do to one another."

Similarly, Susan Washburn claims that the commercialization of sex in our society leads to completely unrealistic expectations of what sex should be:

When sex is treated as a commodity, the consumerist credo, "more is better" is extended to sexual interactions. If one partner is good, two are better, and an orgy the ultimate sexual experience; if one orgasm is good, a Chinese firecracker string of multiple orgasms is better. . . . We collect sexual experiences with the same compulsiveness that marks our accumulation of material goods. We're afraid that we won't get ours before the supply runs out. (Washburn, 1981, p. 230)

To be sure, sexual experiences in early adulthood are often warm, exciting, gratifying, and untroubled. Even casual sex can serve a number of useful purposes, psychologically as well as physically, and there is certainly no reason why having fun is to be frowned on. But the prevailing trend is clearly toward sex in the context of caring relationships, and one place this is particularly evident is in the relatively recent growth of cohabitation — unmarried heterosexual couples living together.

Current estimates suggest that more than two million people

(4 percent of unmarried American adults) are cohabiting, while in Sweden more than 90 percent of the populace has tried this pattern. Although cohabitation is *not* a practice limited to college students, about one-quarter of college students have had the experience and another 50 percent would like to. Actually, about half of the people now cohabiting have previously been married, although the majority of cohabitants appear to be under age thirty-five.

What factors led to the popularity of cohabitation? One psychologist suggests that the women's movement, loosening of restrictive college housing regulations, and the radical political climate of the late 1960s combined with other social changes to promote this practice (Macklin, 1978, pp. 199–200):

> The increase in divorce, and the changing conception of the function of marriage, caused many young single people and divorcees to move cautiously into that state. The increased acceptance of sexuality outside marriage and improved contraception made it easier for nonmarried persons to engage openly and comfortably in a sexual relationship. And the increased emphasis on relationships and personal growth called into question the superficiality of the traditional dating game, and led to a search for styles of relating that allowed for change, growth, and a high degree of total intimacy.

Generally, there are three basic forms of cohabitation — casual or temporary involvement, preparation or testing for marriage, and substitute for or alternative to marriage. Most college students who begin cohabitation see their relationship as affectionate but uncommitted. In one study, however, 96 percent of the students who had cohabited said that they wanted to marry in the future. Often, cohabitation serves as an added step in the courtship process. In fact, given the current trend of delaying the timing of marriage, cohabitation for a period of several years or more may be becoming something of a cultural fixture.

In their recent study of relationships, Blumstein and Schwartz included 653 cohabiting heterosexual couples who had lived together an average of 2.5 years. They found that while cohabitants had more frequent sex than married couples, as cohabitants stayed together longer their frequency of sexual relations declined. Other key findings included:

1. As with married couples, when there were problems in non-sexual areas of the cohabitants' lives, their sex lives suffered.
2. Cohabiting women initiate sex more often than married women, but in older cohabiting couples the male often resents this.
3. Only about one-third of people who are cohabiting have sex with other people outside their relationship.

Blumstein and Schwartz believe that it will be difficult for cohabitation to become a viable permanent institution as long as the traditional marriage model exists.

According to Cherlin, however, not only are people in America now becoming more tolerant of cohabitation, there is some evidence that it is being institutionalized as part of the family system. "Far from being a threat to the primacy of marriage," Cherlin says, "cohabitation is becoming more and more like the first stage of marriage." Since even expert sociologists can't agree, it will probably take more studies to discover the trends.

What effects does cohabitation have on the participants? Generally, cohabitants have given very positive ratings to their experience, describing it as "maturing," "fostering emotional growth and personal understanding," and "improving skills in heterosexual relationships." The great majority of college students surveyed who have had prior experience with cohabitation say they would never marry without living with the person first. Although critics are concerned that cohabitation leads to erosion of the family and a reduced marriage rate, there is little reason to believe that many people are permanently substituting living-together arrangements for marital relationships. Although there is not enough current data, it is possible that cohabitation may actually lower the divorce rate by allowing couples a closer premarital look at each other.

In contrast to single life, marriage, which will be discussed in greater detail in chapters 9 and 13, presents different sexual development patterns. For better or for worse, a majority of young adults eventually marry, embarking on a voyage that can itself create certain sexual difficulties. As the novelty of early marital bliss dissolves in the process of learning to live with one another's quirks and habits, as early dreams of conquering the world give way to a more practical focus on details of everyday

life, sex is likely to become less exciting and sometimes less gratifying for one or both partners. Reflecting this tendency, the
frequency of sexual activity generally declines in the early years
of marriage. Parenthood leads to less privacy, more demands,
and even exhaustion. It is hard to get excited about sex if you
have been running after a two-year-old all day long, just as sex
is apt to lose its attraction if you have worked a fourteen-hour
day at the office.

Pregnancy itself may influence a couple's sexual activity, although it has no uniform effect on sexual feelings or function.
For some women, pregnancy is a time of heightened sexual
awareness and sensual pleasure, while for others no changes are
noticed or sexual feelings decline. Some couples find that late in
pregnancy the awkwardness of a bulging belly and concern
about the baby lead to voluntary abstention from sex. For others,
adjustments in sexual positions or the use of non-coital sex play
solves these problems fairly easily.

There is marked variation in patterns of sexual behavior in
the first trimester of pregnancy. Not surprisingly, women with
morning sickness and high levels of fatigue report losing interest
in sex and a lower frequency of sexual activity, but other women
experience just the opposite effects. In the second trimester,
however, 80 percent of pregnant women note heightened sexuality both in terms of desire and physical response. In the last
trimester, there is typically a pronounced drop in the frequency
of intercourse. While many women think this happens because
they are less physically attractive, husbands generally deny this
explanation and instead voice concern about injuring the fetus
or their wife.

A few practical guidelines about sex during pregnancy are in
order. Women who have a history of previous miscarriages or
whose pregnancy is in danger of miscarriage should abstain
from any type of sexual activity that might result in their having
orgasms, since contractions of the uterus during orgasm could
be risky. If vaginal or uterine bleeding occurs during pregnancy,
it is also wise to avoid all forms of sexual activity until receiving
a medical okay. Air blown forcefully into the vagina during oral-
genital contact can be dangerous for the pregnant woman, if it
causes air embolism (air bubbles in the bloodstream). Cunnilingus *without* air blown into the vagina is not risky. If the mem-

branes have ruptured, intercourse or cunnilingus should be prohibited because of the danger of fetal infection. *Aside from these few cautionary notes, sex during pregnancy is quite safe for the fetus and the mother.*

A few doctors have voiced concern about orgasms causing premature delivery. While it is possible that uterine contractions resulting from orgasm late in the third trimester may start labor in a small number of pregnant women (or that the prostaglandins contained in semen may trigger the same response), a recent report documents no statistical correlation between orgasm or intercourse and premature birth. In fact, being orgasmic by intercourse or masturbation was associated with lower rates of prematurity.

For most young married adults, sex is no longer the frantic secret activity of adolescence or the stylistic *tour de force* of singlehood. While sexual pleasure is not sacrificed or lost, it is balanced with other needs and responsibilities — an important developmental task in this phase of the life cycle. Those who do not succeed in this process of integration are more likely to become sexually dissatisfied. As a result, they may turn to extramarital sex, professional counseling, or divorce. Each of these paths seems well traveled at the present time.

Some couples fulfill the American dream of marital bliss by staying together, raising children, remaining faithful, and loving one another all along the way. Others live out a modified version of this script in which love disappears but the other features are retained. Still others modify the script in different ways: no children, no faithfulness, or no bliss. The appearance to outsiders and the reality within the relationship do not always match.

One concrete bit of evidence that bliss is often missing from marriages can be found in the divorce rate in our society. Since the early 1960s, the annual number of divorces has climbed steadily, reflecting the change in attitudes toward divorce (less stigmatization and sense of failure) and in divorce laws (which make it possible to obtain "no fault" divorces in most states). Recent projections based on statistics gathered by the U.S. Census Bureau suggest that approximately one out of five couples marrying now will divorce before their fifth anniversary, while a third won't make it through a decade of marriage. Four out of ten will divorce before their fifteenth anniversary. Thus, as you might infer from these projections, the majority of divorces involve spouses who have not yet reached middle age.

It is not clear how frequently sexual dissatisfaction is a primary cause of divorce, but marriage counselors are well aware that sexual problems are common in couples whose marriages are troubled. Whether the sexual difficulties precede and contribute to other marital problems, or whether the reverse is more typically true, is simply not known.

Relatively little has been written about sexuality in the aftermath of divorce. It might be expected that the new sense of personal freedom that would follow a divorce would lead to increased rates of sexual activity, and in fact one study of 367 divorced singles has partially confirmed this. The same study, however, found that 27 percent of the men and 36 percent of the women reported decreased amounts of sexual activity. Similarly, a separate survey found that lack of sexual desire was more prevalent among divorced men than among those who were married. To understand why this divergence occurs, we should realize that the reality of life after divorce is not always one of freedom and happiness. These comments illustrate what may be going on:

> A *thirty-two-year-old woman:* Sure I wanted to have new sexual experiences, but someone had to take care of my kids, and the guys I met were all pretty grubby, and I spent a lot of time feeling sorry for myself. Looking back, I guess I wasn't very good company.

> A *thirty-three-year-old man:* It wasn't at all what I thought it was going to be. I sort of expected to find being single again a blessing, sexually speaking. I had visions of myself living out fantasies from *Playboy,* but it didn't really work like that at all. At the beginning, I was so nervous and embarrassed I couldn't function properly; then, I got an infection from a girl I really was beginning to like. Being single is no cup of tea.

> A *twenty-six-year-old woman:* Every guy who asked me out expected that because I had been married, I'd be happy to have sex with him on our first date. It just wasn't comfortable for me, but I don't think many of them cared.

The recently divorced man or woman may be delighted with a new-found sexual freedom, but the initial thrill is likely to be dampened by one of several difficulties. Returning to the rituals of dating and courtship may be annoying or embarrassing, especially to those who have been out of circulation for a decade or longer. Self-consciousness about sex with someone who is almost a stranger may be distressing. Concerns about the adequacy of sexual performance and personal attractiveness may be combined with remorse or guilt over lack of commitment and morality. To all of this, we must also add that there is frequently a sense that time is running out: "I'd better get a partner while I can."

Statistics show that about five out of six men and about three out of four women remarry after a divorce — about half within three years — apparently deciding that a second marriage will give them a chance to do things differently. Often, this is more illusion than reality: statistical projections suggest that more than half of *second* marriages for women born between 1945 and 1954 are expected to fail, with men trailing closely behind with a projected 40 percent second marriage divorce rate. In addition, there is no evidence that divorced persons who remarry have an increased sense of well-being compared to those who remain single.

The problems encountered in second marriages are apt to be complex, including, for example, how to deal with children of earlier marriages, the financial strain of alimony, and relations between the new spouse and the "ex." In addition, many of the same problems that plagued a first marriage tend to reappear in a second one: selfishness, alcohol abuse, lack of communication, and other similar problems don't disappear easily.

One research survey of couples in second marriages suggests that the same sexual problems often reappear as well. These problems are not always apparent before the remarriage since sexual problems, such as low sexual desire, may be temporarily overcome by the excitement of a new romance and by the special attentions that each partner pays to the other. When the romance dies down to a low flickering flame and sex is attempted not in romantic circumstances, but after bathing three kids, washing the dishes, paying the bills, and arguing about who is going to take the garbage out, it is no surprise that sexual feelings are not at an ecstatic peak. In addition, some couples may decide to marry even when they realize that a sexual problem is present. They may either hope that the problem will disappear with time — which is generally unlikely — or they may adopt a "who cares" attitude that deliberately deemphasizes the role of sex even before the marriage begins.

Despite the fact that divorce doesn't always provide as neat a solution to problems as many people think it will, there are certainly many instances in which divorce is the soundest option available. Knowing when to put an end to an unpleasant or painful relationship is often a key step toward creating the chance for a new start.

By their mid- to late thirties, most married people have moved beyond the demands of early parenthood and have acquired a sense of maturity and security in their own world. Individuals who have chosen not to marry have typically worked through whatever uncertainty may have surrounded this life-style choice at an earlier age. For both men and women, this is a time of becoming one's own person and making decisions about future life directions. Women with teenage children begin to look ahead toward establishing life goals outside the home. Others who have postponed childbearing face final decisions about whether to proceed with having a family. Although this might seem to be, at last, the attainment of equilibrium and stability, it is often just the lull before the storm.

MIDDLE ADULTHOOD

At around forty, people enter a period of transition from their younger years to what has traditionally been called "middle age." It is a time when the visions and energy of youth begin to give way to hard realities and when most people first confront their own mortality and sense that time is running out. Some feel alarmed because their physique has folded and their youth "has flown the coop," never to be regained (despite attempts at fancy diets, jogging, hair dye, and face-lifts). For most people, life must be reappraised in terms of goals, accomplishments, and experiences. As a result, a midlife crisis sometimes begins to take shape.

From a sexual perspective, the male seems to be particularly vulnerable to the midlife crisis. Since rumor has it that after age forty a man is "over the hill" sexually, many men begin to check their sexual performance for signs of wear and tear. This is well-described in the following:

> He notes that it takes more time to become aroused. Where it used to be a matter of seconds and a mere glance at the orbs of flesh colliding beneath a pair of tennis shorts, he may take minutes or more to reach erection as he gets older. He also notices, correctly, that he is slower on the comeback. In the sweet agonies of teen age he may have walked about with an erection all day, seldom completely losing it even after he made love or masturbated — a virtual prisoner of his hormones and tight-fitting pants. But now each sexual act has a definite beginning and end, and it may be a matter of hours or all day before he can reach erection again. Comparisons, stinging comparisons . . . he is not the boy he once was. (Sheehy, 1976, p. 305)

Once a man begins to question his sexual capabilities, the odds are that he will experience difficulty getting or keeping an erection. This, of course, "proves" the correctness of the underlying concern, and a vicious cycle is set in motion.

Some men turn to younger sexual partners to heat up their passion and others succeed, at least to a certain degree, in blaming the problem on their wives. But the middle-aged male is in a precarious position of sexual vulnerability. If he goes to a physician to discuss the problem, he may be told (as several of our patients have been), "At your age, you shouldn't worry about it anymore."

The woman's midlife crisis is less apt to include concern about her sexual capacity. For women who have chiefly been mothers, it is a time for an emerging identity, a freeing of the inner self as children reach a relative stage of greater independence. It is a bittersweet time in which a woman who has not established a career or nonfamily interests may mourn the passing of her offspring into their own maturity and may simultaneously look at available options for redirecting her talents and energy. As children leave home, the "empty-nest syndrome" may strike, causing depression and listlessness as the woman tries to deal with too much unstructured time and few sources of rewarding or interesting activities. Since these vulnerable feelings may be followed by or coincide with the menopausal years, it may be a particularly trying period for such women. Interestingly, recent research suggests that it is not only women who may be affected by the empty-nest syndrome. Roberts and Lewis point out that men are also sometimes distressed by their children's departure from home, "discovering that their marriages and friendships had become empty shells about the same time that loved children were leaving." Of course, having children leave home can have positive effects on a marriage, too — for instance, it gives couples a chance to focus on their own interaction and can create opportunities for freer, more relaxed sex as well.

There are more and more variations on the midlife transition today. Career-oriented women who postponed motherhood (and time off from work) until their mid-thirties may be anxious to reconfirm their work identity, but may have trouble finding a job or getting back on the "fast track." Other mothers who refused to restructure their lives around childrearing and continued to work outside the home — either by choice or by economic necessity — may feel considerable guilt or exhaustion. Still other women find their lives complicated by divorce and face childrearing as a single parent; alternatively, if they remarry a man with children of his own, the family interactions may become particularly complex.

Another component of the midlife crisis for both sexes that has been generally overlooked is the phenomenon of sexual burnout, which may affect as many as 20 percent of people in this stage of the life cycle. Unlike occupational burnout, which occurs largely in reaction to a person's intense investment of time and energy in his or her work coupled with chronic, unrelenting emotional pressure, sexual burnout stems from tedium and satiation with the same sexual routines. More than simply boredom, which can certainly be one of its precursors, sexual burnout is typically marked by a sense of physical depletion, emotional emptiness, and a negative sexual self-concept. Caught in the throes of sexual burnout, the middle-aged adult develops a feeling of sexual helplessness and hopelessness, as though nothing can be done to rekindle erotic passion or pleasure. Sexual burnout occurs not only in married couples but also in singles who have previously been very active sexually.

Sexual burnout is not the same as the sexual lack of interest that often accompanies depression. Depression is typically marked by disturbances in sleep and appetite and by a general loss of pleasure in all or almost all usual activities or pastimes. People experiencing sexual burnout do not have these symptoms. While the prognosis for sexual burnout is generally good, since most people recover spontaneously from this syndrome over time, about 10 percent of those affected remain sexually inactive on a relatively permanent basis. Perhaps this indicates that they are happier with celibacy than with cultural stereotypes that prescribe sexual participation as obligatory for a "healthy" adulthood, or perhaps it shows their inability to overcome the negative effects of the sexual burnout syndrome.

Of course, not everyone experiences a full-blown midlife crisis, and for some, the forties and fifties are a time of happiness and sexual satisfaction. One woman told us, "I'd never want to be twenty-one again — being forty-five is more fun!"

Most men have "discovered" their sexuality in a joyous way by their twenties, but — at least in past generations — a sizable number of women did not awaken from their socially programmed sexual dormancy until their thirties or forties. Given the traditional limits set on female sexual behavior and feelings ("nice girls don't . . .") and given the traditional division of marital responsibilities (domestic and childbearing duties are "female," career orientation is "male"), this pattern should come as no surprise. As a result, many women undergo a process of sexual self-discovery in mid-adulthood, perhaps including being orgasmic for the first time. The woman in this phase of the life cycle is just as likely as her mate to seek out extramarital sexual opportunities, although this fact is not written about extensively.

In a survey of 160 midlife women, Lillian Rubin found that the most characteristic pattern was one of *improving* sexuality. She noted that although these women had to overcome sexual inexperience and cultural prohibitions against female sexuality when they were younger (and then had to deal with raising young children), by midlife women were better able to relate sex to their own wishes and needs rather than participating in sexual activity principally to please their partners. Thus, Rubin found that midlife women take the sexual initiative more often than they previously had done (although, paradoxically, some of these women became concerned with putting sexual pressure on their husbands or partners by initiating sex too often).

One intriguing finding has been noted about sex differences in the psychology of middle adulthood. Among those forty to fifty-five, men are apt to exhibit a strong sense of self-confidence and control and typically engage in behavior geared to show their power and proficiency, while women tend to be more dependent, passive, and lacking in confidence. By the late fifties, however, a decided shift occurs. Men seem to move away from their need to demonstrate power and mastery and begin to show more concern for emotional sensitivity and interpersonal relations. At the same time, women frequently begin to show more self-confidence and assertiveness, in effect reversing the earlier roles that had been observed. While the implications of this observation are not entirely clear, it appears to lend support to the notion that some postmenopausal women become more sexually assertive than they had been previously, while men may become more interested in sharing tenderness and affection as they become less preoccupied with career concerns.

A few additional aspects of middle adulthood deserve mention. Some people who were content with being single as young adults begin to yearn for the commitment and long-term companionship that marriage can offer. These individuals may have a difficult time finding a potential mate who fits their expectations. For many in this situation, an increasingly acceptable alternative to the singles bar scene has been the "personal classified" ad section of a newspaper or magazine. It is intriguing to note that a substantial number of these ads run by women in their forties mention marriage as the objective, whereas in a recent review of more than a thousand personal classifieds we uncovered only two males in their forties advertising for a potential spouse. What we found, instead, were many ads from men in mid-adulthood looking for women twenty-five to forty years old to share "fun times" together. Since a recent survey found that ads placed by women get an average of 49 responses, while ads placed by men only draw 15 replies, it is unclear what the exact meaning of this finding is.

While society has long accepted the practice of middle-aged men seeking younger women in an effort to find rejuvenation, friends and family and society in general often view a middle-aged woman's romance with a younger man as improper or shocking. Today, perhaps in part because of higher divorce rates and a larger number of women gaining status and independence through career pursuits, it is no longer rare to see a middle-aged woman dating a man five to ten years or more her junior. There is, however, one typical difference between these relations and those of the middle-aged (or older) male with a younger female partner. The physical attractiveness of the older male seems to be of only minor importance to his desirability compared to his status or power, while the middle-aged woman generally needs to be relatively good-looking in order to attract a younger male.

Contrary to popular misconceptions, there has not been a sudden upsurge in the divorce rate for people in mid-adulthood. In fact, most of the boring or unhappy marriages that survive to middle age will continue to old age rather than be dissolved by a midlife crisis. For those caught in such marital inertia, there is apt to be a decline in sexual ardor at home, which can express itself in a variety of ways. Decreased sexual desire frequently first appears at this time, as earlier pretenses of enjoying sex together are shed in favor of more behavioral honesty. Sexual dysfunctions can also appear as a reflection of marital stresses or of the pent-up hostility of one spouse for another. Extramarital involvement is another sexual possibility that can provide a sense of escape from a stagnant relationship along with the excitement of getting to know a new partner on intimate terms; this topic is discussed more fully in chapter 13.

A recent study conducted by the sex therapist Barry McCarthy of middle-adult couples engaged in marital or sex therapy found that spouses placed little emphasis on committing time or psychological energy to their marital or sexual relations. In this sample, 80 percent of marriages had experienced an extramarital affair. The likelihood of stress in the marital relationship was somewhat higher when the wife was having an affair. McCarthy observed that this might be because the women in his sample tended to become more emotionally involved in their affairs than did the men, who generally focused on the sexual interaction.

Another revealing aspect of this study had to do with husband-wife differences in expectations about sex:

Men's expectations tended to be extreme, either extremely high in terms of frequent, high-intensity sex (the "every sex is dynamite" myth) or an extremely low expectation — after having sex with someone for six months there is just no excitement (the "sex is not for marriage" myth). The woman's views ran the gamut from romanticism to seeing sex as a duty. Many women had an underlying disappointment about their husband's attitudes toward marital sexuality and with their skills as lovers. Yet, the women were reluctant to initiate discussions about sexuality and even more hesitant to experiment sexually and explore with their spouses.

The sexual problems of mid-adulthood are not restricted to heterosexuals, although little attention has been given to the middle-aged homosexual population in America. Here, too, a wide variety of patterns is seen. Gay men in their forties may find that it becomes more difficult to attract younger partners on physical grounds alone. As a result, some homosexual males turn to paid male prostitutes, others become relatively celibate, and still others form long-term relationships that provide companionship and emotional support as well as sexual opportunity. Gay men who were previously in heterosexual marriages — including many who were fathers — often decide to divorce and change to an exclusively homosexual life.

It has been noted that some gay males who have put an emphasis on conquests, techniques, and the ideals of the youth culture develop much anxiety about aging. These men sometimes resort to face-lifts or hair transplants to retain the illusion of their youthfulness (as some heterosexual men do, too) and they often develop depression or become alcoholic as a reflection of the negative sense of self they experience as their aging becomes more apparent. However, most homosexual men do not encounter such difficulties; generally, they have the same types of problems as they grow older that heterosexuals have.

Lesbians are often able to make an easier transition into middle-adulthood because of the fact that more of them are in lasting, one-to-one relationships. Nevertheless, while concern over physical attractiveness may not be quite as strong an issue as it is for some gay males, there can be intense jealousy that arises in these relationships "whenever the relationship is threatened by other women who might intrude and attempt to disrupt the coupling" (Levy, 1981, p. 125).

THE MENOPAUSE

With aging, all women reach an end to their fertility. First, there is a gradual decline in female reproductive capacity from age thirty on, reflecting both a drop in fertility and a higher rate of miscarriages. In addition, abnormalities of the menstrual cycle become more frequent over thirty-five as the aging ovaries respond less efficiently to LH and FSH from the pituitary gland. After age forty, the frequency of ovulation generally begins to decrease, and around age forty-eight to fifty-two, menstrual flow stops entirely in a process called the *menopause*.* However, since deciding when the menopause has occurred can only be done retrospectively — by convention, after one year without further menstrual flow — women who are sexually active at this stage of their lives should continue to practice birth control until it is certain they cannot become pregnant.

The timing of the menopause and the symptoms that accompany it vary greatly from one woman to another. Although the ovaries stop producing all but a minute amount of estrogen, and ovarian progesterone production ceases entirely, small amounts of these hormones are still present because of continued activity of the adrenal glands. LH and FSH levels typically become elevated after menopause.

Symptoms

Although about 80 percent of women experience symptoms due to their changing hormone levels, only a minority of menopausal women seek treatment for symptom relief. This is probably because the majority of symptoms are of a relatively minor nature and tend to disappear with time.

The most common symptom in the menopause is the *hot flash*,

---

* The term "menopause" correctly refers only to a woman's last natural menstrual period. Many people incorrectly use the word "menopause" to refer to a time more correctly called the *climacteric* or the *perimenopausal years:* that is, the several years immediately before and after the menopause.

which affects 75 to 80 percent of menopausal women. Typically, the hot flash appears suddenly as a feeling of warmth over the upper part of the body (very much like a generalized blushing) and is accompanied by reddening, sweating, and, occasionally, dizziness. In some women, hot flashes are infrequent (once a week or less) but others have them every few hours. Hot flashes may last just a few seconds and be quite mild, or they may last for 15 minutes or more in the most severe cases (experienced by less than 10 percent of women). One particularly disturbing feature of the hot flash is that it occurs more often during sleep than in the daytime, in which case it is liable to awaken the woman abruptly and contribute to insomnia.

Current evidence suggests that hot flashes are due to a malfunction of temperature control mechanisms in the hypothalamus. Although estrogen deficiency seems to be a necessary condition for hot flashes to occur, and estrogen therapy effectively combats this symptom, hot flashes generally disappear spontaneously within a few years after the menopause even without treatment. In approximately 20 percent of affected women hot flashes persist for at least five years beyond the onset of this time. Since this symptom is sometimes severe enough to interfere with everyday functioning and there is no test that can predict when hot flashes will disappear spontaneously, deciding whether to obtain treatment or not is very much a subjective decision for the woman.

Other changes also reflect prolonged estrogen deficiency. Lowered levels of circulating estrogen predispose women to shrinking and thinning of the vagina, a loss of tissue elasticity, and lessened vaginal lubrication during sexual arousal, all of which may sometimes lead to painful intercourse. Other physical changes that may occur in the postmenopausal years include thinning of the breasts and the vulva and loss of mineral content in bones, resulting in a more brittle structure (a condition that is called *osteoporosis*).

Although there has been considerable controversy in the past about the risks and benefits of estrogen replacement therapy (E.R.T.) in the menopause and postmenopausal years, strong scientific evidence shows that the symptoms we have discussed can be significantly alleviated by its use. In fact, E.R.T. plays a preventive role in slowing the occurrence of osteoporosis, rather than just alleviating symptoms once they occur. Because there is also considerable evidence that E.R.T. increases the risk of cancer of the uterus, and an unsubstantiated but realistic concern that it may increase the risk of breast cancer, caution is certainly in order. The consensus of medical opinion seems to favor E.R.T. for several different reasons:

1. Adding a progestin to the latter part of the estrogen cycle materially reduces the increased risk of cancer of the uterus that arises from estrogen use.
2. Osteoporosis has now been recognized as a disorder of great seriousness, since it often leads to hip fractures in elderly women, and in 20 to 30 percent of cases, these women die due to the fracture or its complications;
3. E.R.T. may provide protection against certain forms of heart disease.

Most authorities caution that E.R.T. should not be used indiscriminately and that it should be employed in the smallest effective dose for the shortest period of time compatible with the therapeutic need. However, since the need for prevention of osteoporosis is lifelong, some experts (ourselves included) prefer to use E.R.T. on a long-term basis as long as there are no specific contraindications to its use or adverse effects in the individual taking it.

Psychological Aspects

In the past, just about every problem that could befall a woman was inaccurately attributed to the menopause. In eighteenth- and nineteenth-century Europe, for instance, physicians thought women decayed at menopause, and "nervous irritability" was diagnosed in nine out of ten menopausal women. Mistakenly blaming emotional instability, depression, and other psychological problems on the hormonal changes of the menopausal years has not just been a practice of ancient history; it is a relatively common error in modern times that continues even today.

In 1981, Ballinger found no evidence for an increased rate of depression or major psychiatric disorders in the years following the menopause and pointed out that "emotional symptoms at this time of life, as at any other time, are influenced by a multitude of environmental and personality factors." Others contend that menopausal discomfort has been greatly exaggerated, that the menopause may actually be an adaptive, positive event, and that "the mythology surrounding the menopause is based less on the reality of the female experience than on the sexist interpretations of the female experience" by male physicians (Alington-MacKinnon and Troll, 1981). Bart and Grossman also point out that much of the research on the menopause is methodologically flawed, particularly the portion related to psychological aspects of the menopausal years. They suggest that a woman's response to the menopause is in large part a function of her "premenopausal personality and life patterns" and note that menopausal depression (when it occurs) is far more a result of lack of meaningful roles for women and poor self-esteem at this stage in their lives than a reflection of hormonal changes.

While there is some disagreement about the impact of the menopause on female sexuality, several studies suggest there is characteristically a decline in sexual interest and possibly a loss of female orgasmic responsivity in the immediate postmenopausal years. In 1979, Hallstrom studied 800 women in Sweden and found considerable evidence to support this premise; for example, the prevalence of weak or absent sexual interest increased progressively from ages thirty-eight to fifty-four, and a declining capacity for orgasm was reported for these same ages. However, Hallstrom also found that some postmenopausal women showed increased sexual interest and capacity for orgasm. On the other hand, several reports have noted that sexual interest may increase in the postmenopausal years. The discrepancy in these findings may be the result of inadequate research design that does not take into account factors such as the health status of study subjects and their spouses or sexual partners.

LATE ADULTHOOD

In America, sex is generally regarded as something for the young, healthy, and attractive. Thinking of an elderly couple engaging in sexual relations usually provokes discomfort. The idea of sexual partners in a nursing home seems shocking and immoral to most people. Despite these cultural myths, the psychological need for intimacy, excitement, and pleasure does not disappear in old age, and there is nothing in the biology of aging that automatically shuts down sexual function.

Biological Considerations

Aging alone does not diminish female sexual interest or the potential of the woman to be sexually responsive if her general health is good. Specific physiological changes do occur, however, in the sexual response cycle of postmenopausal women. These changes do not appear abruptly or in exactly the same fashion in each woman.

Typically, there is little or no increase in breast size accompanying sexual arousal, although breast sensitivity to stimulation continues. The sex flush occurs less often and less extensively than at younger ages, but this change has absolutely no effect on sexual feelings or functioning. Less muscle tension develops during sexual arousal, particularly in the plateau phase, which is not surprising since this corresponds to the usual decrease in muscle size and strength that occurs with aging. This reduced muscular tension may account (at least in part) for the reduced intensity of orgasm that is sometimes experienced by women in late adulthood.

While clitoral response is not affected by aging, vaginal function changes in two different ways. First, reduced elasticity in the walls of the vagina leads to less expansion during sexual arousal. Second, vaginal lubrication generally begins more slowly than at younger ages and vaginal dryness may create some problems as the quantities of lubrication are somewhat reduced. This condition can be overcome if it causes discomfort either by estrogen replacement therapy or by the use of an artificial lubricant such as K-Y Jelly.

Recent research has shown that the decrease in vaginal lubrication in postmenopausal women is the direct result of diminished vaginal blood flow that, in turn, is caused by low estrogen. In another recent investigation, Leiblum and co-workers found that sexually active postmenopausal women had less shrinkage of the vagina and higher levels of androgens and pituitary gonadotropins (LH and FSH) than sexually inactive women. This

suggests that regular sexual activity may provide at least some
protection against the physiologic changes of aging in relation
to female sexual anatomy.

The normal pattern of reproductive aging in men is quite
different from women because there is no definite end to male
fertility. Although sperm production slows down after age forty,
it continues into the eighties and nineties. Similarly, while testosterone production declines gradually from age fifty-five or sixty
on, there is usually no major drop in sex hormone levels in men
as there is in women.

About 5 percent of men over sixty experience a condition
called the *male climacteric,* which resembles the female menopause in some ways. (Using the term "male menopause" to describe the male climacteric is incorrect since men do not have
menstrual periods.) This condition is marked by some or all of
the following features: weakness, tiredness, poor appetite, decreased sexual desire, reduction or loss of potency, irritability,
and impaired ability to concentrate. These changes occur because of low testosterone production, and they can be reversed
or improved by testosterone injections. It should be stressed that
most men do not have a recognizable climacteric as they age.

The physiology of male sexual response is affected by aging
in a number of ways. The following changes have been noted in
men over fifty-five:

1. It usually takes a longer time and more direct stimulation
   for the penis to become erect;
2. Erections tend to be less firm, on average, than at earlier
   ages;
3. The testes elevate only partway up to the perineum, and do
   so more slowly than in younger men;
4. The amount of semen is reduced, and the intensity of ejaculation is lessened;
5. There is usually less physical need to ejaculate;
6. The refractory period — the time interval after ejaculation
   when the male is unable to ejaculate again — becomes
   longer.

In addition, the sex flush usually does not occur in aging men,
and muscle tension during sexual arousal is reduced, as in

women, since muscle mass and strength generally decrease with aging.

Although the changes in male sexual physiology described above do not usually occur abruptly or represent an impairment of function, men who are uninformed about these patterns may be frightened into thinking something is wrong with them. In other instances, a man's partner may be the one to become alarmed. For example, while many men find that they enjoy sex in their later years without ejaculating at every opportunity, partners who don't realize this may think it reflects poorly on their attractiveness or skill as lovers.

Some men have completely unrealistic expectations about what their sex lives *should* be as they age. While they wouldn't expect to run a mile as fast at age sixty-five as they did at age twenty-five (or to recuperate from their exertion as quickly), they expect to get rock-hard erections instantly in all sexual situations and are worried when they can't make love twice in one evening. The aging male, by misinterpreting these changes, is particularly vulnerable to performance anxiety.

Psychosocial Considerations

In part, our cultural negativism about sex and romance in the geriatric years is a reflection of an attitude called *ageism,* a prejudice against people because they are old, that is similar to the more familiar prejudices of racism and sexism in our society. As Butler and Lewis (1976, p. 4) note: "The ageist sees older people in stereotypes: rigid, boring, talkative, senile, old-fashioned in morality and lacking in skills, useless and with little redeeming social value." These same authors observe that ageism in relation to sexuality is the ultimate form of desexualization: "if you are getting old, you're finished."

Ageism is not restricted to heterosexuals, either, as this quotation from *Gay and Gray: The Older Homosexual Male* (Berger, 1982, p. 191) shows:

> Many older gay men believe that younger gays react negatively to them. Most older gays feel that young people sometimes take advantage of them, do not welcome their company in bars, clubs, and bathhouses, do not care to associate or form friendships with them, and think they are dull company.

It is interesting to note that in a recent study, college students estimated that average married couples in their sixties had intercourse less than once a month and expected that married couples in their seventies had intercourse even less often. As we will see, these are underestimates of actual behavior, showing that young adults are susceptible to ageism in their thinking — at least when it comes to sex.

Kinsey and his colleagues were the first to examine systematically the effect of aging on sexual behavior. Although their studies indicated that sex continued well into late adulthood, they also described a general decline in the frequency of sexual activity for both men and women across the adult age range. A number of other studies confirmed these overall trends, with most reports suggesting that the decrease in sexual activity is partly due to diminished health and partly a reflection of cultural attitudes and expectations.

More recently, a longitudinal study done at Duke University has found that patterns of sexual activity actually remain relatively stable over middle and late adulthood, with only a modest decline appearing in most individuals. For example, men who were sixty-six to seventy-one years old at the start of this study had no overall decline in sexual activity scores over the next six years, and a majority of men and women in the fifty-six to sixty-five age bracket at the beginning of the study had stable rates of sexual activity during this same time period. However, in those over age sixty-five, 18 percent of the men and 33 percent of the women completely stopped sexual activity with their partners over a six-year period.

Various studies of sexual behavior in late adulthood indicate that the male's declining interest seems to be the major limiting factor to continued sexual activity. This declining interest, however, may say more about cultural expectations which become self-fulfilling prophecies than anything else. While the pattern of sexual activity among aged people varies considerably, coital frequency in early marriage and the overall quality of sexual activity in early adulthood correlate significantly with the frequency of sexual activity in late adulthood.

It is interesting to note that most researchers who claim to be studying sexual activity rates in the elderly have actually restricted themselves to studying coital behavior, as though other forms of sexual activity "don't count." Many adults over sixty continue to masturbate, although these are almost entirely individuals who masturbated when they were younger. Not surprisingly, elderly people without sexual partners tend to masturbate more frequently. In couples over sixty, other forms of sexual stimulation continue to be utilized (for example, oral sex and manual stimulation); they not only provide some variety, but also can be a source of pleasure and closeness even if the male is unable to obtain or maintain erections. Here is what one seventy-eight-year-old man told us:

> In my mid-sixties, I had a real problem getting it up. Sexual intercourse became impossible and I got very upset over it at first. But my wife just found other ways to excite me, and after a while — when I relaxed more, I guess — my problem disappeared. Now, even though she doesn't always want to have intercourse when I do, we still use our mouths and tongues on each other, and we plan to keep doing it, too.

In late adulthood, the longer life span of women, combined with their tendency to marry men who are their age or slightly older, frequently creates a situation of widowhood. At the same time that fewer men of the same age are available as new sexual partners, younger men tend to focus their attentions on women under forty. Thus, sexual opportunities and social contacts with men are often extremely limited for these mature women who might otherwise be enthusiastic sexual partners.

In the last few years, a new trend is emerging in which not all women in the sixties or older who find themselves widowed or divorced are passively accepting their fate. Some women in this situation are using their ingenuity to find new partners, as shown by the following ad from the personal classified section of a well-known magazine:

> Tall, handsome intelligent man sixty to sixty-five, wanted for attractive, lively sixty-two-year-old widow with fantastic figure and modern values. If you're interested in romance more than golf, send a photo and your phone. Box xxx.

Although a sixty-five-year-old widow (or widower) may have an interest in sexual activity, social pressures may prevent sexual opportunities or belittle their meaning via snide jokes (e.g., only "dirty old men" are interested in sex, and "dirty old women" should "act their age"). As prolonged abstention from sexual activity in old age leads to shrinking of the sexual organs (just as a healthy arm loses strength and coordination if put in a sling for four months), the older adult is truly faced with a sexual dilemma — "use it or lose it."

An often-neglected aspect of aging in our society is that many elderly persons are relegated to nursing homes or other long-term care facilities for health reasons, to "protect" themselves, or for the convenience of their children. Although research on sexual behavior in this population is understandably sparse, a number of authorities have spoken out in favor of making provisions in such facilities (such as making private rooms available) for those who desire sexual interaction. While most nursing home administrators seem opposed to this notion, and some facilities actually insist on segregating men and women even if they're married, both ethical considerations for protecting the rights of the elderly and more practical considerations having to do with maintaining self-esteem and personal well-being lead us to advocate change in this area. However, it should be noted that permitting sexual expression in the institutionalized elderly is not the entire solution to their plight. Additional steps must be taken to overcome loneliness and boredom in the nursing home environment so that these facilities are not unintentional prisons for our older adults.

In the United States, where there is little preparation or education for aging, it is not surprising that many people are uninformed about the physiological changes in their sexual function in their sixties and seventies. They may mistakenly view these normal "slowing down" processes as evidence that loss of function is imminent. Brief preventive counseling during middle age might result in significant change in this area, but changed attitudes toward sex and aging seem even more necessary. What people must recognize is that given good health and the availability of an interested and interesting partner, there is no reason that sexual enjoyment should have to come to an end in late adulthood. Perhaps the ultimate test of whether we have lived through a sexual revolution will be if attitudes toward sexuality in old age are transformed.

CHAPTER EIGHT

Gender Roles

ON A TELEVISION soap opera, a self-confident, smooth-talking businessman seduces a beautiful but not too bright female secretary. A children's book describes a warm, caring, stay-at-home mother while depicting father as an adventuresome traveler. A newspaper advertisement for cigarettes shows a husky young man enthusiastically dousing a shapely, squealing female companion with water, her wet T-shirt clinging to her bust — the headlined caption reads "Refresh Yourself." Each of these messages tells us something about stereotypes and sexism.

In the past twenty-five years, there has been considerable scientific interest in studying differences and similarities between the sexes for a number of reasons. First, various beliefs about sex differences in traits, talents, and temperaments have greatly influenced social, political, and economic systems throughout history. Second, recent trends have threatened age-old distinctions between the sexes. In 1985, for instance, more than half of American women worked outside the home. Unisex fashions in hairstyles, clothing, and jewelry are now popular. Even anatomic status is not fixed in a day where change-of-sex surgery is possible. Third, the women's movement has brought increasing attention to areas of sex discrimination and sexism and has demanded sexual equality.

As a result of these trends, old attitudes toward sex differences, childrearing practices, masculinity and femininity, and what society defines as "appropriate" gender-role behavior have undergone considerable change. Many of today's young adults have been raised in families where a progressive attitude toward gender roles has been taught or where parents struggled to break away from stereotyped thinking. Thus, there is a continuum of types of socialization today that ranges from old, traditional patterns to more modern versions. This chapter will examine these issues and trends as they influence the experience of being male or female.

MASCULINITY AND FEMININITY

Before you read any farther, you might take a few minutes to write out a list of the traits you would use to describe a typical American man and woman. If your descriptions are similar to most other people's, you probably listed characteristics like strong, courageous, self-reliant, competitive, objective, and aggressive for a typical man, while describing a typical woman in terms like intuitive, gentle, dependent, emotional, sensitive, talkative, and loving.

Most people not only believe that men and women differ but share similar beliefs about the ways in which they differ. Beliefs of this sort, held by many people and based on oversimplified evidence or uncritical judgment, are called *stereotypes*. Stereotypes can be harmful because they lead to erroneous judgments and generalizations and can thus affect how people treat one another.

Because many stereotypes about sexuality are based on assumptions about the nature of masculinity and femininity, it is difficult to offer a concise definition of these two terms. In one usage, a "masculine" man or a "feminine" woman is a person who is sexually attractive to members of the opposite sex. Advertisements for clothing and cosmetics constantly remind us of this fact. In another sense, masculinity or femininity refers to the degree to which a person matches cultural expectations of how males and females should behave or look. In the not too distant past, some segments of our society were upset when long hair became fashionable among young men or when women applied for admission to West Point because these patterns did not "fit" prevailing expectations about differences between the sexes. In still another meaning, masculinity and femininity refer to traits measured by standardized psychological tests that compare one person's responses to those of large groups of men and women.

According to traditional assumptions, it is highly desirable for males to be masculine and females to be feminine. If behavior matches cultural expectations, it helps to preserve social equilibrium and allows for a certain amount of stability in the details of everyday living. Conformity to cultural norms presumably indicates "adjustment" and "health," while straying too far from expected behavior patterns indicates abnormality or even disease. Finally, "masculine" men and "feminine" women are relatively predictable and behave in ways that are fairly consistent and complementary. Fortunately (or unfortunately, depending on your viewpoint), it now appears that masculinity and femininity are unlikely to tell us much about personality, sexual preferences, or life-style, and old stereotypes are now giving way to more useful and dynamic scientific views.

The traditional approach to studying masculinity and femininity looked at these traits as opposites. According to this view, if you possess "feminine" characteristics you cannot have "masculine" characteristics and vice versa. It was assumed that people who scored high on certain traits judged as masculine (e.g., independence, competitiveness) would also have a general lack of femininity. As a result, most psychological tests designed to measure masculinity and femininity were set up as a single masculinity-femininity scale. Furthermore, men and women whose masculinity or femininity scores differed substantially from group averages were judged to be less emotionally healthy and less socially adjusted than others with "proper" scores.

Recent research findings have changed this approach. Instead of viewing masculinity and femininity as opposites, various behavioral scientists now look at them as separate characteristics that coexist to some degree in every individual. Thus, a woman who is competitive can be quite feminine in other areas; a man who is tender and loving may also be very masculine. As we discuss the ways in which gender roles are learned and the impact they have on our lives, it will be helpful to keep this viewpoint in mind.

PATTERNS OF GENDER-ROLE SOCIALIZATION

Even before a baby is born, parents are likely to have different attitudes about the sex of their child. In most societies, male children are clearly preferred over female children, and having a son is more often seen as a mark of status and achievement than having a daughter. This preference probably stems from the belief that men are stronger, smarter, braver, and more productive than women, and that "it's a man's world" (certainly true in the past) — meaning that there are more and better educational, occupational, political, and economic opportunities open to males than to females.

Parents often try to guess the sex of their unborn child and may construct elaborate plans and ambitions for the child's life. If the child is thought to be a boy, the parents are likely to think of him as sports-oriented, achievement-oriented, tough, and independent. If the child is thought to be a girl, parents are more apt to envision beauty, grace, sensitivity, artistic talents, and marriage.

This sort of prenatal thinking is one form of stereotyping, as is guessing that the baby will be a boy because "he" kicks a lot inside the uterus. It is not surprising then to find that the earliest interactions between parents and their newborn child are influenced in subtle ways by cultural expectations.

Such cultural expectations continue at the moment of birth, when the announcement of the baby's sex ("It's a boy" or "It's a girl") sets in motion a whole chain of events such as assigning a pink or blue identification bracelet, choosing a name, selecting a wardrobe, and decorating the baby's room, each of which involves making distinctions between males and females.*

As friends, relatives, and parents discuss the newborn's appearance, gender stereotypes are everywhere: "Look at his size — he'll be a football player, I bet." "She has beautiful eyes —

---

* In the song "A Boy Named Sue," written by S. Silverstein and recorded by Johnny Cash, the father reversed usual gender distinctions in name selection in order to achieve the paradoxical effect of improving his son's masculinity. By giving him the name Sue, the father forced the boy to fight frequently to defend himself from ridicule, thus becoming "tough."

she's a real doll." "See how intelligent he looks!" "She's got great legs already! You'll have to work to keep the boys away." Informal banter about the child's future is also likely to be gender-linked: if friends remark, "You better start saving for the wedding," you can bet they are not talking about a baby boy.

A variety of research reports identify ways in which adults react to their stereotypic gender-role expectations about babies. For example, parents of newborn infants describe daughters as softer, smaller, finer-featured, and less active than sons, although no objective differences in appearance or activity level were noted by physicians. In early infancy, boys receive more physical contact from their mothers than girls do, while girls are talked to and looked at more than boys — a difference in treatment that tends to reinforce a female's verbal activities and a male's physical activity. In another study, two groups of young mothers were given the same six-month-old infant dressed either in blue overalls and called Adam or wearing a pink frilly dress and called Beth; the results showed that "Beth" was smiled at more, given a doll to play with more often, and viewed as "sweet" compared to "Adam." Interestingly, although both mothers and fathers behave differently toward unfamiliar infants on the basis of perceived sex the parents are typically unaware of this differential treatment.

Parents respond differently to infant boys and girls in other ways. They react more quickly to the cries of a baby girl than a baby boy and are more likely to allow a baby boy to explore, to move farther away, or to be alone, thus fostering independence. In contrast, the baby girl seems to be unintentionally programmed in the direction of dependency and passivity.

Gender differences in socializing children occur for reasons that are not fully understood at present. Certainly, cultural influences are important, but biological factors may also be involved. For example, boys' higher rates of metabolism, greater caloric intake, and higher rates of activity may prepare them for earlier independence, or parental encouragement of independence may reflect cultural expectations. Furthermore, the different prenatal hormone exposures of males and females may possibly account for behavioral differences in infancy. Often, parents are unaware of how their actions with their children are different depending on the child's sex. Nevertheless, differential socialization seems to occur even in parents who are philosophically committed to the idea of avoiding gender stereotypes.

By age two, a child can determine in a fairly reliable way the gender of other people and can sort clothing into different boxes for boys and girls. However, two-year-olds do not usually apply correct gender labels to their own photographs with any consistency — this ability usually appears at around 2½ years. Core gender identity, the personal sense of being male or female, seems to solidify by age three. This process is probably assisted by the acquisition of verbal skills, which allow children to identify themselves in a new dimension and to test their abilities of gender usage by applying pronouns such as "he" or "she" to other people.

At age two or three, children begin to develop awareness of gender roles, the outward expression of maleness or femaleness, in their families and in the world around them. It might seem that the child forms very sketchy impressions at first — "Mommies don't smoke pipes" or "Daddies don't wear lipstick" — but the toddler's understanding is greater than his or her ability to verbally express it. It is likely that impressions of what is masculine and feminine form across a broad spectrum of behaviors.

The serious business of young childhood is play, so by examining the objects used in play activities we may be able to learn something about gender-role socialization. Walk through the toy department of a large store and you will quickly see the principle of differential socialization at work. Boys' toys are action-oriented (guns, trucks, spaceships, sports equipment) while girls' toys reflect quieter play, often with a domestic theme (dolls, tea sets, "pretend" makeup kits, or miniature vacuum cleaners, ovens, or refrigerators). Where a particular toy is marketed to both girls and boys, the version for girls is usually feminized in certain ways. For instance, a boys' bicycle is described as "rugged, fast, and durable." The girls' model of the same bike has floral designs on the seat and pretty pink tassels on the hand-grips and is described as "petite and safe." A detailed analysis of the content of ninety-six children's rooms showed that boys were given more toy cars and trucks, sports equipment, and military toys, while girls received many more dolls, dollhouses, and domestic toys. Although many boys today play with "E.T." or "Star Wars" dolls or other action-oriented figures, most parents of boys are likely to become concerned if their sons develop a preference for frilly, "feminine" dolls.

Picture books are another important source for learning gender roles. As Weitzman (1975, p. 110), observes, "Through books, children learn about the world outside their immediate environment: they learn what is expected of children of their age." Although in recent years some changes have occurred, an analysis of award-winning books for preschoolers conducted by Weitzman and her co-workers in 1972 showed marked evidence of gender-role bias. First, males were shown much more frequently than females (there were 261 males and 23 females pictured, a ratio of eleven to one). Second, most males were portrayed as active and independent, while most females were presented in passive roles. Third, adult women shown in these books were consistently identified as mothers or wives, while adult men were engaged in a wide variety of occupations and professions. It is no wonder that girls get a strong message that "success" for them is measured in terms of marriage and motherhood. Fortunately, this imbalance is beginning to change today, with many recent books aimed at preschoolers showing women in a more favorable light.

Television is also a powerful force in the gender-role socialization of young children because it provides a window to the rest of the world. The fictionalized world of Saturday morning children's cartoons is filled with gender stereotypes: the heroes are almost all males and females are shown as companions or as "victims" needing to be rescued from the forces of evil. Even award-winning children's shows such as *Sesame Street* have been criticized in the past because women were seldom shown as employed outside the home and male figures predominated. Advertisements geared at preschoolers perpetrate the same patterns: boys are shown as tough, action-oriented people, while girls are portrayed as more domestic, quieter, and refined.

By the time children enter elementary school, gender-role expectations are applied with some unevenness. A seven-year-old girl who likes sports and climbs trees is generally regarded as "cute" and is affectionately, even proudly, called a tomboy. A seven-year-old boy who prefers playing with dolls and jumping rope to throwing a football is labeled a "sissy" and may be the source of great parental consternation. Although child psychiatrists regard tomboyishness in girls as a "normal passing phase," "effeminate" boys are thought by many researchers to require treatment to prevent them from becoming homosexual or having later sexual problems.

Different patterns of gender-linked play continue during the school years and are reinforced firmly by peer group interactions. Schoolyard and neighborhood play is noticed by other boys and girls, and children whose play preferences do not match everyone else's are thought to be "weird" and are often the butt of jokes. Since there is a powerful motivation to be like everyone else in order to have friendship and group acceptance, this teasing can have a negative influence on a child's sense of self-esteem.

At this age, boys are generally expected to show their masculinity by demonstrating physical competence and competitive spirit in sports activities, which becomes the primary focus of boyhood play. They are rewarded for bravery and stamina and criticized for showing fear or frustration ("Big boys don't cry"). Girls, on the other hand, although physically more mature than boys at corresponding ages in childhood, have traditionally been steered away from highly competitive sports and sheltered from too much exertion. (Today, this pattern is changing considerably as girls are encouraged to enter competitive swimming, gymnastics, soccer, and Little League baseball just as much as they are encouraged to take ballet or music lessons.) Girls are expected to stay clean and be neat, to avoid fighting, and to avoid dangerous activities ("Be a lady"). Young girls often seem to be programmed to cry to show hurt or frustration and find that crying (at least in the presence of adults) often elicits comforting. Thus, males are encouraged to solve problems in an active, independent way, whereas females are more likely to be shown that *their* best way of solving problems is to act helpless and to rely on someone else to take care of them.

Even for the children of relatively "liberated" parents, sexism sometimes inadvertently looms:

> Take my friend Irene, a vice-president of a Fortune 500 company, who at a recent dinner party bemoaned the stiff resistance of male executives to women in senior management. Not ten minutes later, she proudly regaled us with tales of her eight-year-

old son who struts around the house shouting, "Boys are the best, boys are the best."

In Irene's mind, forty- or fifty-year-old executives practice sexist oppression. But when her Jonathan shuts girls out, he is cute, natural ("It's the age," she told me), and turning out to be a "real boy." (Rommel, 1984, p. 32)

There are also, of course, instances where parents react differently to a child's seemingly sexist behavior. In one case, a mother who encountered her eight-year-old son telling his friends that girls are poor athletes took her son on successive weekends to watch the U.C.L.A. women's basketball team and to a women's weight-lifting contest. The boy apparently gained a different perspective on female athletic capacity, because he was seen soon thereafter playing softball with a nine-year-old girl from down the street.

While these sorts of situations are of concern to some parents who want to raise their children in the nonsexist fashion, other well-meaning parents feel that since many young girls "shut boys out" and believe that "girls are best," this is not really sexist at all. They point out that while these responses aren't appropriate for adults, such attitudes foster self-esteem in children.

Much of the child's time is spent in school, where gender-role stereotypes still exist in many classrooms. Elementary school readers show many more male figures than females. History lessons portray a view of the world as male-dominated; in the few instances when women are mentioned, they are usually in a subservient or domestic role (recall how Betsy Ross served the cause of the American Revolution by sewing). Girls are usually assigned different classroom "chores" than boys are (for example, boys might be asked to carry a stack of books, while girls are asked to "straighten up the room"), and teachers often assign activities to boys and girls based on their presumptions about gender-role preferences. In one school, third-grade girls were asked to draw a mural while the boys were asked to build a fort. A girl who said she would rather work on the fort was told by her teacher, "That's not a job for young ladies."

School-age children are also exposed to obvious gender-role stereotypes on television. From commercials children learn that most women are housewives concerned about important decisions like which detergent to use, which soap does not leave a bathtub ring, and which brand of toilet paper is softest. Men, on the other hand, are concerned about health issues ("Four out of five doctors recommend . . ."), economics, automobiles, or recreation (most beer commercials play upon themes of masculinity, for example). With a few notable exceptions, the lawyers, doctors, and detectives on TV are all men, and women — even when cast in adventurous occupations — are shown as emotional, romantic sex objects who cannot make up their minds. It is no wonder that stereotypes about masculinity and femininity continue: children are exposed to them so widely that they come to believe they are true. Supporting this observation, a recent study found that children who watched television more than twenty-five hours per week had more stereotyped gender-role perceptions than age-matched children who watched less than ten hours of television weekly.

Adhering to gender-appropriate roles is even more important during adolescence than at younger ages. What was earlier seen as rehearsal or play is now perceived as the real thing. The rules are more complicated, the penalties for being "different" are harsher, and future success seems to hinge on the outcome.

Adolescent boys have three basic rules to follow in relation to gender roles. First, succeed at athletics. Second, become interested in girls and sex. Third, do not show signs of "feminine" interests or traits. Teenage boys who disregard these rules too obviously are likely to be ridiculed and ostracized, while those who follow them closely are far more likely to be popular and accepted.

The traditional prohibition of feminine traits in male adolescents probably relates to two separate factors. The first is the view of masculinity and femininity as complete opposites that was discussed earlier in the chapter. For a teenage boy to "fit" the male stereotype, he must be achievement-oriented, competitive, rational, independent, self-confident, and so on. If the opposite traits emerge, his masculinity is subject to question. Second, a teenage boy who shows "feminine" interests or traits is often regarded suspiciously as a potential homosexual. In a variation on this theme, in schools where home economics courses were opened to male enrollment, some parents have voiced concern that it would "rob" boys of their masculinity and lead to "sexual deviance." However, in communities where boys take home economics and girls take shop courses, it is remarkable that an easy equilibrium has been reached, with no one "harmed" psychologically by the experience.

The adolescent girl is confronted by a different set of gender-role expectations and different socialization pressures. In keeping with the traditional expectation that a female's ultimate goals are marriage and motherhood rather than career and independence, the prime objective seems to be heterosexual attractiveness and popularity. As a result, the adolescent girl's school experience may push her toward learning domestic or secretarial skills instead of orienting her toward a profession, and the message she frequently gets — from peers and parents — is that academic achievement may lessen her femininity. However, it appears that this pattern is now undergoing considerable change. As it has become more culturally "permissible" for women to enter professions such as medicine and law, or to enter the business world at the management level, more and more teenage females have become comfortable with maintaining a high level of academic success.

For many women, the high value that society places on both achievement and popularity poses a problem. One factor that seems to influence female nonachievement is fear of success: that is, being anxious about social rejection and loss of perceived femininity if success is achieved. This fear is not entirely irrational, as studies show that in adulthood, men often seem to be threatened by a woman who is more successful than they are, resulting in lower rates of marriage for high-achieving women. Interestingly, a recent report notes that females who are masculine sex-typed have lower fear-of-success scores than those who are feminine sex-typed.

Female adolescents also get mixed messages about the relationship between femininity and sexuality. While the traditional message about sexual behavior has been "nice girls don't," or should feel guilty if they do, the primary allure of femininity is sexual, and the "proof" of femininity is sexual desirability. But if femininity is to be valued, why not be sexually active? The dilemma lies partly in the cultural double standard that sanctions varied male sexual experience but regards the female with more than one partner as promiscuous.

To be certain, the traditional gender-role stereotypes related to sexual behavior have been set aside by many adolescents. Teenage girls are much more apt to ask boys out today than they were twenty-five years ago and often take the initiative in sexual activity. This is seen as a major relief by some adolescent males, who feel freed from the burden of having to be the sexual expert, but is frightening to others who feel more comfortable with traditional sexual scripts. As one seventeen-year-old boy put it, "I don't like the feeling of not being in control. What am I supposed to do if a girl wants to make love and I'm not in the mood?"

In many ways, the old "quarterback-cheerleader" idea of masculine and feminine gender roles during adolescence has broadened into newer, more complex, and less clearly defined patterns. Athletic, educational, and career aspirations have become less compartmentalized, styles of dress have been altered, and many colleges that were previously restricted to one sex have now become coeducational. Nevertheless, it is important to realize that the influence of traditional gender-role attitudes continues to affect today's adolescents, showing that the present is still very much the product of the past.

Before proceeding any further with our overview of gender-role socialization, two points must be made. First, our discussion has deliberately highlighted common denominators of this process while ignoring many sources of variability. To believe that children in Beverly Hills, Detroit, and rural Vermont are exposed to identical messages about gender roles is obviously incorrect. Differences in religion, socioeconomic status, family philosophies, and ethnic heritage all influence the socialization process: for instance, researchers have found that gender-role distinctions are sharper in the lower class than in the middle or upper classes. Second, to think that gender roles are entirely shaped in childhood or adolescence implies that adults cannot change. In recent years, however, many young adults have moved away from the traditional gender-role distinctions with which they were brought up and have chosen alternate patterns with which they can live more comfortably. How this trend will ultimately affect future generations is not known.

Despite differences in upbringing and changing attitudes, our culture's gender-role stereotypes usually come into full bloom in the adult years, although the patterns change a little. For men, although heterosexual experience and attractiveness continue as important proofs of masculinity, strength and physical competence (as in hunting, fighting, or sports) are no longer as important as they once were. Occupational achievement, measured by job status and financial success, has become the yardstick of contemporary masculinity for middle- and upper-class America.

For women, marriage and motherhood remain the central goals of our cultural expectations, although this stereotype is now beginning to change significantly. As more and more women join the work force, as more and more women are divorced, as more and more people choose childless marriages, the notion that femininity and achievement are antithetical is slowly beginning to crumble away.

Marriage is a fascinating social institution in which gender roles play out in some unexpected ways. Tavris and Offir (1977, p. 220) observe:

> The irony is that marriage, which many men consider a trap, does them a world of good, while the relentless pressure on them to be breadwinners causes undue strain and conflict. Exactly the reverse is true for women. Marriage, which they yearn for from childhood, may prove hazardous to their health, while the optional opportunities of work help keep them sane and satisfied.

There is a substantial body of research that shows that married men are physically and mentally healthier than single men, but married women have higher rates of mental and physical problems than single women. Gove (1979, pp. 39–40) suggests several aspects of gender roles and marriage that conspire to cause such problems:

1. Women usually have their "wife-mother" role as their only source of gratification, whereas most men have two sources of gratification — worker and household head.
2. Many women find raising children and household work to be frustrating and many others are unhappy with the low status of their "wife-mother" role.
3. The relatively unstructured and invisible role of the housewife is a breeding ground for worry and boredom.
4. Even when a married woman works outside the home, she is generally expected to do most of the housework (and thus

is under greater strain than her husband) and typically has a low-status, lower-paying job and must contend with sex discrimination at work.

5. The expectations confronting married women are diffuse and unclear; uncertainty and lack of control over the future often conspire to create problems and low self-esteem.

Fortunately, there are some positive indications that change is not just on the horizon but actually here in our midst today. Dual-career families are becoming common and more and more men are willingly participating in ordinary household tasks that were previously regarded as strictly "women's work." A small but growing number of men are staying home to be house-husbands while their wives pursue outside careers.

Adult gender roles hinge on areas other than marriage, of course. It is fascinating to see how the same status inconsistency found in many marriages also applies to situations outside the home. In the business world, very few companies have substantial numbers of women as executives (and the secretarial pool is unlikely to have many men). Although a young man who is successful in business is pegged as a "boy wonder," a young woman who achieves corporate success is sometimes accused of having "slept her way to the top." Medical schools and law schools only began to admit sizable numbers of women in the last decade, and even then it took some prodding from the federal government. Furthermore, changes in admissions policies do not necessarily reflect an open-armed embrace. As one woman attending medical school put it:

> From the beginning, I could notice great astonishment that I was attractive *and* bright. My teachers seemed to think that only ugly women have brains. Then, there was a constant sense of being singled out for "cruel and unusual punishment." From the anatomy labs to the hospital wards, the female medical students were gleefully given the dirtiest assignments and made the butt of jokes. I never did understand why a woman physician examining a penis is so different from a male physician doing a pelvic exam, but this seemed to be a constant source of humor.

Complicating matters even more, and showing how widespread sexism remains in our society, is the indisputable fact that many women who enter even the most prestigious professions

"National Women's Institute
Washington, D.C. 20015

Gentlemen:"

are also subjected to sexual harassment. For instance, a recent
report found that 25 percent of female students and faculty at
Harvard Medical School had encountered varying forms of sexual harassment ranging from leering, sexually oriented remarks
to instances of unwanted touching and requests for sexual fa-
vors. At Atlanta's old-line law firm of King & Spaulding, where
a former female associate had filed a sex discrimination lawsuit
to protest against being denied partnership, a summer outing
for law students working at the firm featured a bathing suit
contest for women at which one male partner proclaimed, "She
has the body we'd like to see more of" (*Wall Street Journal,* Dec.
20, 1983, p. 1).

Not only do women have difficulty gaining access to nontraditional occupations, they are also frequently penalized by lower
salaries than those for men and face more obstacles to advancing
on the job. Furthermore, when women are successful in their
achievements at work, the results are more likely to be attributed
to luck than to skill, dedication, or effort. Another form of prejudice that women often have to overcome is shown in a recent

research study that had 360 college students — half of them male, half of them female — evaluate academic articles that were presented as written by either "John T. McKay" or "Joan T. McKay." Although the same articles were used for the evaluations, with only the first name of the author varying, the articles supposedly written by a male were more favorably evaluated by both sexes than the articles supposedly written by a female.

Clearly, sex discrimination is a problem of today's world that will not disappear overnight. However, there are certainly signs of changing times as women now enter "male" occupations like welding and making telephone repairs and as men increasingly infiltrate traditionally "female" occupations. Now that we've had our first female astronaut and our first female vice-presidential nominee from a major political party, we may finally be entering a new era of sex-role expectations.

ANDROGYNY

While reading this chapter, you may have decided that your own personality reflects certain traits socially labeled as "masculine" and others viewed as "feminine." If so, you are like many other people; relatively few individuals are 100 percent one or the other.

In the recent past, as psychologists have discarded some older assumptions about the nature of masculinity and femininity, the concept of *androgyny* has attracted considerable attention. Androgyny refers to the combined presence of both stereotypical feminine *and* masculine characteristics in one person. The word itself comes from two Greek roots: *andro-*, meaning male, and *gyn-*, meaning female.

Just what does it mean to say a person is androgynous? There is no firm agreement on this point among researchers. First, masculine and feminine traits could coexist but be expressed at different times. Kaplan and Sedney (1980, pp. 7–8) explain this dualistic model of androgyny as follows: "She or he might disagree forcefully and assertively with a colleague on a major issue of program development, but act comfortably and caringly toward that same person's distress over a personal problem." In other words, he or she acts typically male, then female. Or, feminine and masculine traits may exist in a fully integrated way within a person. Instead of alternating between feminine and masculine characteristics, the individual blends the two together. For example, an androgynous woman may initiate sexual activity (traditionally regarded as a "masculine" role) but do so in a style that is warm and sensitive (traditionally viewed as "feminine" traits). Thus, becoming androgynous does not imply losing the qualities associated with one's gender and taking on those associated with the opposite sex. It involves developing those opposite-sex qualities that already exist within us and manifesting them in ways determined by our own-sex qualities.

Several recent studies by psychologists show that about one-third of college and high school students are androgynous. Spence and Helmreich found that androgynous individuals display more self-esteem, achievement orientation, and social competence than people who are strong in either masculinity or femininity, or those who have low scores in both areas. Furthermore, Bem's research has shown that androgynous individuals seem to have more flexible behavior than people with more traditional masculine or feminine patterns. Likewise, there is evidence that androgynous females may have fewer psychological problems than masculine- or feminine-stereotyped persons.

However, androgyny may also have some disadvantages. A recent study of college assistant professors found that being androgynous was associated with greater personal satisfaction but an increased amount of work stress. Other researchers found that masculine males, rather than androgynous males, showed better overall emotional adjustment. Androgynous males had more drinking problems, while masculine males were more creative, less introverted, more politically aware, and felt more in control of their behavior. Furthermore, androgyny does not necessarily lead to more effective behavior or problem-solving. In fact, a recent study of 236 college students found that androgyny does not help a person to be more versatile or adaptable; instead, it was found that for both sexes, the presence of "masculine personality characteristics, rather than the integration of masculinity and femininity, appears to be critical" (Lee and Scheurer, 1983, p. 304).

Confusing the issue even more is the fact that depending on how one measures self-esteem, different results may be obtained in studies of androgyny. With this research still in its infancy, it is too early to know if androgyny is a desirable goal for the future or a potential source of trouble. However, it does provoke many interesting questions about "traditional" male/female roles.

THE PSYCHOLOGY OF
SEX DIFFERENCES

The controversy over the psychological differences between the sexes has long been steeped in myth. Some of the early researchers who determined which sex possessed "superior" skills are now being accused of using sloppy methods and male bias to ensure the results affirmed male superiority. Times have changed, and a landmark study published in 1974 dispels many of these myths by concluding that the sexes are more alike than different.

Eleanor Maccoby and Carol Jacklin spent three years compiling and reviewing over 2,000 books and articles on sex differences in children to find which beliefs were backed by hard evidence and which had insufficient experimental support. Their study did not try to explain why these differences and similarities exist; Maccoby and Jacklin wanted only to describe the present state of the research — without bias.

To rate your own biases, take the following true/false quiz:

T   F   1. Girls are more social than boys.

T   F   2. Boys have higher self-esteem than girls.

T   F   3. Girls are better than boys at simple, repetitive tasks.

T   F   4. Boys have greater mathematical and visual-spatial abilities than girls.

T   F   5. Boys are more analytical than girls.

T   F   6. Girls have greater verbal ability than boys.

T   F   7. Boys have a stronger motivation to achieve.

T   F   8. Girls are less aggressive than boys.

T   F   9. Girls can be persuaded more easily than boys.

T   F   10. Girls are more alert to auditory stimulation; boys are more alert to visual stimulation.

The answers, based on the Maccoby and Jacklin research, are surprising.

Question 1 — There is no evidence to suggest that girls are more social than boys. In early childhood, both sexes choose to play in groups with equal frequency, and neither sex is more willing to play alone. Boys do not prefer inanimate objects over playmates, and at certain ages boys spend more time with playmates than girls do.

Question 2 — Psychological tests show that girls and boys are very similar in self-esteem throughout childhood and adolescence, but they pick different areas in which they feel they have greatest self-confidence. Girls believe they have more social competence; boys see themselves as dominant and powerful. These beliefs have no experimental support and probably spring from early social learning rather than an accurate assessment of their own abilities.

Questions 3 and 4 — Both sexes perform equally well at simple, repetitive tasks. Boys excel in mathematical ability from about age twelve and have an increased ability to perceive relationships among objects in space. Boys, for example, are better able to mentally rotate a picture of an object and to correctly describe its hidden side. This difference is not evident until adolescence, suggesting it may arise either from environmental factors (perhaps boys are given more opportunities to perfect this skill) or from hormonal influences.

While some authorities have questioned whether male mathematical superiority is a by-product of cultural expectations, the evidence seems to be somewhat divided. A study by researchers at the University of Chicago found no sign of sex differences in the ability to solve geometry problems among 1,366 tenth-graders. Two other reports suggest, however, that biological influences may be important. In one study, researchers found that men with severe lifetime androgen deficiency had impaired spatial abilities compared to men with normal hormone levels or men who developed androgen deficiencies after puberty. In another study, researchers have found additional evidence linking exceptional mathematical talent to hormonal status, suggesting that prenatal programming of the brain by androgens may predispose to later sex differences that may be magnified at the time of puberty by rising levels of the male hormone. It is important to realize, however, that the sexes overlap considerably in their mathematical abilities, so this minor sex difference should not be used to counsel boys or girls in regard to courses or careers.

Question 5 — Boys are not more analytical than girls. To analyze, one must be able to recognize the important information in a situation uninfluenced by context or surroundings. Boys and girls are equally likely to respond to the unimportant elements in analyzing a problem.

Question 6 — Girls' verbal abilities mature more rapidly than boys'. Boys and girls remain about equal from infancy to early adolescence, but in high school and possibly beyond, females take the lead. Girls score higher on tests requiring the understanding of complex language, creative writing, analogies, fluency, and spelling. As with the boys' greater mathematical abilities, the girls' later increase in verbal ability may result more from socialization encouraging girls to perfect language skills.

Question 7 — Boys and girls can be equally motivated to achieve but by different factors. Girls are motivated to achieve when neither competition nor social comparison is stressed. Boys need direct appeals to their ego and a sense of competition to reach the girls' level of motivation.

Question 8 — Girls are less aggressive than boys, a difference exhibited as early as age two, when social play begins. Boys are more aggressive physically and engage in mock fighting and verbal forcefulness. Their aggression is usually directed at other male playmates rather than at the less aggressive girls. There is no evidence, however, that parents encourage boys to be more aggressive than girls — they actually discourage aggression in both sexes.

Question 9 — Boys and girls are equally likely to be persuaded by others and to imitate the behavior of people around them. Both sexes are equally affected by social pressure to conform. The only verifiable difference is that girls are slightly more likely to adapt their own judgments to those of the group, while boys are able to accept peer-group values without changing their own values even when the two conflict.

Question 10 — Male and female infants respond alike to aspects of their environment that require hearing and sight. They are similarly skilled in identifying speech patterns, various noises, objects, shapes, and distances. This equality persists throughout adulthood.

An alternative approach for detecting differences between the sexes is the "brain-based" study. Neurologists and psychologists measure brain size and use tests such as the electroencephalograph (the "brain wave" test) to measure the brain's electrical response to stimuli. These studies reduce the possibility of experimenter bias because they do not rely as heavily on interpretations of observed behavior as other studies we have discussed.

Major findings of these brain-based studies (published after the Maccoby and Jacklin research) point toward a neurological basis for some sex differences. Dianne McGuinness and Karl Pribram, using an approach similar to Maccoby and Jacklin's, summarized the findings of a majority of these studies. They found that women have more sensitive taste, touch, and hearing. A woman's hearing in the higher ranges is so much better than a man's that a sound at 85 decibels seems twice as loud to her as to him. McGuinness and Pribram also concluded that women have better manual dexterity and fine coordination, are more interested in people, and are more attentive to sounds as infants. Since evidence is beginning to emerge showing some structural anatomic differences between the brains of men and women, it seems that more brain-based studies will be required to clarify the current controversies in this area.

THE TRANSSEXUAL
PHENOMENON

In 1953 the world was startled to learn about Christine Jorgensen, an American ex-Marine who underwent surgery in Denmark to convert his anatomical appearance from male to female. Since then transsexualism has achieved considerable notoriety. Jan Morris's autobiography, *Conundrum,* provides some fascinating details into her own transsexual odyssey. Renée Richards, an accomplished eye doctor and tennis player as a male, provoked quite a stir when she insisted on joining the women's pro tennis circuit as a converted female.

Transsexual individuals persistently feel an incongruity between their anatomical sex and gender identity. They frequently describe their dilemma as "being trapped in the wrong body." Their psychological sense of existence as male or female (their gender identity) does not match the appearance of their genitals and secondary sex characteristics. Looking and being biologically male, the male transsexual wishes to change to female anatomy and live as a woman. Conversely, looking and being biologically female, the female transsexual wishes to change to male anatomy and live as a man.

Precise statistics on the prevalence of this gender identity variation are not available, but one estimate suggests the figure as one in 100,000 for male transsexuals and one in 130,000 for female transsexuals. Among persons who contact gender identity clinics and request change-of-sex surgery, there are many more men than women. Although there has been considerable speculation about the possible cause(s) of transsexualism, there is little agreement on this matter among researchers in the field. Both biological and psychological factors have been suggested as causes.

In the best defined cases of transsexualism, the person has a lifelong sense of being psychologically at odds with his or her sexual anatomy. Typically, this psychological discomfort is partially (but only temporarily) relieved by pretending to be a member of the opposite, desired sex. Many transsexuals describe having had great interest in cross-dressing (i.e., wearing clothes of the "other" sex) during childhood or adolescence. Transsexuals, however, should not be confused with *transvestites,* who cross-dress to become sexually aroused but usually do not want a permanent change of anatomy or appearance. In at least some cases, discovery of transsexual impulses does not occur until adulthood.

Psychotherapy has been generally unsuccessful in resolving the transsexual's basic distress of feeling trapped in the wrong body. As a result, those judged to be authentic transsexuals have been treated in programs designed to lead to change-of-sex surgery — in effect, redoing the body to match the mind. Since such surgery is irreversible, responsible practitioners take a cautious approach and require a one- to two-year trial period beyond the initial evaluation during which the transsexual patient lives in a cross-gender role. During this time, the transsexual begins living openly as a person of the opposite sex, adopting hairstyles, clothing, and mannerisms of that sex, and also assuming a name that "matches" the new gender.

The transsexual male is given estrogens on a daily basis to produce a certain degree of anatomic feminization: breast growth occurs, skin texture becomes softer, and muscularity decreases, for example. However, treatment with estrogens does not remove facial or body hair (electrolysis is required) or raise voice pitch (some male-to-female transsexuals take voice lessons to learn to speak in a more feminine fashion). Estrogen therapy also reduces the frequency of erections and causes the prostate gland and seminal vesicles to shrink.

Transsexual women are treated with testosterone to suppress menstruation, increase facial and body hair growth, and deepen the voice. Surgery is required to reduce breast size. For both male and female transsexuals, hormone treatments are given throughout the trial period of cross-dressing and adjusting to a new set of gender roles. At the same time, the patient's progress is periodically evaluated by a psychiatrist or psychologist. Attention is also directed to achieving legal recognition of the sex change and to personal matters, such as family or religious counseling.

If all goes fairly smoothly in the trial period and the transsexual is judged to be psychologically stable and able to adjust socially to the conversion, the final stage of treatment is surgery to change the sexual anatomy. At present, it is much simpler to perform male-to-female conversion surgery than the reverse. The male-to-female operation requires removing the penis and testes and creating an artificial vagina and female-appearing external genitals. The more difficult female-to-male procedure involves creating a "penis" from a tube made from abdominal skin or from tissue from the vaginal lips and perineum. While the artificial vagina created in the male-to-female transsexual often looks authentic and may allow a fairly full range of sexual response (e.g., vaginal lubrication and orgasm have both been claimed but not scientifically verified), female-to-male transsexual surgery creates an artificial penis that cannot become erect or feel tactile sensation.

In female-to-male transsexual surgery, it is sometimes possible to attain a degree of sexual function by implanting a mechanical inflatable device inside the penis to produce an artificial erection. Experience with this method is limited at the present time, and in any event, ejaculation is not possible. Many female-to-male transsexuals choose to have hormone therapy and surgical removal of their breasts and uterus but do not opt for an artificial penis.

Transsexual surgery is not a cure for this disorder but is only a procedure that may foster a sense of emotional well-being. Recently, the wisdom of surgery for transsexuals has been questioned by researchers from Johns Hopkins University who claimed to find no significant psychological benefits in patients who had undergone such operations compared to those who did not. The matter is unresolved at present, although several prominent medical centers stopped doing transsexual surgery in 1980 because of the lack of solid evidence that the surgery is beneficial.

GENDER ROLES AND SEXUAL BEHAVIOR

The gender roles in most societies have a strong impact on sexual attitudes and behavior. For example, in America it is still widely thought that males are innately more interested in sex than females, that males characteristically assume an active role in sex while females are characteristically passive, and that male sexual arousal occurs quickly and automatically, while females require sweet talk and special handling and even then have only a precarious degree of arousability. Each of these stereotypes has some behavioral consequences: in general, men try to measure up to the cultural expectations and women often accept the notion of being second-class citizens from a sexual viewpoint.

By looking at a culture that has a very different set of expectations about sexual interaction, we can see how limiting these stereotypes are. In Mangaia, a tiny Polynesian island in the South Pacific, the cultural message is that sexual pleasure is for everyone. As a result: "Less than one out of a hundred girls, and even fewer boys — if, indeed, there are any exceptions in either sex — have *not* had substantial sexual experience prior to marriage" (Marshall, 1971, p. 117). Female sexual passivity is frowned on among Mangaians, and sexual intimacy does not require prior establishment of personal affection. Girls are expected to learn to be orgasmic at a young age, and although their first sexual experiences are likely to be with boys of their own age, older and more experienced partners soon become desirable because they can give more sexual pleasure. One particularly interesting observation: "upon hearing that some American and European women cannot or do not achieve the climax, the Mangaian immediately asks (with real concern) whether this inability will not injure the married woman's health" (Marshall, 1971, p. 162). On Mangaia, all women learn to be orgasmic.

We have previously mentioned the double standard that applies to sexual behavior in our society. According to the traditional double standard, males are permitted to have premarital sexual experience while females are expected to remain virgins. After marriage, while fidelity is "officially" expected, it is acknowledged that men might roam and women are expected to remain faithful. In recent years, the double standard has undergone some subtle changes. Many teenagers have discarded the belief that female virginity is necessary or desirable, but it now seems that a young woman must wait for a "serious relationship" to have intercourse, while young men are not so strongly saddled with this expectation.

The double standard also assigns responsibility for being the sexual "expert" to the male. The male is expected to initiate sexual interaction, to control the timing and tempo, to select the proper activities to bring about his partner's arousal, and to bring his partner to orgasm. While this version of the double standard (the idea that sex is something a man does "for" a woman) may be an improvement on the older belief that "good" women had no sexual feelings (in this view, sex was something a man did "to" a woman for his own release), it hardly encourages flexibility and sharing.

The double standard and its variations can create a number of sexual problems. The female, for example, may develop a narrow view of sexual interaction. Feeling that she must prevent the male from "trying to get all he can," her potential pleasure is decreased by her need to set limits. The male, on the other hand, may feel compelled to prove his masculinity by making sexual advances even when he is not particularly in the mood or attracted to his companion.

It now appears that the sexual double standard, like many other gender-role stereotypes, is beginning to be replaced by concepts of equal opportunity and mutual interaction. What a great many men and women are learning is that they cannot achieve the pleasure they both want until they realize that sex is not something a man does to or for a woman but something a man and woman do together as *equal participants.*

The woman who honors her sexuality learns that she can, when she chooses, express openly the full range of her excitement and involvement — the delight of wanting and being wanted, touching and being touched, seeing and being seen, hearing words and uttering them, of fragrances and textures, silence and sounds. The man who appreciates her as a partner can enjoy letting go of responsibility for her satisfaction and can savor her varying moods and desires in conjunction with his own.

The responsiveness of both partners is based on acceptance of each other as vulnerable human beings with unique needs, expectations, and capabilities. Both can express their creative impulses without fear of violating the gender-role expectations of ladyhood or chivalry. Emotional needs, which vary with the mood, time, and place, are not labeled "masculine" and "feminine." Each partner can appreciate the other's sexual urges. If their sexual needs conflict at times, they can gently negotiate a solution — not as representatives of two different sexes but as two separate partners united by a mutual concern.

Sexual emancipation grows out of a sense of self-respect and personal freedom. If you are nothing to yourself, you have nothing to give and expect nothing in return. Sexually perhaps you might consider yourself useful, as an object is useful, but that is all. Before a true partnership is possible, both individuals must have pride in themselves and feel happy in being male or female.

At least half the potential pleasure of the sexual experience comes from how a partner responds. If there is virtually no reaction, or at best passive acceptance, the emotional current steadily weakens and eventually flickers and goes dead. However, with an actively involved partner, one individual's spontaneous feelings, spontaneously communicated, stimulate the other and heighten his or her tensions, impelling that person to act on his or her own impulses. Whatever she gives him returns to her and whatever he gives her comes back to him.

The relationship between the sexes is often conceived in terms of a misleading image: two on a seesaw. Power is the pivot, and if one sex goes up, the other must come down. What women gain, men lose. But the sexual relationship itself shows the analogy to be false. What men and women achieve together benefits both — the quality of life, as it is *individually* experienced, can be greatly expanded by a fully shared *partnership*.

CHAPTER NINE

Loving and Being Loved

WHEN William James wrote his classic *Principles of Psychology* in 1890, he devoted only two pages to "love." While noting the connection between love and "sexual impulses," James observed, "These details are a little unpleasant to discuss." D. H. Lawrence, the English novelist, was much less timid in dealing with this topic. In *Lady Chatterley's Lover* (1926), he suggested that love depends on being uninhibited in all respects, as illustrated in this bit of dialogue between Lady Chatterley and Mellors, her lover:

> "But what do you believe in," she insisted.
> "I don't know."
> "Nothing, like all the men I've ever known," she said.
> They were both silent. Then he roused himself and said: "Yes, I do believe in something. I believe in being warm-hearted. I believe especially in being warm-hearted in love, in fucking with a warm heart. I believe if men could fuck with warm hearts, and the women take it warmheartedly everything would come out all right. It's all this cold-hearted fucking that is death and idiocy."

Until very recently, the topic of love was more in the province of writers, poets, and philosophers than in the minds of psychologists and scientists. Even though it has been said that "love makes the world go round," few sexologists (including ourselves) have addressed this subject in any detail. Nevertheless, we have all felt love in one way or another. Many of us have dreamed of it, struggled with it, or basked in its radiant pleasures. It is also safe to say that most of us have been confused by it too. In this

chapter, we will focus our attention on the complicated relationships between love, sex, and marriage in an effort to reduce at
least some of this confusion.

WHAT IS LOVE?

Trying to define love is a difficult task. Besides loving a spouse
or boyfriend or girlfriend, people can love their children, parents, siblings, pets, country, or God, as well as rainbows, chocolate sundaes, or the Boston Red Sox. Although the English
language has only one word to apply to each of these situations,
there are clearly different meanings involved.

When we talk about person-to-person love, the simplest definition may be one given by Robert Heinlein in the book *Stranger
in a Strange Land:* "Love is that condition in which the happiness
of another person is essential to your own." This is certainly the
love that Shakespeare described in *Romeo and Juliet,* that popular
singers celebrate, and that led Edward VIII to abdicate the
throne of England to marry the woman in his life.

In any type of love, the element of caring about the loved
person is essential. Unless genuine caring is present, what looks
like love may be just one form of desire. For example, a teenage
boy may tell his girlfriend "I love you" just to convince her to
have sex with him. In other cases, the desire to gain wealth,
status, or power may lead a person to pretend to love someone
to reach these goals.

Because sexual desire and love may both be passionate and
all-consuming, it may be difficult to distinguish between them in
terms of intensity. The key feature is the substance behind the
feeling. Generally, sexual desire is narrowly focused and rather
easily discharged while love is a more complex and constant
emotion. In pure, unadulterated sexual desire, the elements of
caring and respect are minimal, perhaps present as an after-
thought, but not a central part of the feeling. The desire to know
the other person is defined in only a physical or sensual way, not
in a spiritual one. This end is easily satisfied. While love may
include a passionate yearning for sexual union, respect for the
loved one is a primary concern. Without respect and caring, our
attraction for another person can only be an imitation of love.

Respect allows us to value a loved one's identity and integrity and thus prevents us from selfishly exploiting them.

The importance of caring and respect was central to the thinking of Erich Fromm, whose classic book, *The Art of Loving* (1956), influenced all subsequent study of this subject. Fromm believed that people can achieve a meaningful type of love only if they have first reached a state of self-realization (being secure in one's own identity). Thus, Fromm defined mature love as "union under the condition of preserving one's integrity, one's individuality," and noted that the paradox of love is that "beings become one and yet remain two." In speaking about the respect inherent in all love, Fromm suggested that a lover must feel, "I want the loved person to grow and unfold for his own sake, and in his own ways, and not for the purpose of serving me."

Fromm's insistence that people must be self-realized before having a "meaningful" type of love overlooks that love itself can be a way of attaining self-realization. We believe that people have a great capacity to learn about themselves from a love relationship, although we also agree with the psychologist Nathaniel Branden's observation that love cannot be a substitute for personal identity.

Peele and Brodsky, authors of a book called *Love and Addiction*, have an interesting viewpoint on what happens when respect and caring are missing from a love relationship. They believe that some relationships of this variety serve the same needs that can lead people to alcohol abuse or drug addiction. The resulting "love" is really a dependency relationship:

> When a person goes to another with the aim of filling a void in himself, the relationship quickly becomes the center of his or her life. It offers him a solace that contrasts sharply with what he finds everywhere else, so he returns to it more and more, until he needs it to get through each day of his otherwise stressful and unpleasant existence. When a constant exposure to something is necessary in order to make life bearable, an addiction has been brought about, however romantic the trappings. The ever-present danger of withdrawal creates an ever-present craving. (Peele and Brodsky, 1976, p.70)

Peele and Brodsky suggest specific criteria for distinguishing between love as a healthy relationship with growth potential versus love as a form of addiction:

1. Does each lover have a secure belief in his or her own value?
2. Are the lovers improved by the relationship? By some measure outside of the relationship are they better, stronger, more attractive, more accomplished, or more sensitive individuals? Do they value the relationship for this very reason?
3. Do the lovers maintain serious interests outside the relationship, including other meaningful personal relationships?
4. Is the relationship integrated into, rather than being set off from, the totality of the lovers' lives?
5. Are the lovers beyond being possessive or jealous of each other's growth and expansion of interests?
6. Are the lovers also friends? Would they seek each other out if they should cease to be primary partners?

These questions are not listed to suggest that there is only one "right" way to love. While most people in love probably can not answer "yes" to all six questions, thinking about these issues may give you some ideas for present or future relationships.

As a practical matter, it is often difficult to draw a line between loving and liking. Although various researchers have tried to measure love, we agree with the observation "The only real difference between liking and loving is the depth of our feelings and the degree of our involvement with the other person" (Walster and Walster, 1978, p. 9).

ROMANTIC LOVE

The great loves of fiction and verse have been romantic loves marked by a whirlwind of emotions from passion to jealousy to anguish. In romantic love, unlike any other type of love, we immerse ourselves almost completely in another person. When Chaucer wrote that "love is blind," he was acknowledging that the intensity of romantic love distorts our objectivity. In our craving for our loved one, we may overlook flaws, magnify strengths, and lose all sense of proportion.

The puzzles and paradoxes of romantic love are many. We will address them by first discussing some psychological theories about the nature of romantic love and then presenting a conceptual model of the romantic love cycle.

It should be no surprise that there is little agreement among psychologists on a valid definition of romantic love. Branden (1980) says that it is "a passionate spiritual-emotional-sexual attachment between a man and a woman that reflects a high regard for the value of each other's person." Since we believe that romantic love is not restricted to heterosexual relationships, this definition is too restrictive. Fromm did not define romantic love specifically, but it appears that he used the term "erotic love" to mean the same thing. In contrast, other definitions of romantic love do not include a sexual component as a requirement.

Recently, psychologist Dorothy Tennov coined the word *"limerence"* to describe the particularly powerful form of romantic love in which a person is said to be "love-struck" or "head-over-heels" in love. Limerence is marked by preoccupation with thoughts of the loved one and the certain knowledge that only this person can satisfy your needs. The limerent lover's mood depends almost totally on the actions of the loved one; that person's every gesture or word is doted on in hope of approval and in fear of rejection.

Limerence, like other forms of romantic love, is an affliction as well as a joy because it is almost completely outside rational control. The consuming emotional ups and downs of limerence can interfere with other relationships, reduce the capacity for work or study, and disturb a person's peace of mind. According to Tennov, many people never experience limerence (although they may experience love), while other people pass through a series of limerent episodes.

According to a love survey conducted by sociologist John Alan Lee there are six primary types of romantic love. Lee assigned Greek and Latin names to these categories, pointing out that many love relationships are composites of two or more of these patterns. *Eros* describes a love based on physical attraction — an intense, sexual magnetism. Erotic love, according to Lee, is often quick to ignite and quick to flicker out and infrequently turns into a deep lasting relationship. *Ludus* refers to a playful, casual variety of love. Ludic lovers are apt to engage in gamesmanship and usually do not show high levels of commitment to each other. They may date several partners and preserve their options by avoiding dependence on their lover. In this form of love, sex is more "fun and games" than intimacy or commitment.

Storge is warmth and affection that slowly and imperceptibly turns into "love without fever, tumult, or folly." Storge emerges from friendships, but it has no identifiable starting point where people realize they fell in love. Storge is a solid, stable type of love that can withstand crises, but it lacks dramatic passion. *Mania,* in contrast, is a stormy, topsy-turvy kind of love. Mania connotes madness and agitation, and the manic lover is driven by powerful urges — the "need for attention and affection from the beloved is insatiable." The manic lover is either climbing a mountain of ectasy or sliding into a valley of despair. Manic love is like a roller-coaster ride: the thrilling dizziness and ups and downs usually come to an abrupt and rapid ending. *Pragma* is a more level-headed practical love. The pragmatic lover is searching for the proper match with a mental checklist of desirable features for the loved one. Once a likely candidate is found, if there is some degree of mutual agreement, pragmatic love may develop into more intense feelings. *Agape,* Lee's final category of love, is based on the traditional "Christian" view of love as undemanding, patient, kind, and everpresent. It is interesting that Lee admits to never having found an unqualified example of agape, so this type of love remains an ideal more than a reality.

Lee believes that the most satisfying love relationships are between lovers who share "the same approach to loving, the same definition of love." A manic lover who demands intensity and commitment, for example, would be tormented by the ludic lover's playful approach to love. While Lee endorses the idea that two lovers of the same type are most likely to be compatible, there is no guarantee that a person's "love-style" is unchangeable and the same in all relationships. The style of any love relationship probably emerges out of the personalities, needs, and prior experiences of the lovers. People may even avoid a style of love that did not work out for them previously — after all, we can learn from our mistakes.

Research on romantic love has not been exhaustive, but we will summarize the current data and present our clinical observations in terms of a "romantic love cycle." The model shown in Figure 17 does not imply that all romantic loves are identical or that each romantic love passes through the phases of this cycle in a predictable, sequential fashion. Instead, it gives us a way to organize our thinking about romantic love.

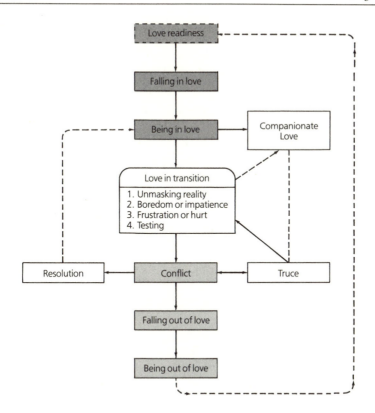

FIGURE 17.   The Romantic Love Cycle

   Although there is no certifiable proof that people can be in a
state of love readiness, we believe such a state of mind exists.
Love readiness does not always result in falling in love, but it
does seem to increase its probability.

   Love readiness consists of several different elements. First,
love is seen as something that is desirable and rewarding rather
than as troublesome or encumbering. People who look at romantic love as a sign of weakness or as a distraction from career
development are unlikely to permit themselves to fall in love (as
far as anyone can control such emotions). But people who believe that love is ennobling and brings out their best may actively
search for a suitable love object. Second, there is a longing for
interpersonal intimacy and companionship. This longing may
be motivated by loneliness, jealousy for someone else's love relationship, or a desire to replace a past love. Third, sexual frus-

tration often contributes to the state of love readiness. This frustration may result from sexual deprivation or from a wish for sex as part of a passionate, committed relationship. Casual sex may be accessible but less fulfilling. Finally, love readiness may reflect the hopefulness people have about being loved; if this is so, the frequency of romantic love may decline as people get older, as Dorothy Tennov has noted, because their expectations of having their feelings returned are reduced.

Some people always seem to be in a state of love readiness but never succeed in getting any farther. While they may eventually make concessions and enter into relationships that don't really qualify as romantic love, their mental quest doesn't seem to end. Other people never experience love readiness or pass beyond it after a brief time. In trying to evaluate love readiness, remember that no one has ever measured it, and that it's not unusual for people who don't match the description given here to find themselves falling in love. As we said before, love is neither rational nor completely predictable.

What triggers love is still the subject of guesswork. But the process of experiencing romantic love begins with a stage of falling in love that ranges from the instantaneous "love at first sight" to a gradual process that requires months or years of development.

A sudden flash of love is unlikely to occur unless a person is in a state of love readiness. In real life, instantaneous love is the exception, not the rule, and falling in love is a process that can start out in many ways. Dating provides an opportunity for discovering if two people like each other, are compatible, and can meet one another's needs. Being physically near a person is another pathway: by close association, you can come to some preliminary conclusions about a person's desirability without announcing even the possibility of a romantic interest. Friendships sometimes blossom into love, although it may be difficult to pinpoint exactly when the falling in love occurs. The trusting atmosphere of an intimate friendship may make passion seem out of place, and love, if it develops, may be a low-keyed rather than fiery emotion.

Two aspects of falling in love are particularly likely to ignite passion — the excitement of getting to know someone intimately and the excitement of sex. Both types of arousal intensify the push toward love by positive reinforcement.

Being in love, like falling in love, can occur whether or not love is returned. If there is no indication that reciprocal love might develop, the probability of a person's staying in love begins to decrease rapidly. But a person who has reached the "being" phase of romantic love is usually inventive, hopeful, and willing to accept even the flimsiest signs of reciprocation.

The romantic lover at this stage may be head-over-heels in love (caught in the grips of limerence) or in a more tranquil, self-satisfied, secure, and objective state. Sexual attraction is almost invariably strong, although it may not lead to action because of shyness, sexual problems, or moral constraints. In some situations, people may try to defuse their sexual impulses by masturbation or strenuous exercise either to keep their love "pure" or if their loved one shows no interest in a sexual relationship. Whatever form being in love assumes, it is usually a passing phase, lasting an average of one to two years, according to many researchers. Most of the time, romantic love either changes into another form of love, called companionate love, or gradually dissolves because of conflicts, boredom, or lack of interest.

The transitional phase of romantic love is a pivotal time. Here, the initial excitement of getting to know someone and the passion of a new sexual relationship begin to lessen and the thrill is going if not gone. Lovers begin to notice imperfections in each other that were previously unobserved or ignored, and boredom or impatience begins to set in. Frustration occurs when love does not measure up to our fantasies, when we realize that all our problems are not "cured," or when we discover that the ecstasy cannot go on forever without intermissions.

Characteristically, lovers in this transitional phase begin to test one another — and the presumed strength of their love — in various ways. Each lover is likely to try to force or trick the other into becoming what they were thought to be or what he or she would like them to become. Power struggles and competitive strivings emerge. Testing becomes a means of making a rational decision about the future of the relationship: "Do I want to stay with this person, or should I get away now, while I can?" Jealousy may rear its ugly head, anger may erupt, and conflict is almost unavoidable.

The transition stage of love is basically a time for testing reality. In a sense, love pulls its head down from the clouds, and the conflicts and doubts that arise may lead to a stage of falling out of love or measures may be taken to push the relationship into a temporary state of truce. The truce may either lead things back to the "love in transition" stage (with the probability being high that further conflicts will occur), or may lead directly to a companionate love relationship.

If, on the other hand, given the ingredients of motivation, flexibility, cooperation, and a little bit of luck, the conflicts are resolved, the relationship returns to the "being in love" stage. If this occurs, the new version of the relationship may actually be stronger, strengthened by the ability to survive its conflicts successfully. Mutual trust is no longer just a matter of faith but a by-product of experience.

Just as people falling in love delight in learning everything about their partners and revealing much about themselves, people falling out of love are less open, intimate, and interested in their partners. Concern for the partner's happiness becomes a second priority rather than a guiding light and eventually becomes an incidental thought. Communications may be strained because the two lovers are no longer "on the same wavelength," and whatever troubles occur at this point in the relationship hardly seem worth the effort to overcome.

Love relationships come apart in different ways, most of which are painful. Only about 15 percent of love relationships end by mutual consent, according to the researchers Hill, Rubin, and Peplau. Many times one person pulls out of a love relationship while the other is still "in love." In this situation, the falling-out-of-love stage occurs at different times for the two lovers. The heartbreak and sorrow of the deserted lover are sometimes very similar to a grief reaction, passing through a period of tearful mourning and shock followed by a time of persistent, haunting memories before there is a return to happiness. At other times, the jilted lover becomes angry, vengeful, or determined to avoid future love at any cost.

Once having fallen out of love, some people quickly revert to a state of love readiness, no worse for wear and perhaps even wiser and wealthier from their love experience. There is a kernel of truth to the notion that a person "on the rebound" may be more open (and more vulnerable) to a new love relationship. On the other hand, there often seems to be a "refractory period" early in the "being out of love" phase during which it simply is not possible to fall in love again.

COMPANIONATE LOVE

It is rare for the passion and excitement of a romantic love relationship to last for more than a few years. Usually, romance is replaced (except for occasional brief flickers) by another kind of love, which comes to a new state of equilibrium. This is called *companionate love,* which can be looked at as a steadier love based on sharing, affection, trust, involvement, and togetherness rather than passion.

Companionate love is not just a sorry substitute for romantic love, although it can deteriorate into drabness and routine if not sustained by continued caring and respect. Many companionate love relationships include an exciting, satisfying sexual side, and in many ways the partners may find that their pleasure in each other increases. Companionate love is less turbulent and more predictable than romantic love, so many people find it to be a soothing, secure kind of relationship.

Companionate love is most characteristic of marriage and other long-term committed relationships. Because it is less possessive and consuming than romantic love, it allows two people to carry on their lives — working, raising children, having hobbies, relaxing with friends — with a minimum of interference. It is a reality-based and steadier love, as opposed to romantic love, which is all too often based merely on ideals and fantasies.

RESEARCHERS ON LOVE

No one is completely certain about what causes the feeling of love. Although the image of Cupid practicing his archer's skills is clearly not a satisfactory explanation, no good alternative has really been proposed. Much of the research on love has addressed this issue from the perspective of social psychology, looking at interpersonal attraction as a possible source for some answers. Let's see what findings from these studies might apply.

Physical appearance seems to be an important element in determining how attracted one person is to another. Nursery school children, teenagers, and adults all regard good-looking people with more favor than their less attractive peers. People tend to be more socially responsive and more willing to provide help to attractive individuals. There is fairly solid evidence that despite sayings like "You can't tell a book by its cover", or "beauty is only skin deep," the physical attractiveness of females has a strong influence on their dating frequency — a finding that is far less true for males.

In a thought-provoking experiment, Dion, Berscheid, and Walster found that physical attractiveness also influences our expectations about other people's personalities and behavior. On the basis of photographs, both men and women rated good-looking people as more sexually warm and responsive, interesting, poised, sociable, kind, strong, and outgoing than less attractive people. The group judged as physically attractive was also seen as more likely to attain high occupational status, to make better husbands or wives, and to have happier marriages than the less attractive group. And a recent study involving college-age couples matched by computer found that both sexes reacted most positively to their dates if they were good-looking; intelligence, personality, and social charm had little to do with the romantic chemistry.

Physical attractiveness seems to be more important for females than for males in affecting interpersonal relationships. A man's occupational status or financial success is generally more important in rating his "social desirability." Physicians, lawyers, and other professionals are consistently rated by women as more desirable dates or marriage partners than men with low-status occupations such as janitor or waiter.

Here are a few other salient findings by "love researchers":

1. The notion of "love at first sight" may in many cases be only a myth that fulfills our need for instant acceptance and rationalizes our feelings of sexual arousal by giving them a

"dignified" label, but in some cases it proves to be real. Unfortunately, only the passage of time will show which is which.

2. The old bit of folklore that men prefer "hard-to-get" women just is not true.

3. College-age women fall in love more frequently than men of college age, but college-age men fall in love more quickly.

4. Men hang on longer in a dying love affair than women, and women end more romances than men.

So far we have discussed love as though it is solely a product of the mind. Some scientists suggest that there may be a biological component as well. Two types of evidence can be cited in favor of this idea.

First, evolutionary biologists point out that reproductive success may be at least partly linked to love. Hundreds of centuries ago, successful reproduction hinged on two factors: (1) genetic diversity to ensure the health of offspring, and (2) the man's closeness to this sexual partner during pregnancy and the infancy of their newborn child to provide protection and food, and to help in childrearing. Love might have created more stable attachments than sexual attraction could accomplish by itself. Love also drew genetically unrelated persons together to engage in mating, thus diversifying the gene pool and contributing to the survival of the species.

A second type of evidence that may shed light on the biological side of love also relates to sex. In 1964 the psychologist Stanley Schachter devised a theory based on the physiological responses that accompany many emotions — a pounding heart, sweaty palms, heavy breathing, and so on. According to Schachter, how we distinguish between love, anger, jealousy, nervousness, and other emotions is not based on our bodily reaction only (since the responses may be identical) but rather on the way we interpret or label what we are experiencing. In Schachter's sense, love is a matter of physiological arousal interpreted in a certain way. Recent studies have shown, however, that Schachter's theories are only partly correct. It now appears that there are some specific differences in the reactions of the autonomic nervous system to various types of emotions, raising the possibility that the feeling we call love may be accompanied by a unique set of physiological responses.

One plausible explanation is suggested by the psychiatrist Dr. Michael Liebowitz, who contends that the excitement and arousal of romantic love are a direct result of surging levels of two neurotransmitters, dopamine and norepinephrine, that carry chemical messages that bridge the gap between nerve cells in the brain. In his 1983 book, *The Chemistry of Love*, Liebowitz argues that these neurotransmitters are activated by visual cues — noticing someone who fits our ideal of attractiveness, for example — and then bathe the pleasure center of the brain in a sea of chemical messages. Liebowitz also believes that intense, transcendent love experiences may involve a separate neurotransmitter called serotonin, which can produce an almost psychedelic high, while companionate love may rely more on the brain's production of narcotic-like substances called endorphins that give a sense of tranquility.

Although research on love has become more fashionable in recent years, the following observation by a prominent psychologist still applies. "So far as love or affection is concerned, psychologists have failed in their mission. The little we know about love does not transcend simple observation, and the little we write about it has been written better by poets and novelists" (Harlow, 1958, p. 673).

LOVE AND SEX

The relationship between love and sex in our society is quite complicated. Traditionally, females were taught that love is a requirement for sex, while males were urged to obtain sexual experience whether or not love was present. Gradually in the 1960s and 1970s, premarital sex became more acceptable for females, at first if they were engaged to be married and later if they were involved in a "significant relationship" (one usually defined by love). Today, although restrictions have loosened even more for some, many heterosexual couples still need a statement of love before they feel morally comfortable with the idea of "going all the way."

It is tempting to categorize sex involving people who are not in love as casual sex, distinguishing it from relational sex. But people who do not love one another can have a strong relationship, and lovers can sometimes have casual sex in the sense that there's not much thinking about it or interpersonal communication going on. Sex can be mechanical, impersonal, and hurried whether or not two people love each other.

There is nothing inherently bad about impersonal sex if it is clearly consented to by all parties. Under certain circumstances and for some people, impersonal sex may be enjoyable in its own right. Others are offended or distressed by impersonal sex and could never consider participating in a group sex scene or having sex with a stranger or prostitute.

Some people enjoy a more personal, intimate brand of sex, hoping that it may develop into love. There are no guarantees that this will happen, so if this is the only reason for sexual participation, these people are liable to feel disappointed, cheated, or angry. It is probably easier for those who engage in casual sex to view sex without love as an experience valued for its own pleasures and unique returns. If nothing more is expected, disappointment is less likely.

Some moralists would be happier if it could be proven that sex without love does not work. There is no evidence, however, that sex is always or usually better if you are in love. We have worked with hundreds of loving, committed relationships where the sexual interaction was in shambles and with hundreds of people who deeply enjoyed sex without being in love.

Another aspect of this topic deserves mention. A love relationship, unless it's on the rocks, offers some protection against being used sexually. As we pointed out earlier, love is marked by caring and respect that buffers this risk. Sex without love is probably more likely to create misunderstandings and involves a greater risk of being used for the purposes of your partner without much regard for your own feelings and needs. It is harder to gauge the trust in a nonloving relationship, so "proceed at your own risk."

There are a variety of circumstances in which a love relationship exists without sex. Parent-child love, brotherly love, and so-called platonic (friendly, nonsexual) love are some obvious examples. There are also forms of romantic and companionate love in which there is no sexual component. In the purest forms of love without sex, both partners agree to abstain from physical intimacies. They may make this choice because of religious beliefs, a lack of interest in sex, or a desire to wait until marriage. This comment from a twenty-five-year-old man shows another motive for abstaining from sex:

> My partner and I became so caught up in our sexual activity that the rest of our relationship was neglected. After we decided not to have sex, we found more time and energy to love each other in ways that meant more to us.

There are also unavoidable circumstances that may prevent or limit sexual interaction in a love relationship, such as serious illness or geographic separation. Sometimes the decision is one-sided rather than mutual, in which case one lover may be left in a sexually frustrating situation.

Many people are troubled by questions about what's "right" and "wrong" sexually. Adolescents and young adults are often uncertain about the pros and cons of premarital sex. Others are puzzled about the morality and propriety of extramarital sex. When love is added into the equation, even more complex questions arise. Does sexual attraction to someone other than your lover mean that love is dying? Does "extrarelational" sex undermine the quality of love or trust in a primary love relationship? Should the presence or absence of love be the major determinant of our sexual decisions?

There is no simple formula for answering such questions. Each person approaches sexual decision-making from a framework of personal values, beliefs, and experiences that tip the balance one way or another. Some love relationships will wither and die if sex is "off limits." In other relationships, premature leaps toward the bedroom may jeopardize the foundation of love. For one person, abstaining from premarital sex is a matter of deep moral conviction, with potentially negative consequences if these convictions are set aside. For someone else, the same act of abstention prevents them from learning that there is poor sexual compatability with their future mate, creating a difficult problem to be faced early in marriage.

The dimensions of sexual decision-making can be appreciated better if we recognize some of the sources of potential conflict. It is easy to see how our society encourages us to adhere to a

BLOOM COUNTY   By Berke Breathed

particular set of sexual values. Our religious beliefs — and the strength of these beliefs — may also guide us. But at the personal level of decision-making, we must deal with the tensions between our various needs. For instance, there is built-in conflict between the following pairs of values: commitment versus freedom, privacy versus intimacy, sexual novelty versus permanence, independence versus sexual fidelity. A person's sexual decisions are outgrowths of how she or he judges the relative importance of such personal values.

Sexual decisions are unfortunately sometimes made on the basis of guilt, ignorance, or impulse. This creates a different kind of decision-making process that is more likely to be second-guessed later on. Since we must live with our sexual decisions, we should take an active role in making sexual choices instead of just letting the choices "happen."

LOVE AND MARRIAGE

To discuss love without discussing its relationship to marriage would be like discussing free enterprise with no mention of money. Love and marriage are certainly not synonymous, but they are linked in important ways.

While marriage, unlike love, can be defined in legal terms, its psychosocial dimension is most closely related to love. We do not believe that marriage is the only way to go, the "best" choice, or a perfect solution to life's problems. We recognize that some couples who live together without a marriage certificate are more meaningfully wedded, more committed to each other, than other couples whose marriage is legally sanctioned. But we agree with this idea:

> For all the books and preaching and counseling and psychological knowledge we have today, for all our new ideals of freedom and our emphasis on personal growth and fulfillment, marriage is still basically two people trying to love each other and answer each other's needs. (N. O'Neill, 1978, p. 4)

Our culture is unique in the emphasis it places on love *before* marriage. Unlike societies in which marriages are traditionally arranged by parents, with courtship following tightly controlled etiquette, we are led to believe that "love conquers all" and live our lives accordingly. In India, China, Japan, and parts of Africa and the Arab world, arranged marriages are common. Often, these matches preserve social and economic order and create a stable setting for family living. The partners in an "arranged" marriage — who may be "pledged" to each other during childhood — are not expected to begin their marriage because of love. Instead, devotion and responsibility are expected to grow as the marriage develops, and love may or may not occur.

Paradoxically, in America, where our strong sense of freedom and democracy carries over into more selection processes, we have an extraordinarily high rate of divorce. This may tell us something about the lack of love education we receive while growing up. We are expected to recognize love and to select a marriage mate " 'til death do us part" largely on the basis of love. The odds are strong, however, that most of us have gotten more training in learning how to drive than in learning how to love.

A person caught in the throes of romantic love is apt to be drawn to thoughts of marriage like a moth to light. The desire to be one with your loved one, especially when the love is mutually felt, fits neatly into our expectations of marriage as a form of intimate, lasting romance. But the irrational nature of romantic love often causes us to overlook potential problems and to minimize what will happen when and if the passion dies down.

This doesn't mean that you must adopt a scientific, calculating analysis of likely marriage candidates. Despite the success some computer dating services claim in finding the "perfect" match, no one has succeeded in devising a foolproof formula for marital success. Like many things in life, the process of selecting a mate is largely a matter of common sense combined with an element of luck.

At the commonsense level, it helps to realize that men and women tend to be happiest in equitable relationships. People feel most comfortable when they are getting from a relationship what they believe they deserve: too much or too little (inequity) often leads to discomfort and dissatisfaction. Research on "equity theory" shows that men and women generally marry someone with similar physical attractiveness, intelligence, and attributes. Marrying someone with a markedly different socio-economic, educational, or cultural background is likely to be riskier than marrying someone more closely matching your own characteristics, but this is where personal choice comes into play. A general set of statistics is not an infallible guide, and sometimes you're better off following your heart than your head. Since people tend to change over the years, even the most impressive statistics may be invalid ten years later.

One way of assessing someone's marriage potential is to stay in a long-term relationship to see what happens. After the initial luster of love wears off you can see how problems are solved, how interactions change in times of stress and with the passage of time, and if boredom or dissent becomes commonplace. You can be fairly certain that premarital problems are likely to intensify rather than disappear, once the honeymoon is over. One way of accomplishing this type of assessment is by living together, a route chosen by many young adults today.

One last note on mate selection. While sex is not the most important ingredient in most marriages, it does help to know if you and your spouse-to-be are sexually compatible or have considerable difficulty getting things together sexually. This doesn't mean that you should rush into a sexual relationship (again, your own values must be applied here) or that you need to have a performance checklist (a tactic that could boomerang in unexpected ways), but you do need to decide how to deal with the sexual side of your relationship and to think about how important sex is (or isn't) to *you*.

Marriage is rarely a faithful replay of the fairy tale ending "and they all lived happily ever after." Living in a marriage and making it work is no easy matter. It's easier to be loving when you aren't awakened by kids at 3:00 A.M., when you're not squabbling with the in-laws, or when your sexual advances aren't repelled by a headful of curlers, a facial mask, cigar breath, or Monday night football.

In the real world, few marriages maintain a perpetually loving relationship. Even companionate love shifts its intensity from time to time as a married couple reacts to the ordinary stresses and strains of life together. At any given moment, a husband and wife may dislike each other or even hate each other, yet still spring back into a love relationship.

Researchers have found that marital dissatisfaction tends to increase the longer people have been married. While it is possible to speculate on its many causes — lessened sexual interest, the responsibilities of parenting and occupation, failure to share time together, poor communications, and changes in personal attractiveness, to name just a few — many marriages suffer from a kind of benign neglect that relegates the relationship to a low priority, thus removing the very elements that usually sustain love.

To maintain love or help it grow, marriage partners must invest emotions and energy on a continuing basis. Spouses who succeed in "giving" to one another — in communications, physical warmth, shared interests, and shared responsibilities — are also likely to succeed in staying in love.

While some marriages succeed quite well in keeping love alive, others evolve into business relationships or a sort of "roommate" status in which love fades away completely. Partners in marriages that become strained over time may stay together to protect the children or to adhere to a philosophy that simply will not allow for separation or divorce. Although these marriages may be loveless, they are not necessarily "bad." Even good marriages are susceptible to a disappearance of love.

In an essay called "The Future of Marriage," Morton Hunt (1977) observed: "Formal promises to love are promises no one can keep, for love is not an act of will; and legal bonds have no power to keep love alive when it is dying." The nature of marriage in the 1980s may have changed considerably from earlier times, but the nature of love has not. A major challenge facing any marriage is to preserve the spark of love — a task that requires hard work and creativity.

CHAPTER TEN

Intimacy and Communication Skills

THE SEARCH for intimacy is a familiar part of our lives, yet finding and sustaining satisfying intimate relationships seem to be difficult undertakings for many people. This might be surprising at first, since the benefits we derive from sharing warm, trusting relationships — like enjoyment, acceptance, comfort, support, and companionship — have an obviously self-fulfilling role in our lives. Yet despite the fact that most people agree that intimacy is desirable, there is no clearcut path to follow for establishing an intimate relationship, and preserving intimacy, or helping it grow, seems to be a major problem today if we consider current divorce statistics or the tens of thousands of couples seeking marriage counseling.

Understanding the nature of intimacy — what it is, how to achieve and sustain it — while also recognizing the potential pitfalls and problems of intimacy is the subject of this chapter. In addition, since effective communications are, in many ways, essential to developing and maintaining intimacy, we will also discuss communication skills as they apply to intimate relationships.

INTIMACY AND
INTIMATE RELATIONS

The word "intimacy" comes from the Latin *intimus*, which means innermost or deepest. Intimacy can be defined as a process in which two caring people share as freely as possible in the exchange of feelings, thoughts, and actions. As we will use the term in this discussion, intimacy is generally marked by a mutual sense of acceptance, commitment, tenderness, and trust. This definition allows us to see that intimacy is not precisely the same as romance or even strong affection. Friends may work or play together and enjoy each other's company but have little or no exchange of their inner thoughts and feelings, and a person may fall in love with someone without expressing that feeling to the loved one.

In much the same way, intimacy as a temporary condition (in contrast to an ongoing process) is sometimes situation bound: it is defined by a particular set of circumstances that leads to openness with no commitment or tenderness necessarily present. For example, two people sitting next to each other on a long flight might engage in a very personal conversation, but if they later meet accidently at a cocktail party, they might be quite uncomfortable about their disclosures. Similarly, two people who have engaged in casual sex together without any real exchange of feelings may have "been intimate" sexually but they have not experienced the sharing and caring of intimacy as we have defined it.

Thus, we can see that there is much variety and complexity in intimate relations; they are not all alike by any means. In particular, we can expect that intimacy without romance or sexual interaction (where the status is agreeable to both persons involved) is apt to be quite different from the intimacy that accompanies love or romance or sexual passion. But before we go on to examine the nature of interpersonal intimacy in more detail, it is helpful to take a brief look at another component of intimacy that undoubtedly influences a person's relationships with others: intimacy with self.

A number of psychologists have stressed that a person's ability to form intimate relationships with others depends in part on having a firm sense of self based on realistic self-knowledge and a reasonable degree of self-acceptance. Such self-awareness helps us identify our needs and feelings and thus enables us to share them with others. Self-acceptance is also an important building block for interpersonal intimacy because it allows people to be themselves without pretending to be something other than who and what they are.

People who don't like themselves very much or who feel ashamed of who they are often have a difficult time establishing and maintaining intimacy because they are preoccupied with trying to prove themselves to others or with trying to gain recognition or respect. Even if they are succesful in these efforts, their feelings about themselves usually don't change in a lasting way. Others who are anxious or depressed about themselves may deal with these feelings in ways that block self-awareness: by using drugs (including alcohol) for escape, by sitting passively in front of a television set to distract themselves from themselves, or by becoming immersed in their work. Still others who are unhappy with themselves try to find personal satisfaction in relationships in which someone else cares for them, protects them, provides for them, or entertains them, but this is often only a short-term solution.

This does not mean that a person must be totally happy with him- or herself in order to be capable of intimacy with others. When we look within ourselves, we may not always like what we see. Generally, we separate what we like from what we don't like and use this process to try to change. If we are honest in our self-appraisals, the intimate knowledge we develop helps us relate to others. At the same time, a person who *never* looks inward (whether out of fear, laziness, or self-hatred) has such distorted self-perceptions that it is unlikely he or she can contribute fully to a relationship with someone else.

One useful caution should be added here: it is important to retain our sense of self even while involved in an intimate relation with another and not to become so preoccupied with a relationship that we lose touch with our sense of self. An intimate relationship that absorbs most of your time and emotional energy can be exhilarating, but it can also leave you little time for knowing yourself. Such an all-absorbing relationship can be draining or damaging rather than fulfilling. In contrast, intimate relations that enhance your self-acceptance and self-knowledge are likely to be positive elements in your life.

People view interpersonal intimacy in a number of different ways, as these remarks indicate:

A twenty-two-year-old single woman: What I really want most from a relationship is the privilege of being honest all the time. I don't know if that's possible, though.

A twenty-nine-year-old married man: To me, intimacy means sharing. You share the good and the bad, the joys and pain, and through all the sharing you know your partner cares about you.

A thirty-five-year-old single woman: It comes as a shock to some people to realize that lesbians can have committed, lasting relationships, but that's exactly what Sallie and I have. To us, that commitment is a lifetime bond.

Because the joys of intimacy are numerous and varied, it is a mistake to think of it as a single, unchanging condition. Not only does intimacy exist in various degrees of intensity and in different types of relationships — between friends, between lovers, between family members, and so on — intimacy as a process fluctuates within any given relationship over time. This is partly because each partner's expectations and hopes influence how they evaluate what they're getting from the relationship. People who feel that an intimate relationship is consistently unfair or one-sided generally are more likely to end the relationship (or look for another to take its place). In contrast, those who view their relationships as equitable and balanced are likely to be happiest and stay together for longer times. In addition, the intensity of intimacy in a particular relationship is influenced by external circumstances, such as geographic separation or work pressures, which can temporarily divert a person's energy and attention from the relationship to other aspects of his or her life.

In order to better understand the process of intimacy, we will examine its basic components: caring, sharing, trust, commitment, honesty, empathy, and tenderness. However, it is important to recognize that these components do not usually exist separately from each other but instead are blended in a unique amalgam in which each strengthens and solidifies the others.

Caring is an attitude or feeling you have for another person which is generally related to the intensity of your positive feelings toward them. Although you might have positive feelings about someone with whom you have no personal involvement — on the basis of good looks alone, for instance, you might feel positively about a person sitting across the table from you in the library — the mutual caring characteristic of intimacy occurs only when two people share and interact together.

The sharing of thoughts, feelings, and experiences that accompanies the growth of intimacy in order to learn about each other requires spending time together without the ordinary barriers with which people protect their privacy. Thus, one of the key steps in developing an intimate relationship is *self-disclosure,* the willingness to tell another person what you're thinking and feeling. Because there is no certainty that the other person will be interested in what you have to say, and because it takes some time to establish the trustworthiness of the other person, most people begin the process of self-disclosure gradually. Instead of revealing their fondest dreams and deepest fears all at once, people generally develop personal openness in a relationship as they find it reciprocated and as they see signs of the other person's continuing interest.

This process of intimate sharing is not limited to superficial or pleasant things alone, but should extend across a broad spectrum. According to noted psychologist Carl Rogers, saying " 'I want to share myself and my feelings with you, even when they are not all positive,' almost guarantees a constructive process to communications." Sharing uncertainties, worries, and other personal problems with an intimate partner is essential for the growth of intimacy.

Although sharing thoughts and feelings is important to intimacy, it is important that experiences be shared also. Research has shown that people who share mutually rewarding experiences are most likely to develop and sustain a warm, caring relationship. Such shared experiences can include many things. Sharing hard times and good times, sharing childrearing or a dual career, sharing recreational activities, or sharing in planning for the future are all examples of how intimate couples interact.

Sharing experience does not necessarily mean that intimate partners should do everything together. Although such a system might work well for a few couples, most people would find total sharing difficult. This is partly because a particular activity or experience doesn't always provide equal rewards to each partner. You might love jogging, for instance, while your partner

prefers to play bridge. Forcing each other into activities that are
not mutually enjoyable just for the sake of togetherness is unwise. Furthermore, constant and complete sharing is not necessarily a measure of a couple's intimacy. Indeed, maintaining an
identity independent of an intimate relationship is also important to the longevity of the relationship. Pursuing individual
interests and maintaining a circle of friends gives a person a
chance to process the feelings generated in intimate interactions.
Such efforts help to prevent partners from becoming psychologically overloaded with too much one-to-one togetherness. In
addition, such independence allows people to bring new experiences and thoughts to their primary relationship, which can also
help it grow.

The process of self-disclosure we mentioned above doesn't
occur in a vacuum but depends on the degree to which you trust
the person to whom you are making disclosures about yourself.
Thus, trust is another necessary ingredient for intimacy, and,
like caring and sharing, trust develops over time. While people
trying to form an intimate relationship usually have to make
some initial assumptions about trusting each other, trust solidifies when a partner's behavior matches his or her words. If he
promises to help and be there and his behavior confirms those
words, she comes to trust him. If she promises never to laugh at
his personal secrets and she doesn't, he comes to trust her. Once
trust grows strong, two people are able to share even more information about their thoughts and feelings without fear that
this will be used against them in some way.

Another component of intimacy, commitment, is generally an
outgrowth of the caring, sharing, and trust that develop in the
early stages of an intimate relationship. Commitment requires
both partners to work willingly to maintain their intimacy
through periods of crisis, boredom, frustration, and fatigue, as
well as through times of joy, prosperity, and excitement. Intimacy that surfaces only when life is on the upswing is a fleeting,
unreliable form of closeness rather than the more encompassing
interaction most people would like it to be. Here, too, Carl Rogers has nicely captured the essence of the commitment of intimacy: "We each commit ourselves to working together on the
changing process of our present relationship, because our rela-

tionship is currently enriching our love and our life and we wish it to grow."

Realistically, it is important to recognize that the degree of one's commitment to an intimate relationship may change over time. Those who pledge themselves to each other "forever and ever" on the basis of a passionate relationship that has lasted only a few weeks may find that as they get to know each other better their desire to stay together lessens. Even couples who have shared years of satisfying intimacy may find that they later grow apart or develop problems that undermine their relationship. Thus, commitment should be regarded as an attitude that states current intentions without being an irrevocable guarantee of the future. Nevertheless, commitment that is backed up by a willingness to work to overcome problems that might develop in a relationship is an important ingredient for a durable future.

Honesty is another necessary part of intimacy, although total honesty in the sense of full self-disclosure is not necessarily good for a relationship. Too much honesty can be devastating to any relationship if it is not tempered with an understanding of how a given message might affect one's partner. But there is a decided difference between keeping some things private — that is, setting limits on the self-disclosure that occurs — and deceit. When deliberate deception occurs in friendships or romances, it generally undermines the quality of information exchange that can occur, and therefore, undermines intimacy. Indeed, the presence of deceit in a relationship usually is a warning sign that manipulation of one form or another is occurring.

If a person engages in deceit in an intimate relationship, even with the best of intentions, the discovery of the deceit almost inevitably leads to a loss of trust. Thus, telling lies (a sin of commission) is usually more harmful to intimacy than keeping something private (a sin of omission). This means that if your partner asks about something that you don't feel you can discuss truthfully, you can always say "I don't want to discuss this particular subject" without violating your partner's trust. Of course, putting too many topics "off limits" can lead your partner to wonder what you're hiding and may also result in your partner's pulling away from openness, too; as in most aspects of intimate relations, partners tend to parallel each other's behavior.

Empathy is the ability to understand and relate to another person's feelings and point of view. In order for self-disclosures to occur between intimate partners, each must feel that he or she is listened to, and understood (or at least accepted) by the other person. Such empathy enables each person in an intimate relationship to act in ways that support and help the other and to avoid or limit destructive, irritating, or alienating attitudes.

One of the most neglected aspects of intimacy is the expression of tenderness, which can be achieved either by spoken messages or by physical contact (e.g., hugging, cuddling, holding hands), as well as by direct behavior. This ingredient of intimacy often seems to be particularly difficult for men who have been socialized to be purely rational, action-oriented beings: they seem bewildered by tenderness, or afraid that it is an "unmanly" way of acting. On the other hand, some of these men are able to be physically tender, but lack comfort or familiarity with the verbal side of tenderness. Both components — verbal and physical tenderness — are usually necessary for romantic intimacy. In fact, much of the time that people complain about intimacy disappearing from their relationships they are actually acknowledging that the amount of tenderness they receive from their partners has noticeably declined. Thus, paying attention to ways to express your tenderness to your partner, by both words and actions, is one of the best ways of keeping intimacy fresh and satisfying over time.

Finally, it is important to recognize that unless intimate partners are willing to set aside many of the ordinary defenses they use in everyday life, it can hardly be said that the intimate relation is a special one. It is hard to be intimate with a person who continually denies the reality of his or her inner feelings (for example, someone who always pretends that everything's great). It is equally difficult to have satisfying intimacy with a person whose behavior is based on pretense (for instance, a partner who is always looking for "status" and ways of impressing others). On the other hand, people who are able to relinquish such defenses in favor of being themselves, authentically and spontaneously, are apt to find intimacy more rewarding.

Sex Differences in Intimacy

A twenty-eight-year-old woman: Whenever I go out with a guy, it seems like all he's interested in is sex. We sleep together once or twice and suddenly he disappears. I think all the men I know are afraid of a really intimate relationship.

A twenty-five-year-old man: It's sad to hear women condemning men for being unwilling to get into close, loving relationships. Many of my friends value intimacy highly, although admittedly it's not easy to find.

The two opinions quoted above about sex differences in intimacy highlight a much-debated topic. Currently, there aren't any reliable data on whether men and women have different levels or types of motivations for intimacy. Thus, the best we can do is review current research evidence on sex differences in particular aspects of intimate behavior, such as self-disclosure.

A number of studies show that women seem more adept at self-disclosure than men, and that girls and women disclose more intimate information to their friends than boys or men do. In addition, girls tend to have more intimate friendships than boys do, and women show a higher correlation between friendship and intimate disclosures than men do. Furthermore, women have an easier time building deep, loyal, noncompetitive friendships with other women than men do with other men.

However, the research evidence does not uniformly support the view that there are major sex differences in self-disclosure. Rubin and his co-workers, who conducted a study of 231 dating college couples in 1980, found few differences in the levels of self-disclosure that men and women made to each other. Fifty-seven percent of each sex had made full disclosure of their previous sexual experiences to their current partner, 73 percent of the men and 74 percent of the women had fully disclosed their feelings about their sexual relationship together, and 48 percent of the men and 46 percent of the women had given their partner their honest views on the future of the relationship. Although some differences were found (e.g., women revealed more about their greatest fears, their feelings toward their parents, and their feelings about their closest friends, while men revealed more about the things they were proudest of, the things they liked best about their partners, and their political views), the researchers noted that overall, their sample of college students generally adhered to a norm of "full and equal disclosure." Other studies have also found that men confide more in their girlfriends than in anyone else and that sex differences in self-disclosure are minimal.

Other research indicates that intimacy is somewhat easier for women than men and/or that intimacy is more rewarding to or ingrained in women. For example, lesbians are more likely to pair off in intimate relationships than gay men. Similarly, sex therapists have noted that fear of intimacy is relatively common in men but less frequent in women. Furthermore, men seem to want "instant intimacy" more often than women, an attitude that indicates a fundamental misperception of how intimacy actually develops.

How can we explain such differences? First, we should realize that the existing research focuses on intimacy in only a limited way, particularly emphasizing verbal self-disclosure. This approach necessarily avoids a more comprehensive view of intimacy as an ongoing experience in which time together, physical contact, and shared activities may outweigh the importance of the verbal exchanges that occur. Thus, it is possible that with more sophisticated studies, male-female intimacy differences would prove to be minor or nonexistent. However, it may be that early differences in the socialization of males and females in our society account for later differences in intimacy skills. Generally, females in our culture have been socialized to show their feelings, while males have been taught to keep their feelings hidden and to show no signs of weakness or fear. (As Kate Millett succinctly put it, "Women express, men repress.") In addition, females tend to be touched more during infancy and early childhood than males, something that might lead to later sex differences in intimacy. Similarly, the competitive, aggressive behaviors that are generally encouraged in males in our society do not, in turn, encourage intimacy, while the nurturance and sensitivity usually encouraged in females do enhance intimate behavior.

Whatever differences in intimacy preparation exist because of childhood socialization, men are certainly fully capable of intimacy; some of them simply seem to need a while to learn how to find it. In fact, the author Gail Sheehy has noted that men seem to become increasingly concerned with intimacy from age forty on, although many men certainly develop a great deal of intimacy at much earlier ages. Perhaps the real dilemma of the sex-differences-in-intimacy problem has been aptly described by Rubenstein and Shaver in their 1982 book, *In Search of Intimacy*, who point out that although "men and women need intimacy to the same degree . . . fewer women than men get their needs met, despite women's expertise, because so many men are intimacy-takers rather than givers."

Intimacy Problems

Although most people readily express a need for intimacy in their lives, it often seems to be elusive. In this section, we will examine several different types of problems people have with intimacy, including common barriers to intimacy, fear of intimacy, and pseudo-intimacy.

Some people seem to be able to forge close relationships easily, while others have a difficult time getting past the "social acquaintance" stage. The fortunate few who can comfortably develop closeness and rapport with others in a seemingly effortless way are a distinct minority. Most of us have to work at developing intimacy, and most of us, at one time or another, find that our intimacy overtures are ignored or rejected. Here is a list of common reasons for difficulty initiating or maintaining intimate relations.

1. *Shyness.* People whose shyness causes them to avoid social interactions or to isolate themselves in social settings are unwittingly restricting their opportunities for intimacy. Paradoxically, shy people often long for intimacy and companionship in their lives, but they seem unwilling or unable to take the risks necessary to overcome their shyness.

2. *Aggressiveness.* People who behave aggressively often scare others away or cause them to adopt a defensive posture. The typical concern seems to be "I'll be overpowered by this person," and few people look for relationships in which they'll be dominated by someone else. Toning down aggressive language and behavior can improve a person's chances for intimacy.

3. *Self-centeredness.* Being preoccupied with oneself commonly turns others off. We all know people who insist on being center-stage all the time, who ignore the needs of others (not out of malice but because of lack of awareness), who monopolize conversations, and who are generally unwilling to do what a partner wants unless it coincides with their own needs. These people frequently initiate intimacy by telling others a great deal about themselves, but they tend to have a more difficult time maintaining long-term relationships.

4. *Selfishness.* Going beyond self-centeredness, selfishness is apt to be far more damaging to the development of genuine intimacy. Selfish people are often manipulative and try to gain a tactical advantage over others to get their own way. The selfish person doesn't care much about what's best for the relationship or best for the other person; instead, he or she seeks to exert control for personal gain.

5. *Lack of empathy.* The person who is unwilling or unable to accept and understand another's views, thoughts, or feelings has a difficult time in intimate relationships. Often, these people seem to have difficulty listening: either they block out what their partner says or they fail to internalize the message and look at the situation from the partner's point of view. Empathetic people do not just sympathize with the feelings and needs of others, they try to respond to these feelings and needs as well.

6. *Conflicting or unrealistic expectations.* Many people are so idealistic about intimacy that they expect the impossible, creating a situation that frequently leads to disappointment, frustration, or, possibly, to giving up. In other intimate relationships, the partners' goals may be so different that the relationship fails. For instance, if one person is looking primarily for companionship and entertainment in a friendship, while the other is looking for a deeply philosophical, intellectual relationship, they are not likely to find a pleasing intimacy together.

Needless to say, this is not a complete list of all possible barriers to intimacy. There are other conditions, such as depression, drug abuse, or severe physical illness, that may make intimacy extremely difficult even when the other ingredients seem to be in place. But it is also important to realize that intimacy is often extraordinarily resilient, making its own way in the face of un-

BLOOM COUNTY   By Berke Breathed

foreseen obstacles. Perhaps that's one reason why so many of us are concerned with finding and keeping intimacy in our lives.

Another common problem people have in interpersonal relations is the fear of intimacy. People with such a fear are typically anxious about intimacy because of distrust, fear of rejection, and fear of losing control. In addition, many people who fear intimacy have negative self-images; they believe that they have nothing of value to bring to an intimate relationship and doubt the judgment of someone who seems interested in them because they consider themselves unworthy and uninteresting.

People who are untrusting and fear rejection sometimes avoid forming intimate relations entirely, preferring to have many superficial relationships instead of a relationship that calls for taking risks and making a commitment to someone else. These people guard themselves from hurt, but they also isolate themselves emotionally. Others enter into intimate relations but protect themselves by regulating the degree of closeness. Whenever the relationship threatens to become too intimate, they pick a fight, become distracted, or bury themselves in work; in short, they construct a buffer against the demands of the relationship and thus calm their fears by keeping the intimacy under control. Helen Kaplan notes that sometimes *both* partners in a relationship have intimacy conflicts.

> Such couples long for closeness with each other, but when they achieve a certain point of contact they become anxious. Then one or the other will behave in such a manner as to create distance. When distance reaches a certain point anxiety and longing for

closeness will be evoked in the couple. They miss each other and move closer to each other again — but not too close. Then the see-saw will move in the other direction. (Kaplan, 1979, p. 184)

In some cases, fear of intimacy is a lifelong condition. Sometimes such a fear reflects traumatic relations with parents during early childhood; in other cases it develops after a painful experience in an intimate relationship in which a person was not only hurt but intensely disappointed. While most of us survive the breakups of intimate relationships none the worse for wear, this isn't true in all cases — and if the emotional scars are thick enough, the fear of intimacy is most understandable.

A different type of intimacy problem can be identified by the term "pseudo-intimacy." It is possible to distinguish between genuine intimacy, which is a positive, self-enhancing process, and pseudo-intimacy, which is more pretense than openness, more manipulation than sharing. The latter form of intimacy is marked by the following features:

1. One person looks to the other to meet most or all of his or her needs rather than taking the responsibility to meet these needs.
2. There is a big gap between what is said and what is done.
3. Mutual trust is missing from the relationship or has been deliberately and repeatedly violated by one of the partners.
4. The commitment in the relationship is either one-sided or illusory.
5. One person in the relationship persistently acts in a selfish fashion and shows little interest in giving.
6. Communication is one-sided (one partner monopolizes the talking or has little to say).
7. One or both partners order each other about and criticize each other for not following these demands.
8. Conflicts and arguments consume much of the time and energy of the partners with little or no resolution of key issues typically occurring.

This doesn't mean that genuine intimacy is present only if there is perfect tranquility and affection in a relationship. Commitment to a relationship and caring about one's partner do not ensure that intimacy always produces positive, happy feelings, or mutual agreement on all issues. People who are very much in love, for instance, can have moments when they hate each other just as people who feel tenderness toward each other sometimes act in cruel ways. This variation of feelings doesn't mean that a couple doesn't have a meaningful intimacy, it simply shows that intimate relationships are highly complex. And in the final analysis, it is exactly this complexity that gives intimacy its greatest value — the strength that bonds us together, one to one, in a unique relationship of mutual giving and getting.

COMMUNICATIONS

As we said in the introduction to this chapter, it is through effective communication that intimacy is established and can grow. Thus, understanding how to communicate effectively is a cornerstone of interpersonal and sexual relations, yet few of us are taught the skills of intimate communications. In schools we learn to write essays and term papers and sometimes even learn the fundamentals of how to speak before a group or to debate, but when it comes to developing intimate communication skills, we are left alone. The following discussion provides some practical, commonsense suggestions for developing your ability to communicate effectively in personal relationships.

Communication usually begins with the intent to convey information to someone else. The sender must convert the intent into an actual message that is presented to the intended recipient. The message may be verbal (words, sounds) or nonverbal (consisting of a look, a touch, or an action). The recipient must not only receive the message but also understand and interpret its meaning. At each one of these seemingly simple steps, things can and do go wrong.

In many cases, the sender doesn't succeed in saying what he or she really means. Sometimes, for example, people can't find the right words to convey what they're feeling, or what they need, so the messages they send are inaccurate. Even if the message has been accurately formulated, something may go wrong in the sending process so it's never received at all or is received in a garbled fashion. How often has someone missed the main point of your message, and after you explained yourself (perhaps with some exasperation) said, "Oh, that wasn't what I thought you said."

Next, the receiver may not be turned on and so may miss the message (that is, a person may not be listening to what you're saying), or he might hear what he would like or expect to hear, rather than what is actually said.

Possibly the single greatest source of communication trouble, however, is in the way messages are interpreted by those who receive them.

> *Man:* I told you earlier this evening I didn't want to make love tonight.
> *Woman:* I thought you just meant you didn't want to *then,* I didn't realize you meant for the whole night.

Although the seemingly simple art of communication can often be difficult and complex, there are steps we can take to ensure that our messages are sent as clearly as possible *and* that we are open to receiving messages as efficiently as we can. We examine these steps in the following sections.

Sending Signals Clearly

Effective communication begins with the message sent from one person to another. If an unclear message goes out, even an attentive listener is likely to be confused and forced to guess about the intended meaning. There seem to be three main reasons for this lack of clarity.

1. *Not saying what you mean.* When people aren't able to find the right words to express their feelings, they may not be fully in touch with their feelings. People may also avoid saying what they really mean so they won't hurt someone they care about, so they won't be embarrsssed, or so they won't risk being rejected.
2. *Sending mixed messages.* Mixed messages carry contradictory meanings. This can happen when body language or a person's tone of voice contradicts the spoken words. For instance, if someone says "That's lovely," but grimaces while speaking, the listener is apt to be confused. Likewise, a person who says "I am *NOT* upset" in a forced, slowly articu-

lated voice is indicating just the opposite. Mixed messages also occur when there's an inconsistency in the content of a message, as when one part of the message negates the other: "I love it when you're rough with me, but I wish you'd be more gentle," or "I really don't want to worry you, but I think I may be pregnant" are examples of this type of problem.

3. *Not being specific.* Vague statements leave a listener frustrated and wondering "What did he/she mean?" For instance, being told "We should really have more romance in our lives" by your partner might lead you to ask yourself: "Does this mean there's something wrong with our relationship? Am I being criticized? Should I be doing something new? Is my partner unhappy? What does my partner want?" A more specific statement such as "I'd love it if you would read me some love poems once in a while to help me feel romantic," wouldn't leave those loose ends.

Clarity in communications can be enhanced in a number of different ways. Here are some general suggestions to think about:

1. Think through what you want to say and how you'll say it particularly if it's an important or emotionally charged message.
2. Let your partner know what your priorities are; try not to crowd in so many requests and instructions that it's difficult to grasp your key points.
3. Be concise. Long-winded discussions are more likely to confuse than clarify. On the other hand, being concise doesn't mean being simplistic or superficial. Don't leave out important information about your feelings or desires in order to be brief.
4. Don't talk *at* your partner. Give him or her a chance to respond and interact.
5. Try not to begin communications by criticizing or blaming your partner. Starting on a negative note puts your partner on the defensive and makes objective listening difficult.
6. Don't be afraid to put what you need to say in a letter if you're having trouble saying it face-to-face. Writing it down

shows that you cared enough to take the time to say it carefully.

7. Ask for feedback from your partner to be sure you've been
   understood and to get his or her reactions.

Nonverbal Communications

After a lovemaking session one night, Cathy withdrew into a
stubborn silence. When George asked her what was wrong, she
said, "Nothing at all," but the firm set of her lips and the way she
rolled away to avoid his touch told George how to interpret these
words — that in fact something *was* bothering her. With some
patience and encouragement, George was finally able to find out
what had upset Cathy. She hadn't had an orgasm, and she felt
the reason was that he had stopped stroking her clitoris too soon.

As this example shows, the nonverbal side of communication is
often at least as important as the words that are spoken. In fact,
one psychologist suggests that of the total feeling expressed by a
spoken message, only 7 percent is verbal feeling, 38 percent is
vocal feeling, and 55 percent is conveyed by facial expression
(Mehrabian, 1972). Posture and positioning (body language)
also are powerful forms of nonverbal messages, sometimes saying "Keep away" and sometimes inviting intimacy and closeness.
Sitting in a relaxed fashion sprawled out next to your partner
usually conveys a sense of comfort and warmth, while sitting
rigidly on the edge of your chair at a deliberate distance from
your partner usually conveys a sense of withdrawal, annoyance,
or preoccupation. Unspoken messages can also be powerfully
transmitted by touch, which can suggest an attitude of caring
and accessibility.

It's important to recognize that inconsistencies between nonverbal cues and verbal content are usually resolved in favor of
the former; in this sense, nonverbal messages are more "powerful" than spoken words alone. For this reason it's useful to communicate in ways that maintain consistency between the verbal
and nonverbal messages you send to your partner, taking care
to avoid sending mixed messages by saying one thing with your
words and something different with your body language or
vocal tone. Thus, one way to improve the chances of communicating effectively is to be aware of your own nonverbal language

— an aspect of communicating to which many people never pay attention. It also helps to actually practice ways of sending positive nonverbal messages that express trust, commitment, and caring rather than suspicion, rejection, or impatience. You can do this by yourself, with the aid of a mirror or tape recorder, or with your partner's help. Together you can discuss the nonverbal communication patterns in your relationship and see how they can be improved.

Not surprisingly, nonverbal messages apply in a special way to sexual interactions. At times, they indicate displeasure or resentment. For instance, if your partner's body tenses up whenever you stimulate the genital area with your tongue, you may begin to think that he or she is uncomfortable with this caress no matter what is said. Likewise, if your partner usually moans with passion as you make love together, the sudden absence of such sounds may make you feel as if you were doing something wrong. At other times, nonverbal messages convey a sense of pleasure, involvement, warmth, or similar feelings. In addition, nonverbal communications during sex can help your partner see what you like without breaking the mood by words. And taking your partner's hand and guiding it on your body, or showing your partner exactly how you'd like to be touched, can be a true gift of sexual intimacy.

Although touch can be used as an effective means of nonverbal communication in a variety of ways, intimate partners often seem to talk too much and touch too little, missing many opportunities to convey feelings of tenderness or affection to each other. In many situations, a long, tight hug says more about the way people feel about each other than a ten-minute dialogue. Likewise, stroking a partner's hair or face, or leisurely kissing, or performing a sensual massage can convey a sense of caring and pleasure that goes beyond words. On the other hand, if people confine their touching to sexual situations, they compartmentalize the physical side of their interaction, sometimes making sex seem like a bartered commodity used to attain closeness.

Vulnerability and Trust

Communicating in an intimate relationship differs in certain ways from communicating with other people in your life. This is

partly because partners in a truly committed, intimate relationship can make the very basic assumption that neither one of
them deliberately intends to hurt the other, an assumption that
can't always be made in our dealings with the rest of the world.
While this doesn't mean that emotional hurts will never occur, it
does provide a safety net of trust and support that allows each
person to become uniquely vulnerable in an intimate, caring
relationship.

The willingness to risk being vulnerable, which is at the essential core of intimacy, and the trust that makes it possible encourage people to say what they're feeling or thinking. They feel free
to reveal things about themselves — including fears, shortcomings, and failures — without worrying that this information will
be used against them at any time. Thus, while trust and vulnerability are not methods of communication, they are necessary
preconditions for intimate communications to occur.

"I" Language

One of the most direct ways to communicate clearly and to avoid
mind-reading games in a relationship is to use a highly effective
style of communicating called "I" language. The basic premise
of this approach is that a person should take responsibility for
him- or herself, since no one knows better than the individual
what he or she is feeling or needs at any given moment. By
beginning as many statements as possible with the pronoun "I,"
a person takes responsibility for his or her self-expression. "I"
sentences tell what you feel, what you need, or what you want.
"I'd love it if we could cuddle and kiss," "I'm feeling restless
now," or "I wish we could spend more time talking to each
other" are examples of "I" language. To some people, "I" language sounds selfish because we're taught from an early age that
it's not polite to talk about ourselves excessively. Intimacy, however, requires that a person open up and express his or her
feelings without beating around the bush.

In contrast, sentences that begin with "you" are apt to be demanding or accusatory, provoking defensiveness in the other
person: "You don't kiss me much anymore" or "You don't spend
enough time talking to me" have a very different tone from the
"I" messages previously listed.

"We" sentences are potentially problematical because they compel one person to speak for both. This requires that the person make assumptions or guesses about his (or her) partner's moods, preferences, and needs — and while a charming sense of togetherness may result when the "we" assumptions prove accurate, on many occasions they are actually annoyingly off-target. "We" messages can also encourage imbalanced communications: one partner may monopolize the talking by speaking for both almost all the time. In this situation, the less assertive partner won't say much of anything and is liable to submerge his or her requirements beneath the flood of directives from the outspoken partner. Such lopsided communication is not conducive to close, intimate relations.

"I" language, then, provides an excellent means for one partner to put his or her emotional cards on the table in intimate dialogues instead of coyly fencing around. This openness, in turn, invites the other person to speak openly as well. Consider the following contrast in styles and content:

Without "I" language

Eileen: What do you want to do tonight?
John: Oh, I don't know. What would you like to do?
Eileen: Well, I wanted to do something we'd both enjoy.
John: Don't you have any ideas? [The conversation is apt to continue in this wheel-spinning mode for awhile, since Eileen and John are trying not to pressure or offend each other by making the first suggestion.]

With "I" language

Eileen: I'm feeling a bit lazy tonight, so I'd like to stay home and relax.
John: I was kind of looking forward to getting out — I thought maybe we'd go dancing.
Eileen: I don't think I'd really enjoy that the way I'm feeling. I'm just too tired to handle that right now.
John: Well, there's a ball game on TV I can watch, so maybe we'll get out this weekend.

If John and Eileen hadn't agreed on what to do, "I" language could be used to reach a negotiated solution that would be mutually satisfactory. Here's an example of how it might be achieved:

> *Eileen:* I really wasn't planning to go out tonight, I was hoping to stay in and take it easy. Would that bother you?
> *John:* Well, I don't have my mind set on dancing, but I certainly wanted to get out and do *something*.
> *Eileen:* I guess I can handle something that doesn't involve exerting much energy. How about a movie? How does that sound to you?
> *John:* Oh, that'd be great. I wanted to see that new Woody Allen flick and we could catch the nine o'clock show.
> *Eileen:* That's fine with me. I'll take a nap if you make dinner, and then we can get going.
> *John:* Sounds good to me. I'll wake you up in a half hour, okay?

When a satisfactory compromise can't be found via this sort of negotiation, the next step is for both partners to examine the relative strength of their different needs. Some couples find that this is easiest if each person rates the intensity of his or her needs on a quantitative scale (for instance, using a scale from -10 to $+10$, where 0 represents a neutral feeling). Other couples compare their needs by discussion that doesn't involve exact quantification. In either case, the basic premise of such negotiations is that it will usually be in the best interests of the relationship to go in the direction of the person whose need is greatest, as long as the other person isn't hurt by this process. It is also possible to negotiate so that each partner does something separately from the other; being in an intimate relationship doesn't mean always doing things together.

A word of caution about "I" language: Phony "I" sentences, such as ones that begin "I think that you . . . ," or "I feel that you . . . ," are really just "you" sentences camouflaged by the addition of the "I think" or "I feel." These should be avoided since it's not the grammatical construction of the sentence that's the key; the essence of "I" language is to speak for yourself without accusing or blaming.

The most complete, functional "I" messages should not simply announce what or how you're feeling (especially if it's negative) but should go on to say what you think you need to try to maintain (or change) the feeling. This prevents your partner from the often frustrating task of having to conjure up a remedy for whatever ails you. In sexual situations, this principle is particularly true. Instead of saying, "I don't like it when you dive for my crotch," you might rephrase the message to say "I really get uncomfortable when I feel sex is rushed and hurried, but a slow leisurely tempo turns me on." This type of communication avoids the trap of sending half a message — what you *don't* like — by completing the message and saying what you'd prefer. Here again, by assuming responsibility for stating your own needs and preferences, you relieve your partner of having to figure out what will please you.

It is also important to realize that "I" language is not the only way of communicating effectively in an intimate relationship. In fact, since intimacy generally produces a sense of thinking about a partnership as "we" or "us" rather than simply "you" and "me" (Hatfield, 1982), there is nothing wrong with using language that emphasizes this viewpoint. Similarly, "you" sentences that offer positive rather than critical content — for example, "You're so kind and sensitive" — are certainly welcome in any relationship. Thus, "I" language should be seen as a potential way of achieving clarity in intimate communication instead of as the only correct way of communicating with your partner.

Expressing Affection

While it might seem that expressing affection in an intimate relationship ought to be the easiest thing in the world, marriage counselors and sex therapists frequently find that even loving couples often neglect this side of their relationship. Although affection is expressed in actions more meaningfully than words, never hearing words of affection can be troubling and can lead people to question whether their partners really care for them.

> I know deep inside me that she really loves me, but she never says it anymore. I feel a little stupid about it, because I can't exactly say to her, "Laura, please tell me you love me" — then I

wouldn't be sure if she really meant it or if she was just saying it to make me happy.

Similarly, when affection is expressed only during sex, and not at any other time, it can lead a person to feel as though it's a limited or conditional affection — in other words, "I love having sex with you" rather than "I love you."

Some couples find creative ways of expressing their feelings for each other. These can be as varied as a note stuck inside a jacket pocket, a poem written by one person for the other, or a quick telephone call that says "I just wanted to say how crazy I am about you." Whatever style is chosen, being sure that your intimate partner knows of your affection (as long as it's real) is an important key to the durability of any relationship.

Expressing Anger

Anger at its raging peak is almost certain to distort communications, and it makes any real dialogue quite difficult. Postponing serious discussions about the source of your anger and its possible solutions until you have simmered down is usually a wise thing to do if at all possible. If not, it's best for both partners to recognize that much of what is said in anger may be regretted later on, since it may not be meant (it may just be said to hurt).

We believe it is important to recognize that anger is usually not a primary emotion. In most cases anger develops from preceding feelings of hurt, resentment, or frustration. If these can be identified and discussed while they're in their early stages, *before* they grow into anger, there is a much better chance of dealing with them successfully and avoiding the harm that anger can produce. However, when anger does occur, it is often better to release it quickly, in small doses, and in appropriate ways, than to let it simmer in continuing resentment and hostility until it boils over or explodes. As the social psychologist Carol Tavris points out, "Couples who are not defeated by rage and the conflicts that cause it know two things: when to keep quiet about trivial angers, for [the] sake of civility, and how to argue about important ones, for the sake of personal autonomy and growth" (1982, pp. 222–223).

Some authorities believe that an outburst of anger can help

people discharge pent-up tensions and can "clear the decks" to allow for a return to a relative state of emotional equilibrium. Others have concluded that anger is generally "constructively motivated" (not intended to hurt someone, but to bring about change) and have conducted research showing that anger is usually beneficial to both the person expressing the anger and the person who is its target. Thus, it is important to realize that getting angry with your partner occasionally is certainly no sign that your relationship is doomed. As Carol Tavris observes: "In the final analysis, managing anger depends on taking responsibility for one's emotions and one's actions: on refusing the temptation, for instance, to remain stuck in blame or fury or silent resentment" (1982, p. 226).

The Art of Listening

Many people have the mistaken notion that being a good listener simply means sitting back in a chair and keeping your mouth closed. But the ability to listen accurately and empathetically is actually a complex process. Here are some specific pointers about what it takes to be an effective listener:

1. *Effective listening requires your undivided attention.* Trying to listen while you're doing something else, like watching TV or reading, tells your partner that you don't think that what he or she has to say is very important. In addition, listening with "half an ear" increases the chance that you'll miss a detail or nuance of your partner's message (verbal or nonverbal) that may be crucial to its overall meaning.

2. *Effective listening is an active rather than passive process.* The best listeners show the speaker that they are involved in the communication process even though they are temporarily silent. This can be done by eye contact, nodding your head, or asking an occasional question to clarify a point without disrupting your partner's message.

3. *Effective listeners are patient in their listening style.* People don't always plunge right in to intimate discussions without first establishing that their partner is receptive and willing to talk. If they feel rushed, they'll either skip the conversation entirely (and probably feel angry about your inaccessibility) or be forced to convey their message in a choppy, incomplete version. The pa-

tient listener realizes that a bit of encouragement early in a conversation can set the stage for a more meaningful dialogue later on. At the same time, patient listeners refrain from the temptation to barge in with their own comments before the other person has completed his or her message.

4. *Effective listeners avoid putting undue emphasis on one word or phrase in a message and wait for the message to be completed before they react to it.* This is particularly true in sexual matters, where many key words (such as "orgasm" or "satisfaction") can trigger an emotional response. One man became so agitated and annoyed when his wife mentioned the frequency of their sexual relations that he missed her actual message — that she was enjoying sex now more than ever before — as he mentally rushed to defend himself from criticism.

5. *Effective listeners pay attention to what the speaker is actually saying instead of approaching conversations with preconceived notions of what might be said.*

6. *Effective listeners are attuned to their partners even when there's been no request for a discussion.* Sometimes the most important communications occur in odd, offhand moments rather than in planned, formal dialogues. Unless you're tuned in to this possibility and receptive to what is being said, you seriously cut down the chances of spontaneous communications, which are often the most valuable.

7. *You don't have to agree in order to listen — in fact, it can be useful to agree to disagree.* The point of being a good listener is to understand what the speaker is saying; this doesn't mean you have to endorse the message. Recognize that your partner is expressing his or her feelings, which may be very different from your own.

In addition to these points, it is also a good idea to realize that the listener's role is not a totally silent one. Most often, intimate communications invite some form of dialogue, with the listener making some acknowledgment of having heard the message, checking out any areas of uncertainty by asking for further clarification, and paraphrasing the overall gist of the message to be sure it has been correctly understood. The following exchange at the end of a longer conversation illustrates how this might be done:

Dan: I hear your concern about our sex life. I don't know quite what to say right now — it's taken me by surprise.
Jane: That's okay, I just want you to think about it a while, not to have an answer tonight.
Dan: Is the major thing that's bothering you that sex has gotten too mechanical for us?
Jane: Well, that's part of it, but I'm also getting a little bored.
Dan: So you want us both to think about how we can get more creativity and tenderness in our sex?
Jane: That's it exactly.

Notice, in this dialogue, that when Dan checks out Jane's meaning and then paraphrases her earlier comments, he doesn't use "I" language. Dan isn't talking about himself here; instead, since he's trying to clarify Jane's "I" messages, he uses sentences that are focused on her rather than himself. Part of the art of listening is deciding when to listen and when to respond.

Talking about Sex

Although communicating about sex doesn't always involve words, letting a partner know what is important or pleasing to us sexually often requires verbal statements. Yet many people are particularly hesitant when it comes to talking about sex, perhaps because of embarrassment, fear of rejection, or concern that talking about it will cause sexual spontaneity to disappear.

As children, most of us were discouraged from saying much about sex and many never learned the terminology to describe their sexual anatomy. So, part of the hesitancy people have in talking about sex with a partner is actually a carryover from these childhood taboos.

Our difficulties in talking about sex with even an intimate partner also are related to the sexual scripts our society has written, particularly the one that casts the male as the sexual expert and the female as the passive, naive participant.

> I'd been dating Larry for three or four months when I finally decided that I had to talk to him about sex. I just wasn't enjoying his style, which was much too fast and rough for me. But every time I tried to bring up the topic, my vocal cords seemed to freeze

and I backed off. Finally, I wrote him a letter that broke the ice, and we worked it out pretty easily.

Nancy and I were having some problems with our sex life about a year after we got married, but I somehow couldn't bring myself to say anything to her about it. I figured that I was supposed to know what was wrong and how to fix it because I was the male — and because I was more experienced. It took us three years before we saw somebody who helped us straighten things out.

Many couples have difficulty talking about sex, so it's not surprising that people frequently put up with awkward or frustrating sexual patterns or don't openly express their desires. Yet talking about sex, like any other form of communication, can be facilitated by some thought and practice. Here are some pointers that may prove useful:

1. *Talk with your partner about how and when it would be most comfortable to discuss sex.* You may be surprised to find that your partner is also hesitant about sexual discussions and that simply bringing the topic out in the open provides you both with a good opportunity for defusing tensions. You may also be able to determine when it will be easiest to talk about sex. Some people prefer avoiding the "instant replay" analysis right after making love, but others feel this is the perfect time for talking since events and feelings are fresh in your minds. Whatever you decide about this, the important thing is to let your partner know that you're interested in feedback about your sexual interaction. Armed with this knowledge, your partner won't be worried that you'll react to anything that's said as though it was criticism.

2. *Consider the possibility of using books or other media sources to initiate discussions.* This approach allows partners to discuss what they've read or seen and relate it to their personal preferences or dislikes. The advantage is that the discussion is more abstract — in effect, a discussion of sexual ideas as much as sexual action — and thus doesn't sound so much like "When you touched me here I didn't feel good." The disadvantages are that the books you read may not fit your own styles and needs and that some of the suggestions you read about may be offensive or uncomfortable.

3. *Use "I" language as much as possible when talking about sex*

together, and try to avoid putting blame on your partner for your own patterns of response (or lack thereof).

4. *Remember that if your partner rejects a type of sexual activity that you think you might enjoy, he or she is not rejecting you as a person.*

5. *Be aware that sexual feelings and preferences change from time to time.* It's very tempting, on hearing that your partner likes to have his or her earlobe licked, to do this automatically every time you make love. That way, you might think you can't be accused of forgetting. The problem is that doing the same things over and over tends to get boring and sometimes becomes downright unpleasant. The other side of this coin is that a partner who doesn't generally like a particular form of sexual stimulation, such as oral sex, may develop a yen for that activity on any given occasion. Be flexible in translating your talks about sex into action; be prepared to change when necessary or advisable.

6. *Don't neglect the nonverbal side of sexual communications, since these messages often speak louder than words.* Don't be afraid of showing your partner just how you like to be touched: with firm or feathery stroking, with vigorous or slow rubbing, with alternating intensity to the touch, or with a consistent pattern held for some time. Since it's often difficult to express your precise preferences in words, put your hand on your partner's and demonstrate. Not only is this a perfect "I" message, it also relieves your partner of the need to guess at what you like.

7. *Don't expect perfection.* Intimate relations can stumble if partners expect that sex should always be a memorable, passionate experience. Realize that just as your mood can change, or your physical feelings ebb and flow, so too can sexual experiences range from ecstatic peaks to fizzled-out fiascos. It isn't necessary to analyze what went wrong whenever sex wasn't superlative; instead, it's useful to talk with your partner to be sure that you both have realistic expectations about sex rather than impossible dreams that can only lead to disappointment.

It's important to realize that talking about sex with your partner isn't something to do once and then put aside. Like all forms of intimate communications, this topic benefits from the ongoing dialogue that permits a couple to learn about each other and resolve confusions or uncertainties over time.

If your partner doesn't communicate very openly about your sexual interaction, and you've tried to draw him or her into discussions a number of times only to be shut out, you need to examine your options. If you are generally happy with your sex life together, there may not be a pressing reason to talk about sex. As long as you feel that you can make your needs known, respecting your partner's silence on the subject can be the wisest course. If, on the other hand your sex life is unsatisfactory (whatever the reason), attempting a candid conversation by stating your feelings and concerns and asking your partner to respond in kind certainly seems in order. Some couples find that a visit to a sex therapist or counselor is helpful if talking about sex together is difficult; the therapist or counselor may be able to pinpoint the source of reluctance for such discussions and suggest ways of solving this type of problem.

CHAPTER ELEVEN

Sexual Fantasy

FANTASY ALLOWS us to escape from the frustrations and limits of our everyday lives. Through fantasy, a person can transform the real world into whatever he or she likes, no matter how briefly or improbably. Although it is only a make-believe excursion of the mind, fantasy can help people find excitement, adventure, self-confidence, and pleasure.

From childhood on, most people have sexual fantasies that serve a variety of functions and elicit a broad range of reactions. Some are pleasant or exhilarating; others are embarrassing, puzzling, or even shocking. In this chapter, we will discuss the functions of sexual fantasies and then provide a classification for describing the most common types of fantasy patterns. For purposes of clarity, we will restrict our use of the term "sexual fantasy" to refer only to wakeful thoughts as distinguished from sleep-associated dreams.

Although every child learns that pretending is an important type of play, sexual fantasies after childhood are usually not thought of as playful. This attitude may exist because sex is usually regarded as a serious matter, even in the imagination. Furthermore, some religious traditions regard a thought as equivalent to an act; thus, a person who has "immoral" sexual daydreams or desires is as sinful as a person who acts upon these impulses. Fantasies have also been viewed as having implications for mental health. Psychoanalysts were the only group for half a century to study fantasy in any depth. They viewed "deviant" sexual fantasies — those portraying anything other than heterosexual acts that led to intercourse — as immature expressions of the sex drive and as blocks to the development of more mature sexuality. Many psychoanalysts also believed that such fantasies were likely to be forerunners of "deviant" sexual behavior.

Generally, imagination, creativity, and playfulness are part of the act of fantasizing. However, if a fantasy becomes a controlling force in a person's life, the play element may be completely eliminated. This situation isn't very different from the person who becomes addicted to gambling (which also begins as a form of play) or the person who gets so caught up in a competitive sport, such as long-distance running, that the playful side of the activity is lost.

At times, it may be difficult to distinguish sexual fantasy from sexual desire. Just as your awareness of hunger and thinking about what kind of food you'd like to eat may blend together, your sexual appetite may merge with thoughts about how sexual satisfaction may be obtained. Although a fantasy may be valued strictly as a piece of fiction as opposed to a preview of an expected reality, this distinction does not always hold. In some cases, a sexual fantasy expresses sexual desire, while in others it *provokes* sexual desire that does not necessarily require the fantasied act for fulfillment.

THE CONTEXT OF
SEXUAL FANTASIES

Sexual fantasies occur in an astonishingly wide variety of settings and circumstances. Sometimes these imaginative interludes are intentionally called forth to pass the time, to enliven a boring experience, or to provide a sense of excitement. At other times, sex fantasies float into our awareness in a seemingly random fashion, perhaps triggered by thoughts or feelings of which we have little awareness.

Among the most common varieties of sexual fantasies are those that can best be described as old familiar stories. The origin of such a fantasy, if it can be traced at all, might have been a book, a movie scene, or an actual experience. The person using this fantasy finds it to be particularly pleasing and comfortable and returns to it again and again. On different occasions, minor variations may be played out in the fantasy, but the fantasizer almost invariably occupies a central role in the story line.

How a particular fantasy comes to be preferred and repeated over and over is not entirely clear. Sometimes the primary fascination with this sort of fantasy lies in its sexual arousal, while at other times the pleasure may be more related to the "director's role" — being able to control the scene, plot, and actors. In many instances, the complexity of this fantasy makes it more suitable for use in solitary situations than during sexual activity with a partner.

In another form of the preferential fantasy pattern the person repeatedly uses a particular *type* of fantasy — group sex, for example — but no characters or story line connect one fantasy to another. The first pattern described is like playing a specific record again and again, while this pattern is more like playing a certain type of record — country and western or classical music — repeatedly.

There are at least two situations in which preferential fantasies may become troublesome. For some people, the repeated and exclusive use of such a fantasy may lead to a situation in which the fantasy becomes necessary for sexual arousal. The person no longer responds sexually to his or her partner since sexual arousal depends on the fantasy alone. Infrequently, preferential fantasies can become obsessions that may interfere with thinking or behavior. Obsessional fantasies will be discussed more fully in chapter 15.

Just as children exercise both their curiosity and creativity when they pretend, people who use sexual fantasy also draw on these elements. The desire to know about something not yet experienced, forbidden, or seemingly unattainable is often a key feature of sexual fantasies. For instance, a married woman who has always been faithful to her husband may fantasize about an extramarital liaison, or a fifteen-year-old boy may fantasize about making love to a woman pictured in a *Playboy* centerfold. In both examples, the fantasy does not necessarily mean that the person wants to actually participate in the fantasied behavior.

Fantasy in the context of curiosity may be highly arousing and intriguing but it may also include scenes that seem outlandish, preposterous, disgusting, fearful, or silly to the fantasizer if examined in a detached, rational moment. This discrepancy is not surprising, because most of us enjoy things under some circumstances that would be unpleasant, embarrassing, or even frightening in others.

The creative side of sexual fantasies is not only linked to curiosity. Just as there is a creative aspect in an artist painting a scene from memory or an author writing a vivid detail about a past event, there is also a creative element in sexual fantasies that draw on memories of past sexual experience. The fantasizer embellishes the memories and molds them into new forms while retaining the essence of the remembered experience. The fantasy can be smoothed out or improved in comparison with the real-life experience; in the world of fantasy, blemishes, fatigue, and distractions disappear while passion mounts and the action is unencumbered by trivial details.

In creating a sexual fantasy, the fantasizer not only programs the action but also orchestrates the emotions of the principal fantasy figures. If a woman wants her fantasy lover to be strong and silent, he is; if she prefers a verbal, gentle partner, she instantly has such a person at her command. If a man wants a passionate, aggressive woman in his sexual imagery, she is immediately there; if he prefers a reserved, unwilling partner to force into submission, this is easily achieved. In this sense, fantasy provides a dimension that is relatively unattainable in real life because even if your partner is willing to try whatever sexual acts you suggest, you have no way of controlling his or her character or emotions.

Many people regard their fantasies as private property and keep them to themselves. In recent years, it has been suggested that sharing fantasies between partners fosters intimacy and understanding. The implication is that not sharing your sexual fantasies may be selfish or immature. Those who believe in sharing fantasies point out that after a long time in a relationship, many couples discover that they each have fantasies about the same activity (for instance, having anal intercourse). Fearing that their partner may be embarrassed, offended, or unwilling to "play," many couples do not share their fantasies. Sometimes the fantasies of two people in a relationship are quite complementary, as when one person wants to be spanked during sexual activity, while his or her partner fantasizes about spanking someone.

People who suggest that keeping fantasies private reflects immaturity do so for a number of reasons, some of which are at least partially incorrect. They believe that being embarrassed or ashamed of one's fantasies is in and of itself a mark of immaturity. They also feel that intimate relationships should have no barriers to communication because open communication is a mark of maturity and commitment. A third reason they give to "prove" that sharing sexual fantasies is best is that sharing fosters a deeper degree of understanding between partners and so is likely to improve the relationship. Finally, they often point out that sexual fantasies are more likely to be kept private by people who are sexually inhibited or "uptight." By bringing these fantasies out into the open, a person can become less inhibited (therefore, more "mature") and may attain more sexual satisfaction.

Such arguments are oversimplified in many ways. To begin with, there is nothing wrong or immature about having private thoughts or feelings. If private fantasies give a distorted view of a person's preferences, sharing the fantasy with a partner may result in misperception of what that person needs or wants instead of better understanding. For instance, if a woman occasionally fantasizes about being raped and enjoys the fantasy, this does not mean that she wants to be raped or would enjoy being raped. The same can be said of a person who fantasizes about robbing a bank: he or she can hardly be said to have a criminal mind or to be "dangerous" because of this type of fantasy.

A partner may not only misunderstand a fantasy, but may also believe that he or she is in some way expected to play it out in real life. Although the partner can say "no," there may be a subtle pressure, whether intentional or unintentional. Having learned that your partner is turned on by a particular sexual fantasy, do you agree to try it, even if it is a bit uncomfortable, in order to be open-minded and sensitive to your partner's needs? What if you mistakenly decide to "try out" the fantasy when your partner doesn't really want to? Sometimes, once a fantasy has been shared with a partner, it provokes jealousy, guilt, or self-doubt. This is particularly true when one partner assumes that the other's fantasy indicates dissatisfaction or a desire to try someone else.

Many people find that after telling a partner about their most highly charged sexual fantasy, the turn-on value of the fantasy fizzles. While this does not always happen — as sometimes the erotic stimulus of the fantasy increases — it is a potential pitfall. Unfortunately, there is no way of knowing beforehand if partners will benefit from sharing details of their sexual fantasies or if problems will result.

Not all sexual fantasies are willfully conjured up or pleasing. Some fantasies recur over and over again despite being unwanted; other fantasies flood into a person's awareness in a frightening fashion, producing inner turmoil, guilt, or conflict. Fantasies of this sort either may result in sexual arousal or may be so distressing that they shut off sexual feelings.

Usually, intruding fantasies that depict sexual situations or conduct that the fantasizer considers abnormal or bizarre (yet also arousing) include some imagined form of punishment or injury as the price to be paid for the sexual indulgence. The punishment within the fantasy may range from physical afflictions (such as venereal disease or cancer) to being discovered by others in the midst of sexual activity, being arrested and jailed, or being deprived of sexual satisfaction via one calamity or another. Other intruding fantasies may result in real-life problems such as avoidance of sexual activity, profound sexual guilt, or sexual dysfunction.

It is not difficult to imagine how distressing it might be for a forty-year-old married woman with conservative religious and sexual values to find that during sexual activity with her husband she repeatedly has fantasies about having sex with a group of men. Similarly, a man who prides himself on his macho image and is strongly antihomosexual may be alarmed to find himself fantasizing about performing oral sex on another man. If distressing fantasies recur regularly, counseling may be required. Psychologists, psychiatrists, or sex therapists can help a person troubled by such fantasy patterns learn how to "switch the channel" (as you would switch from a disturbing TV show to a more pleasant one), or can teach thought-blocking techniques to deal with the situation.

FUNCTIONS OF
SEXUAL FANTASY

Our use of the sexual imagination is quite varied. Fantasies function at many different levels to boost our self-confidence, provide a safety valve for pent-up feelings, increase sexual excitement, or to let us triumph over the forces that prove troublesome in the everyday world, to mention just a few. Some of the most common functions of sexual fantasies will now be described.

We have already said that fantasy and sexual desire often merge together. People with low levels of sexual desire typically have few sexual fantasies and will often benefit from treatment that helps them form positive fantasies. Many times, sexual fantasies are used to induce or enhance sexual arousal, and while fantasies are often combined with masturbation to provide a source of turn-on when a partner is not available, fantasies are also extremely common during sexual activity with someone else.

For some, the use of fantasy provides an initial boost to getting things under way. Others use fantasy to move from a leisurely, low-key sexual level into a more passionate state. One of the most frequent patterns we have encountered is the use of a particularly treasured fantasy to move from the plateau phase of arousal to orgasm. Some men and women report that they are unable to be orgasmic unless they use fantasy in this way.

Sexual fantasies can enhance both the psychological and physiological sides of sexual response in many ways: counteracting boredom, focusing thoughts and feelings (thus avoiding distractions or pressures), boosting our self-image (in our fantasies we can assume our desired physical attributes and need not worry about penis size, breast size, or body weight), and imagining an ideal partner (or partners) who suits all our needs.

Sexual fantasies also provide a safe, protected environment for engaging the imagination and letting our sexual feelings roam. They are safe because they are private and fictional: privacy ensures that fantasies are undiscoverable, while the fictional makeup of our fantasies relieves us of personal accountability.

Another safety feature of fantasies lies in having the "director's role," even to the point of being able to end the fantasy abruptly if it becomes uncomfortable or threatening. Without such safety, the erotic value of most people's fantasies would probably decrease substantially.

If you consider that most sexual fantasies involve situations, partners, and/or behavior that might be judged improper or illegal if they were real, the importance of safety as a backdrop for excitement becomes apparent. A mild-mannered, genteel college professor can fantasize about orgies with the three attractive coeds in the first row of his classroom without risking his tenure or reputation. A young woman lawyer can fantasize about raping one of her clients without jeopardizing her standing before the bar. A teenage boy can construct elaborate sexual fantasies about ravishing his best friend's mother without risking parental punishment or losing a friend. Clearly, the element of safety ensures the appeal and power of these erotic images.

Fantasies of all types function as psychological safety valves that discharge inner tensions or needs in a relatively painless way. In our fantasies, we can get even with others for real or imagined injustices, conquer fears by carefully controlling the action and emotions, and compensate for any personal shortcomings that are troublesome in real life. Consider the following examples from our files:

> *A thirty-two-year-old married woman:* My husband and I had a lousy sex life for years, and it mirrored a lousy relationship. During this time, my sex fantasies almost always involved making it with other men while he was forced to watch me with great humiliation. It was sweet revenge, I guess. . . .
>
> Once we started marriage counseling, things began to improve. We learned how to talk together, and our sex life improved too. The interesting thing was, my fantasies began to change. I guess I no longer had a need to retaliate.

> *A twenty-two-year-old unmarried male medical student:* I've always been very uptight about sex. I suppose one reason is that I'm embarrassed about the size of my penis, which seems very small. In my fantasies, the woman I'm with always remarks on how big my penis is and seems in awe of its power. I found that if I used this fantasy while I was really with someone, I was much less nervous. It sounds silly, but I really felt better about myself.

Since we live in a society with a strong tradition of sexual restrictions and taboos and we learn not to discuss sexual behavior, fantasy often provides an important means of clarifying and dealing with sexual conflicts or confusion.

For most people, fantasy also provides a way to preview an anticipated experience and to prepare themselves for what to expect and how to act. While this function of sexual fantasies may be most prominent during adolescence or with any people who have only limited sexual experience, it is very important. The opportunity to visualize oneself in a certain form of erotic activity — oral-genital sex, for example — allows one to anticipate some problems that may occur. By replaying a scene several times, fantasizers can develop a better idea of how to minimize difficulties and can also partially desensitize themselves to feelings of awkwardness, embarrassment, or surprise. Of course, if and when the fantasy is transformed into fact, the actual event may be considerably different from the imagined one in feelings, tempo, and other details. Nevertheless, a sense of comfort usually results from using fantasy as rehearsal.

Fantasy and Sexual Values

Many people misunderstand the nature of sexual fantasy and think that it expresses an actual desire to participate in or experience a given situation. This is like saying that a person who daydreams about being a war hero wants to go to war or that a person who fantasizes about having children is ready or willing to be a parent. Professionals are not immune to confusing the issues still further, as Lonnie Barbach (1980, p. 119) explains:

> I worked with a group of feminist therapists who argued that it is sexist to derive pleasure from rape fantasies or fantasies that portray male domination. It was difficult for them to separate the sexual pleasure the fantasy provided from its political interpretation. I also knew a lesbian therapist who nearly panicked when she found herself having heterosexual fantasies, fearing that she might be a "latent heterosexual."

In an ongoing study of sexual fantasies being done at the Masters & Johnson Institute, we have found that most women who are aroused by fantasies that portray "unusual" sex practices such as rape, incest, sex with animals, or sadomasochistic sex indicate that they have no interest whatsoever in acting out the fantasy. In contrast, men appear to be somewhat more adventuresome. About two-thirds of the men we have interviewed who have such fantasies declare that they would be willing to try them under the right circumstances.

A landmark study by the psychiatrists Hariton and Singer of the sexual fantasies of married women during intercourse with their husbands stressed that fantasy content does not indicate sexual problems, psychological problems, or personality flaws. In fact, most people with recurrent sexual fantasies feel neither the desire nor the need to act on them in real life.

While most people realize that a fleeting fantasy is not an indepth revelation of the inner psyche, it is tempting to assume that a favorite fantasy theme says important things about our psychological makeup. There are no research data showing that this is true for all people (although it may be true in individual instances). Our sexual and personal values may differ considerably from our fantasy lives, just as an actor's true identity may vary greatly from the dramatic roles he plays.

While we believe that most sex fantasies are natural, creative experiments that help combat loneliness or boredom and that defuse forbidden urges, some experts have different views. In a paper titled "Why We Should *Not* Accept Sexual Fantasies," a California psychologist, Bernard Apfelbaum, describes fantasies as "cut-off parts of us signalling wildly to get back in." He believes that sex fantasies stem from dissatisfaction with reality and have a high potential for creating relationship conflicts. For example, if one partner feels that the other's turn-on comes from a fantasy rather than from personal involvement, an instinctive sense of being ignored intrudes and blocks sexual responsiveness (he assumes, of course, that fantasy and involvement do not mix). Apfelbaum also suggests that having private, unshared fantasies lessens intimacy and trust in a relationship, and says: "Sexual fantasies always offer us precious clues about what needs to be done to strengthen our relationships."

A somewhat different approach is taken by Avodah Offit, a psychoanalyst in New York, who believes that sex therapists overemphasize the acceptability of sex fantasies. Offit thinks that if reality and fantasy are closely matched, this indicates a "well-integrated personality," a sort of psychological togetherness. If fantasy strays too far away from our personal realities, the inconsistencies point to potential personality problems. Finally, Offit regards sex fantasies as "a pale substitute for the complexities of joy and pain which are requisites for loving a real person" (Offit, 1977, p. 201).

Another psychoanalyst-sex researcher, Robert Stoller (1979), believes that sex fantasies are a private pornography that allow us to gain revenge over a previously painful situation. He suggests that there is a flame of hostility at the core of all sex fantasies (and all sexual excitement). Psychiatrist Natalie Shainess (1971) takes an even stronger position. She says that fantasies during intercourse are "symptomatic of sexual difficulty" and "signs of sexual alienation." She also believes that healthy women do not fantasize very much except when they are young and inexperienced, and if fantasies persist, "you can assume there's greater pathology."

Finally, Alan Rapaport, a clinical psychologist, takes the viewpoint that *any* fantasy that occurs during person-to-person sex is debasing because it reduces personal involvement. "If a person is caught up in a private fantasy while making love . . . it interferes with a more sharing and intimate relationship" (Goleman and Bush, 1977).

Fantasy as Fact

Although many people say that they have no wish to transform their sexual fantasies into reality, for some the opposite is true. What motivates a person to lean one way or the other is uncertain, but some of the relevant factors may be: (1) how powerful an erotic turn-on is involved, (2) how receptive, trustworthy, and understanding the partner is perceived to be, (3) how a person feels about himself or herself, and (4) how unusual or bizarre a fantasy appears.

Reliable statistics on how many people act out their sexual fantasies in real life are not available. For some couples, the acting out involves a limited dramatization, playing roles in a carefully controlled way — a rehearsal of the fantasy instead of the entire experience. For example, a woman who fantasizes about being spanked may ask her partner to give her a gentle

spanking that is more symbolic than real, or a man who fantasizes about having sex with a young teenage girl may ask his
partner to dress and act like a thirteen-year-old. In such situations, the fantasy comes to life in the sense of being "in the flesh"
rather than imaginary, but it is still not the real thing. The limited dramatization form of acting out fantasies is particularly
appealing to many people because of its safety and control, but
it is often less psychologically satisfying than the purely imaginary fantasy since it is "only an act."

Some people go farther in transforming a sexual fantasy into
real life. A married couple may respond to an ad in a "swingers'
magazine" to try out a fantasy of switching partners. A person
with fantasies about being tied up (bondage) may convince his
or her partner to do so. In some cases, the fantasy becomes more
fulfilling, more meaningful, and a part of the continuing sexual
relationship. Very often, however, the result is less than expected: sexual fantasies that are tried in real life often turn out
to be disappointing, unexciting, or even unpleasant. Nancy Friday, who has studied male and female sex fantasies for almost a
decade, says: "I think that for every person who has written to
me about the joys of performing their sexual dreams in reality,
there have been three or four who knew in advance that it
wouldn't work, or who tried it and were disappointed" (Friday,
1975, p. 280). Our research also indicates that for many people,
transforming fantasy to fact is unsatisfactory, resulting at times
in a complete loss of the erotic value of the fantasy. A twenty-
two-year-old female college student told us:

> I used to have one particular fantasy that never failed to work. It
> was almost an electric thing, like flipping a switch and then
> "Zowie." I almost always had my best orgasms, and most exciting
> sex, when I flashed this fantasy through my mind. Then, unfor-
> tunately, I decided to try it out with my partner. We were both
> interested in this, I wasn't embarrassed or uptight, but it just
> didn't click together for me. After we had tried it two or three
> times, the fantasy itself became less exciting and less reliable and
> finally just didn't work at all. It was like losing a best friend.

A similar point is made by Karen Shanor in a book called *The
Fantasy Files:*

Often when a fantasy is finally acted out, it does not occur again with any frequency as a fantasy. Only if the acting-out experience is amazingly good does the thought remain prominent. . . . Most of the time reality does not live up to the excitement of the fantasy, and the fantasy is therefore modified or significantly lessened in its importance.

One other aspect of the fantasy-as-reality topic deserves mention. Men have often turned to prostitutes to assist them in living out a sexual fantasy because the prostitute-client situation is apt to be psychologically safe for a number of reasons. First, the prostitute is a sexually experienced person who has probably "seen it all" and is therefore unshockable. Second, the encounter with the prostitute is protected in a social sense — privacy is ensured and the experience is isolated from everyday relationships. Third, the transaction is primarily sexual rather than personal. Finally, a man requesting unusual sexual services from a prostitute risks no loss of self-esteem or personal repute — if he can afford the price, his fantasy is implemented without much ado. In newspapers across the country that carry sexually explicit classified ads, it is not unusual to see escort services (usually fronts for commercial sex) advertising "Try out your fantasies, whatever they may be" or to find female and male consorts marketing a particular fantasy angle, as shown by a poetic prostitute whose forbidding picture, black cape and all, was captioned, "My whips and chains will thrill your veins. Call Mistress _____."

CONTENT OF SEXUAL FANTASY

The range of the erotic imagination is almost limitless. One person fantasizes about animals, another about movie stars, another about enemas, diapers, or South Sea islands. The action may be explicit and detailed or shifting and vague. The stylistic variations of sexual fantasies reflect the richness of the human mind.

In this section, we will briefly examine some of the more common types of sexual fantasies, realizing that many fantasies are difficult to pigeonhole neatly in a classifcation system.

Experimentation

One popular type of fantasy is to visualize experiences that have never been tried in real life. The content may focus on novel circumstances like being the star of a porno movie, having sex in a public place, or being a prostitute. The fantasy may also explore unusual forms of sexual activity such as sex with an animal or incest. In some cases, the thrill of the forbidden is important; in others, the desire for the unique or the untried is more relevant. For this reason, experimentation fantasies are often used to overcome sexual boredom.

Conquest

At the core of all conquest fantasies is an element of power. The power may be expressed in the ability to command, to force, or to seduce someone else into sexual relations. Assuming the imaginary role of ruling monarch or slave-owner can set the stage for commands to be obeyed. Nonphysical force used to coerce someone into sex can be conjured up in roles such as prison warden, school principal, boss, or blackmailer. If physical force is involved, the fantasy is classified as rape or sadomasochism. Of course, the power to seduce others can be written into almost any fantasy script.

The flip side of the conquest fantasy is the idea of being conquered — of being commanded, forced, or seduced by someone else. The variety of roles that can be assumed to orchestrate such a fantasy is immense. The key element is being powerless to resist, whether for reasons of fear, resignation to one's fate, blind obedience, economic difficulties, or protecting another person.

A subtype of the conquest/conquered fantasy is one version of the domination/humiliation theme. Here, power is not only used to obtain sexual activity, but circumstances must be degrading or embarrassing. The victim might be forced by the power-figure to grovel, to wear unusual items of clothing (diapers, see-through clothes, shackles), or to engage in humiliating acts. The turn-on comes more from the debasement than from the sexual activity per se.

Switching Partners

Imagining sex with a different partner is one of the most common varieties of sexual fantasies. Sometimes the imagined partner is a previous partner (an old boyfriend or girlfriend, for example). More often, the fantasized partner is simply someone considered desirable: a friend, a neighbor, a relative, a teacher, a salesclerk, or an imaginary ideal.

Some people are uncomfortable when they have a fantasy of a different partner during sex with their lover or spouse. Guilt feelings can result if the person views such a fantasy as a sign of infidelity. If you have ever felt this way, you might be interested in knowing that replacing an established partner with another person in fantasy is so common that the chances are quite high that you and your own partner have both had this vision.

A subtype of the "switching partners" fantasy is the version that can be called "celebrity sex." Here, the fantasied partner is chosen from the public roster of celebrities: movie stars, television personalities, sports figures, politicians, musicians, artists, authors, and faces in the news. Just as past generations incorporated images of celebrities like Clark Gable, Humphrey Bogart, Marilyn Monroe, and Jayne Mansfield into their sexual fantasies, people continue this practice today.

We find the following story to be particularly intriguing, since it demonstrates a fairly consistent finding about sex fantasies, even those involving "celebrity sex."

> One young woman developed an elaborate fantasy of having intercourse with Mick Jagger, a member of the Rolling Stones. When she began having intercourse regularly as a "groupie" following different bands around, she always reverted, during the act itself, to images of her childhood fantasy of Mick. In the course of her travels as a "groupie," she actually finally encountered the real Mick Jagger. In bed with him at last, she still found it necessary to resort to her *fantasy* Mick Jagger because the real one, after all, was not as prodigiously gifted in bed as she had long fantasized him to be. (Singer, 1980, p. 187)

Group Sex

Another common type of fantasy that is related to the "switching partners" pattern is the group sex scene. The details of group sex fantasies vary from elaborate orgies involving friends to images of ancient Rome or being on a Hollywood movie set where the evenings turn into sexual Olympic games. At the other end of the spectrum are group sex fantasies where a man imagines himself making love to two women or a woman visualizes herself with two men. In some group sex fantasies the action is bisexual, while in other imaginary scripts the action is strictly heterosexual although there may be dozens of arms, legs, and sex organs in motion at any given time. Interestingly, many people who enjoy this type of fantasy tell us they might be willing to try it out if given the right set of circumstances.

Watching

Some people become particularly aroused by fantasizing scenes in which they are watching others engage in sex. In the purest form of this fantasy type, the observer never actually enters the action, although he or she may be either visible or hidden from the sexual participants. In variations on this theme, the observer reaches high levels of sexual excitement from watching and then joins in the physical festivities. "Watching" fantasies are not unusual among married couples, where one person fantasizes a scene in which he or she watches their spouse having sex with someone else. "Watching" fantasies, however, are not equivalent to actual voyeurism, discussed in chapter 15.

Rape

Fantasies about rape are possibly the most misunderstood of all sex fantasies. Some people think that women who fantasize about rape are really yearning for such an event to occur and suggest that the fantasy represents an unrealized wish. This distorted interpretation has no basis in fact. It is more useful to look at rape fantasies as providing reassurance to some women that they are being sexually passive rather than aggressive, since this conforms to our cultural stereotypes about sexual behavior. In addition, rape fantasies absolve the "victim" of personal responsibility for enjoying sex. As Nancy Friday (1973, p. 109) observes: "By putting herself in the hands of her fantasy assailant — by *making* him an assailant — she gets him to do what she wants him to do, while seeming to be forced to do what he wants. Both ways she wins, and all the while she's blameless, at the mercy of a force stronger than herself."

Rape fantasies have a variety of forms. A woman may visualize herself as the rapist or a man may envision himself as the victim or victimizer. Some heterosexual men occasionally have fantasies about being raped homosexually: as in the woman's situation discussed earlier, the man's fantasy frees him of "responsibility" for the homosexual act since he has been forced into it.

Idyllic Encounters

In sharp contrast to rape fantasies are those with strong overtones of romance and tranquility. "Idyllic encounter" scripts usually involve meeting a stranger under near perfect conditions — a quiet garden, a secluded, moonlit beach, a tropical paradise — where instant romantic attraction blossoms forth and an ecstatic sexual interlude takes place. Then the characters usually go on their way, happy but unencumbered.

One famous literary rendition of the idyllic encounter fantasy appeared in Erica Jong's *Fear of Flying*, where it was described as the "zipless fuck":

> The zipless fuck was more than a fuck. It was a platonic ideal. Zipless because when you came together zippers flew away like rose petals, underwear blew off in one breath like dandelion fluff. . . .
>
> For the true, ultimate zipless A-1 fuck, it was necessary that you never get to know the man very well. . . . [A]nother condition was brevity. And anonymity made it even better.
>
> The zipless fuck is absolutely pure. It is free of ulterior motives. There is no power game. The man is not "taking" and the woman is not "giving." . . . No one is trying to prove anything or get anything out of anyone. The zipless fuck is the purest thing there is. And it is rarer than the unicorn.

Sadomasochism

Inflicting pain (sadism) or receiving pain (masochism) may be a source of sexual arousal. Sexual fantasies with sadomasochistic themes invoke images of being beaten, tied, whipped, chained, tickled, teased, handcuffed, contorted, gagged, burned, spanked, or otherwise victimized or doing the victimization. Physical force or pain is vividly present and usually the turn-on value of the fantasy is in direct proportion to the protests of the victim. Here again, as with many other fantasies, there is often no desire to live out the fantasy. One research volunteer told us, after describing a favorite masochistic fantasy in some detail, "I really can't stand physical pain and I don't know why this turns me on."

GENDER DIFFERENCES IN SEXUAL FANTASIES

It used to be thought that men had more frequent sex fantasies than women and that the sexual fantasies of women were "tamer" than those of men. This viewpoint reflected the idea that women were less interested in sex than men but were more interested in interpersonal relations. For example, Barclay (1973) believed that female sexual fantasies usually dwelled on emotional, romantic elements, while male sexual fantasies sounded like pornographic books: explicit in sexual detail, with little regard to emotions. Similarly, John Money suggested that women have only two predominant "core" fantasies — masochistic fantasies and fantasies of soft objects and touch — whereas men have a much larger fantasy repertoire (J. Petersen, 1980). Morton Hunt (1975), who conducted a survey for *Playboy,* found that 75 percent of men and 80 percent of women fantasized about intercourse with a loved person during masturbation. Yet men reported fantasies of intercourse with strangers twice as often as women did, while women reported more frequent masturbatory fantasies about being forced to have sex and homosexual activities.

Our impression is that men and women are more similar than different in their sexual fantasy patterns. The idea that women do not have sex fantasies is now clearly recognized as outmoded. In studies conducted three decades ago, Kinsey and his colleagues found that 64 percent of women who masturbated used erotic fantasies. More recently Crépault and co-workers reported that 94 percent of women used sexual fantasies, which is quite consistent with our finding that 86 percent of 300 women aged eighteen to thirty-five had erotic fantasies at various times. For both men and women, sexual fantasies are more common during masturbation than during intercourse. In addition, a recent study found that 71 percent of men and 72 percent of women used fantasy to enhance sexual arousal. On the other hand, several studies have shown that the sex fantasies of women tend to be more passive than those of men (women tend to visualize their role in the fantasy as having something done to them by someone else, rather than being the active "doer"), perhaps reflecting an innate psychological difference but more likely accounted for by the way in which most females are socialized to see their roles in sexual interaction as essentially following the lead set by males.

The similarity of fantasy content between the sexes has been supported by our continued research findings. At least in recent years, the sexual fantasies of women are quite explicit and sexually detailed. While changing sexual attitudes and the greater availability to women of sexually explicit materials have undoubtedly led to some changes in the acceptability of fantasy in the last decade, we suspect that male-female differences in fantasy patterns were never quite as large as they were thought to be.

CHAPTER TWELVE

Solitary Sexual Behavior

A twenty-eight-year old man: My first clear memory of sexual arousal was at age twelve or thirteen while I was reading a sexy book. I kept the book hidden from my parents, and I marked all the "good parts" with paper clips. I must've read that thing a thousand times.

A thirty-year-old woman: I went to see an X-rated movie with my boyfriend, and I liked it so much I went back to see it again by myself.

A twenty-one-year-old woman: When I was thirteen, I discovered that rubbing my genitals against a pillow was sexually arousing, and I had lots of orgasms that way. I didn't learn that that was masturbation until I got to college.

THE SUBJECT of solitary sexual behavior is surrounded by misconceptions and prohibitions. Most people have been taught that it isn't polite to talk about masturbation, and the activity itself is usually pursued with some fear about the possibility of discovery. As a result, people lack accurate information on the subject, and many are uncertain about how masturbation might affect their physical health or emotional stability.

Despite these concerns resulting from our cultural conditioning, new attitudes toward solitary sex (often called autoerotic activity) in its various forms have been emerging. Instead of labeling masturbation as an inferior or improper type of sexual behavior, many authorities now see it as a normal part of sexual development. The feminist Betty Dodson (1974, p. 55) takes an even stronger position: "Masturbation is our primary sex life. It is *the sexual base*. Everything we do beyond that is simply how we choose to socialize our sex life." The use of erotic readings, pictures, or movies by men and women has become accepted more widely than in the past. In a sense, solitary sexual behavior has come out of the closet: it has even been celebrated as the ultimate source of our sexual self-awareness. In this chapter, we will discuss several pertinent aspects of sexual behavior without a partner.

MASTURBATION

Masturbation can be defined as sexual self-pleasuring that involves some form of direct physical stimulation. Most often, masturbation is done by rubbing, stroking, fondling, squeezing, or otherwise stimulating the genitals, but it can also be carried out by self-stimulation of other body parts such as the breasts, the inner thighs, or the anus. The term "masturbation" refers to the act of self-stimulation without regard to the outcome; that is, sexual self-stimulation need not lead to orgasm to be masturbation.

In this book, we have deliberately used the term "masturbation" to refer to sexual *self*-stimulation. Stimulation received from a partner, although similar in many ways, involves an interactional element that makes it helpful to maintain this distinction. While masturbation can occur as part of sexual activity with a partner, our focus in this chapter is primarily on masturbation as a private act.

As we have already pointed out, masturbation often begins during childhood and occurs commonly in both males and females throughout the life cycle. Masturbation is also found elsewhere in the animal kingdom. Ford and Beach noted that many species of apes and monkeys "form habits of self-stimulation," and other mammals also practice masturbation:

> Sexually excited male porcupines, for example, walk about on three legs while holding one forepaw on the genitals. . . . Male elephants sometimes manipulate their semi-erect penis with the trunk. . . . Male dogs and cats regularly lick the phallic organ,

often showing convulsive pelvic movements which indicate the stimulatory value of the resulting sensations. . . . One [captive] male [dolphin] had a habit of holding his erect penis in the jet of water intake, and other [dolphins] characteristically rubbed the tumescent organ against the floor of the tank. . . . (Ford and Beach, p. 160)

Despite this apparent "naturalness" of masturbation from an evolutionary viewpoint, Ford and Beach noted that most human societies consider masturbation by adults to be undesirable. For an understanding of how a negative attitude toward masturbation developed in our own society, a glimpse back in history is revealing.

The origins of the word "masturbation" are not entirely clear, although it seems to have been coined in Roman times. While it was previously thought that the term derived from the Latin *manus* (hand) and *stupro* (to defile), scholars now believe it has a Greek root, *mezea* (genitals), with the original meaning, "to arouse the genitals."

The ancient Greeks and Romans were relatively silent on the subject of masturbation, although Hippocrates believed that excessive loss of semen caused spinal consumption. Even though the Bible has no clear-cut prohibitions against this sexual activity, both traditional Judaism and Christianity generally regarded masturbation as sinful. (The story of Onan [Genesis 38:9–11], which had been thought of as an edict against masturbation, is now thought by modern scholars to describe coitus interruptus, a very different act. Nevertheless, masturbation was referred to as "onanism" well into the twentieth century.)

The attitude of the Catholic Church has not changed in recent years; in a Vatican *Declaration on Sexual Ethics* (December 29, 1975), it was noted that "masturbation is an intrinsically and seriously disordered act." The Declaration went on to say, "Even if it cannot be proved that Scripture condemns this sin by name, the tradition of the Church has rightly understood it to be condemned in the New Testament when the latter speaks of 'impurity,' 'unchasteness' and other vices contrary to chastity and continence."

Masturbation was sometimes referred to as an "unnatural act" by church leaders because it had no reproductive goal, but later it was described as "self-abuse," "defilement of the flesh," and "self-pollution." In large part, the credit for giving masturbation such a bad reputation belongs to a Swiss physician, S. Tissot (1728–1797), who brought the matter into the scientific arena and transformed masturbation from a simple sin to an illness that had to be cured.

Tissot believed that all sexual activity was dangerous because it forced blood to rush to the head, leaving too little in the rest of the body, so that nerves and other vital tissues slowly degenerated. In keeping with the scientific knowledge of his time, he was certain that this form of nerve damage caused insanity. Tissot was convinced that masturbation was a particularly "dangerous" form of sex because it was convenient and could be started during the vulnerable years of childhood, and because the masturbator's guilt over his or her sinfulness further irritated the nervous system and made it more susceptible to damage.

The "proof" of Tissot's theory could be seen in mental asylums, where patients were either observed in the process of, or openly admitted to, masturbating. By the time Tissot's notions had crossed the Atlantic to America, the average doctor was quite willing to believe that masturbation caused insanity, epilepsy, acne, weight loss, decreased mental capability, weakness, lethargy, and — the ultimate punishment — early death. Benjamin Rush, one of the signers of the Declaration of Independence and a leader in early American medicine, published several pamphlets supporting Tissot's attitude toward masturbation. Fortunately, American physicians rejected Tissot's belief that sexual intercourse between married people was harmful.

Parents searched desperately for ways to keep their children from being stricken. Physicians were happy to oblige; after all, it was the conscientious doctor's duty to put an end to masturbation. Much energy and money was spent on cures ranging from elaborate belts, locks, and cages — to protect the genitals from roving hands — to surgical "cures" which left little for the patient to fondle.

The nineteenth-century medical profession in America attacked masturbation with zest. The battle was fought on two main fronts — diet and physical constraint. Gravies, alcohol, oysters, salt, pepper, fish, jelly, chocolate, ginger, and coffee were forbidden to masturbators (both male and female) as it was thought that they irritated the nerves and increased sexual desire. Dr. W. F. Morgan (1896) told readers of the *New York Medical Times* that cheese, eggs, and asparagus should not be eaten by young men as an evening meal because they increased the chance of nocturnal emissions.

Other doctors blamed tight "britches," the friction of sheets, handling of the genitals during urination, and the touching of children's genitals by nursemaids or parents during bathing. If "irritants" were removed from the diet and tight "britches" removed from the wardrobe but masturbation still continued, drastic steps became necessary. Doctors prescribed such remedies as straightjackets at bed time, wrapping the child in cold, wet sheets to "cool" desire, and tying the hands to the bedposts. The U.S. Patent Office granted several patents to variations of the medieval chastity belt that shielded the genitals from fondling. Parents could padlock their children into these elaborate "genital cages" and tuck away the key. (One particularly torturous version constructed for adolescents and adults consisted of a tube lined with metal spikes into which the penis was inserted. If the penis became erect — it would be pricked.) By the early part of this century, metal mittens were being sold to deter the evil wanderings of little children's hands as well as an alarm that rang in the parents' bedroom if their child's bed was moving.

For those seeking a more permanent solution to their problem (cages, belts, and metal mittens had to be removed for bathing, leaving the wearer vulnerable to temptation), doctors prescribed other treatments: leeches could be applied to the genital area to suck away blood and relieve the congestion that caused sexual desire; cautery (burning of genital tissue by an electric current or hot iron) was believed to deaden the nerves and decrease feeling and desire. The extreme cures — castration and removal of the clitoris — were most popular in the 1850s and 1860s. American medical journals of the mid-1800s also reported that castration was often a successful treatment of insanity. Our modern belief in immediate circumcision of newborn males is partly a carryover from the Victorian conviction that this surgery discouraged masturbation. Foreskin made the penis difficult to wash; removal of the foreskin thus lessened the amount of time the genitals had to be handled.

Slowly, beginning in the early 1900s, the American medical community began to realize that masturbation caused neither acne nor insanity. A few brave doctors even recommended that females masturbate to relieve hysteria and that males masturbate instead of picking up prostitutes (and veneral disease). As recently as 1930, however, a medical authority continued to warn of the dangers of "onanism" which could lurk in activities like rope-climbing, bicycle riding, or running a sewing machine. He argued that "the path leads to imbecility and premature senility," "loss of spirit, weakness of memory, dependency," "apathy," "languor, irritability, headaches, neuralgias, dimness of vision" and so on (Scott, 1930, p. 424).

As the twentieth century neared its midpoint, medical authorities relied on research instead of speculation and slowly moved away from the notion that masturbation caused illness or insanity. For example, the 1940 edition of the respected pediatric textbook, Holt's *Diseases of Infancy and Childhood,* removed its discussion of masturbation from a chapter called "Functional and Nervous Disorders" and rejected the use of surgery, mechanical restraints, threats, or punishment to deal with this type of sexual behavior. The new edition took the position that masturbation caused no physical harm and defined the problems of masturbation as primarily the worry and guilt that a child may feel.

By the time of the Kinsey reports (1948 and 1953), both public and professional thinking about masturbation had shifted significantly from the early part of the century. But carryovers remained: even today, some people half-jokingly believe that masturbation will lead to "hair on the palm of your hand" or misshapen genitals, while others are convinced masturbation causes sterility, sexual dysfunction, fatigue, or memory loss. Unfortunately, some physicians are no better informed on this subject than the public.

A number of studies in the last fifteen years indicate that attitudes toward masturbation have relaxed considerably compared to earlier times. In 1975, Morton Hunt, reporting on the *Playboy* survey of sexual attitudes and behavior, found that only one out of six men or women between the ages of eighteen and thirty-four felt that masturbation is wrong. In the forty-five-and-over age groups he studied, approximately one-third of the women and men viewed masturbation as wrong. In 1974, Arafat and Cotton, reporting on a questionnaire administered to 230

college males and 205 college females, found that most of those
who did not masturbate refrained because of lack of desire. Of
the nonmasturbators, 32 percent of the males versus 14 percent
of females thought of masturbation as a waste of energy, immoral, and producing cheap feelings. Only a small fraction of
those who did not masturbate cited guilt, inhibition, or religious
beliefs as their reasons.

As we mentioned in chapter 6, many teenagers are still troubled by masturbation and continue to have concerns over the
possible effects of masturbation on their health. This belief was
humorously described by the novelist Philip Roth in *Portnoy's
Complaint:*

> It was at the end of my freshman year of high school — and
> freshman year of masturbating — that I discovered on the un-
> derside of my penis, just where the shaft meets the head, a little
> discolored dot that has since been diagnosed as a freckle. Cancer.
> I had given myself *cancer.* All that pulling and tugging at my own
> flesh, all that friction, had given me an incurable disease. And
> not yet fourteen! In bed at night the tears rolled from my eyes.
> "No!" I sobbed. "I don't want to die! Please — no!" But then,
> because I would very shortly be a corpse anyway, I went ahead as
> usual and jerked off into my sock. (Philip Roth, *Portnoy's Com-
> plaint* [New York: Random House, 1969], p. 19)

Today, masturbation is a more accepted form of sexual behavior than it has ever been in the past, but there are still some
lingering doubts. These doubts center on the following sorts of
issues:

1. *Masturbation is sinful.* This, of course, is a matter of moral
or religious conviction that each person must deal with in his or
her own way. Several studies have found that people who are
strongly religious masturbate less often than those who are not
religious or who have less strongly held religious beliefs.

2. *Masturbation is unnatural.* The logic of this statement is
hard to grasp. If naturalness refers to what occurs in nature,
then this statement is incorrect since masturbation has been observed in many animal species. Furthermore, the numerous reports of masturbation during infancy or early childhood also
refute the notion that it is unnatural.

3. *Masturbation may be a part of growing up, but adults who
masturbate are psychologically immature.* Freudian theory generally

supports this viewpoint, suggesting that adult masturbation is a symptom of psychosexual immaturity except when it is used as a substitute for heterosexual intercourse when no partner is available. Today most authorities believe that adult masturbation is a legitimate type of sexual activity in its own right. The tension between these two positions is created by different theories of psychological maturity; yet no studies show that adults who masturbate are less mature than those who do not. Some experts believe that masturbation is "immature" only when it is *exclusively* and *compulsively* practiced even though other outlets are easily available.

4. *Masturbation tends to be habit-forming and may prevent the development of healthy sexual functioning.* Most sexologists and psychotherapists see this claim as a throwback to nineteenth-century thinking. There is mounting evidence that *lack* of masturbatory experience may lead to psychosexual problems such as impotence or anorgasmia and learning about masturbation is a central feature of many sex therapy programs.

In addition, it is now clear that masturbation may have a number of additional benefits. For example, it can provide a viable (and pleasurable) sexual outlet for people without partners, including the elderly. It can also be beneficial to persons whose sex drives are greater than their partners' at a particular moment. Finally, it is often a gratifying way of releasing tension, thus helping a person relax.

Techniques of Masturbation

People use a wide variety of methods of sexual self-pleasuring. For some, a single approach to masturbation, with only minor variations on the theme, is used over and over again. Other people experiment with a number of different masturbatory methods, sometimes selecting a few favorites for consistent use, and sometimes preferring continued inventiveness to repeat performances.

Masters and Johnson noted that no two women they studied had been observed to masturbate in exactly the same way. Even if the general pattern of physical self-stimulation was similar, the timing, tempo, and style of each individual's approach were unique. While men, in general, have less diversity and more "sameness" in their masturbation patterns, individual embellishments or idiosyncrasies exist here too.

A complete catalogue of the varieties of techniques used in masturbation could fill a full-length book and would probably be boring reading. Thus, we will restrict our discussion to common patterns of masturbation with brief mention of a few interesting variations.

The most common form of female masturbation is to stimulate the clitoris, mons, or vaginal lips by stroking, rubbing, or applying pressure by hand. Clitoral stimulation may be accomplished by rubbing or stroking the clitoral shaft or may result from applying pressure to the mons or tugging on the vaginal lips. The clitoral glans, or tip, is rarely rubbed directly during masturbation because of its sensitivity. If clitoral stimulation is concentrated in one area for a long time, or if intense stimulation is applied to one spot, pleasurable sensations may lessen because the area can become partially numb.

Interestingly, only a few women masturbate by inserting a finger or object into the vagina: Kinsey and his associates found that about 20 percent of women used this approach, and Hite found that only 1.5 percent of women masturbated by vaginal

insertion alone. Similarly, only a small percentage of women routinely include breast stimulation as a part of masturbation (Kinsey's group found that 11 percent of women who masturbate incorporated breast play into self-stimulation).

Most women masturbate while lying on their backs, but some prefer a standing or sitting position. Hite found that 5.5 percent of her sample usually applied clitoral/vulval stimulation lying face down, with a hand placed between their legs. Other women prefer to masturbate by rubbing their genitals against an object such as a pillow, chair, bedpost, or doorknob. (A few years ago, one of our research subjects insisted on bringing her own pillow to our offices for a study involving masturbation since she couldn't masturbate without it.) In variations on this theme, the woman may rub her genitals with fur, velvet, silk, or any soft material.

About 3 percent of women usually masturbate by pressing their thighs together rhythmically, and some women prefer to masturbate using some form of water massage of the genital region or perineum. The use of oils or lotions during masturbation is a fairly common practice but is usually only a secondary part of the masturbatory experience.

As women have become more informed and liberated in their attitudes toward masturbation, they have increasingly used hand-held vibrators to enhance sexual sensations. Almost half of the young women in our studies who masturbate have tried a vibrator at least once, and one-quarter of these women prefer the vibrator over other methods of self-stimulation. Vibrators come in many sizes, shapes, and styles. Some are cylindrical or shaped to anatomically resemble a penis, and others have changeable attachments that permit a variety of stimulatory modes. A few are discreetly designed and are sold in fashionable stores without any hint of their possible sexual utility.

The vibrator's intensity helps many women reach orgasm quickly and easily during self-stimulation. However, the "instant orgasm" of the vibrator-induced variety may create problems. If a woman consistently uses intense mechanical means to achieve orgasm quickly, she will not appreciate the various stages of build-up to her release of sexual tension. Her pleasure may actually diminish, leaving her with a sense of restlessness or frustration.

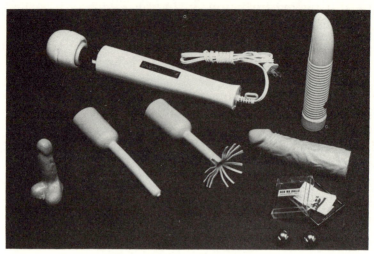

Vibrators, dildos, and ben-wa balls.

The vibrator's intensity (especially when used to induce multiple orgasms) can give ecstatic sexual pleasure to one woman, but it may produce a painful uterine spasm for another. When the vibrator is used to relieve menstrual cramps it may be effective for some women, but it can increase menstrual flow to an uncomfortable level for others or it may even create more severe cramps. If applied to the same favorite area with great frequency, the vibrator may actually deaden the feelings temporarily. There are other highly individual reactions that should be considered before using this instrument freely, such as the presence of genital infections or skin disease.

The sensual high that can be produced with a vibrator is pleasurable, fast, and reliable for those women who are free of emotional or physical problems connected with its use, but the vibrator should be seen objectively *for what it is for each individual:* a toy, a bridge, a crutch — the means to a desired response or a substitute for an absent partner in a time of need. Objectivity is the key, perhaps because overindulgence with a vibrator, like overindulgence in food, can so easily become a way of masking real needs and genuine feelings.

Vibrators are usually applied to the external genitals, but some women prefer to insert the vibrator into the vagina and move it slowly in and out. Other objects may be inserted into the vagina during masturbation, including dildos (artificial penis-shaped objects usually made of rubber), ben wa balls (two marble-sized metal balls that are put inside the vagina and provide stimulation as they roll against each other), and a variety of other objects such as candles, soda bottles, and cucumbers. Among little-used but novel approaches to this type of female masturbation we've encountered are: an electric toothbrush, a dildo made of ice, and a lucky rabbit's foot.

Most males masturbate by rubbing, stroking, or pumping the shaft of the penis with one hand. Scrotal stimulation or direct stimulation of the head of the penis is relatively infrequent, al-

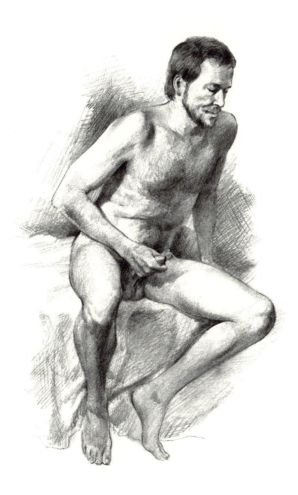

though men sometimes stroke the whole penis in an up-and-down motion. A few men focus their self-stimulation on the raised area (the frenulum) just below the head on the underside of the penis, and a few masturbate primarily by pulling their foreskin back and forth.

The typical male masturbatory episode begins with a comparatively slow, deliberate touch. As sexual arousal increases, the tempo also increases, and by the time of impending orgasm, the stroking motion becomes as rapid as possible. During ejaculation, penile stimulation is variable: some men slow down, other grip the penis firmly, and others stop all stimulation.

During adolescence, some males participate in group masturbation (commonly referred to as "circle-jerks") where there may be a contest to see who can ejaculate most quickly or farthest. Other adolescents (and some older men, too) may be more concerned about catching the ejaculate in tissues, a washcloth, or something else. Because of concern for the "evidence" of masturbation — dried semen leaves a telltale stain — many males masturbate in the bathroom, where a little soap and water will remedy this problem.

A relatively small percentage of males employ some type of friction against an object such as a bed or a pillow as a preferred form of masturbation. Other "hands-off" varieties of male masturbation depend on thrusting the penis into something — the neck of a milk bottle, a cored apple, or modeling clay, for example — in a form of simulated coitus.

Gadgets assisting male masturbation abound and are widely advertised in sex tabloids, magazines, and direct mail catalogues. They include: numerous models of "artificial vaginas" made of rubber or other soft, pliable material; "inflatable life-size dolls" variably equipped with vagina, breasts, open mouth, and anus; suction devices (manually and electrically operated) that promise to deliver the ultimate forms of sexual ecstasy for the male without a partner. These devices can be used with lubricating lotion or cream and may also have features that add vibration or heat to the experience. It should be pointed out that these devices are not always carefully manufactured and may pose some physical risk if they go haywire. Several years ago, some cases of severe penile injury were caused by inserting the penis into a vacuum cleaner hose.

Statistically rare varieties of male masturbation include the two
or three males per thousand who perform oral sex on themselves as well as males who masturbate by inserting objects into
the urethra or the anus. Breast stimulation is rarely included as
a regular feature of male masturbation.

Separating Fact from Fiction

Data about masturbation are a bit tricky to interpret. You may
recall that Kinsey and his colleagues found a wide discrepancy
in the incidence of masturbation between male and female adolescents, but some recent studies suggest that this difference may
be narrowing (see chapter 6). A similar trend may also be occurring in regard to masturbatory behavior in adulthood.

The Kinsey reports stated that 92 percent of the males and 62
percent of the females queried had masturbated at least once in
their lives. More recently, two separate studies came up with very
similar statistics: the *Playboy* survey (Hunt, 1975) found that 94
percent of 982 adult males and 63 percent of 1,044 adult females had masturbated, and Arafat and Cotton's study (1974) of
435 college students found masturbatory experience in 89 percent of males and 61 percent of females.

However, Levin and Levin (1975), summarizing data from a
Redbook questionnaire survey answered by 100,000 women,
found that almost three-quarters of the married women had
masturbated since marriage. Providing additional evidence that
more women seem to have tried masturbation today than in past
decades, Hite reported that 82 percent of her sample of 3,000
women had masturbatory experience.

There are several possible explanations for this rise of female
masturbation:

1. Negative attitudes toward female masturbation seem to
   have softened, although some women continue to feel guilty
   or ashamed of this activity.
2. Women have learned about masturbation at an earlier age
   and in more explicit detail than in the past, primarily
   through the media (books, magazines, movies). As a result,
   masturbation is less likely to be discovered only accidentally.
3. Both men and women have become more aware that sex-

uality is a positive aspect of being female. Acting upon sexual feelings is thus a legitimate activity for women, who are sometimes encouraged to first try masturbation by a sexual partner.

Although the number of females who masturbate seems to be increasing, research data do not suggest that females as a group masturbate as often as males. While there is wide individual variability and some females masturbate several times a day, it appears that males masturbate about twice as often as females.

Kinsey and his colleagues (1948) found that among individuals who masturbate, the average frequency for single sixteen- to twenty-year-old males was 57 times a year, dropping to 42 times a year in the twenty-one- to twenty-five-year-old group. In contrast, the average frequency for single females aged eighteen to twenty-four was about 21 times a year. The *Playboy* survey suggests a contemporary increase in female masturbatory activity: the eighteen- to twenty-four-year-old sample comparable to Kinsey's had an average masturbatory frequency of 37 times a year.

Many people assume that once someone has married, his or her use of masturbation should all but disappear. This generally doesn't happen, though — in the *Playboy* survey, 72 percent of young married husbands masturbated, with an average frequency of about 24 times per year, and 68 percent of young married wives were actively involved in masturbation, averaging approximately 10 times per year. The *Redbook* survey came up with similar findings. Even among older married couples, masturbation continues as a common type of sexual behavior.

Most of the old myths about masturbation causing health problems have now been laid to rest. Physical tolerance for masturbation (or any sexual stimulation) in fact has a built-in safety valve: once the system has reached a point of overload, it temporarily shuts down and does not respond to further stimulation. There is no evidence that masturbation causes physical problems other than the rare cases of genital injury stemming from overly vigorous stimulation.

Nevertheless, a few authorities caution against "excessive" masturbation although they rarely define the term. Men almost always see "excessive" masturbation as somewhat more than their own rate. Very few people we have seen as either research subjects or as patients feel they masturbate too much. Of those who do, the concern is often, "It's excessive because I'm married" or "It's excessive for my age." Rather than count masturbatory episodes, it is probably more useful to consider whether masturbation involves anxiety, conflict, guilt, or an overwhelming compulsiveness. If it does, a person may benefit from professional help, but if masturbation leads to satisfaction and pleasure, it's unlikely to be a problem.

At the opposite end of the spectrum, in the rush to legitimize masturbation, there is often a built-in implication that everyone *should* masturbate. People who have *never* masturbated, while in a statistical minority, should certainly not be made to feel abnormal. People who choose not to masturbate — whether or not they've tried it, whether or not their choice is based on religious conviction, personal preference, or some other consideration — have every right to their decision without any intellectual browbeating by self-proclaimed experts in sexual health. Sexual decisions, in the final analysis, must be personal.

SEXUAL AROUSAL AND SLEEP

We have already mentioned that sexual reflexes function in a rhythmic fashion during sleep (see chapter 3). In addition to having sleep-associated erections or vaginal lubrication, people can experience orgasm during sleep. While it may not be entirely accurate to call this a type of sexual behavior, this form of solitary sex deserves discussion too.

Kinsey and his coworkers found that 83 percent of all males experience nocturnal ejaculation at one time or another, with the highest incidence and frequency of this phenomenon occurring during the late teens. The average frequency of about once a month during this period declines substantially during the twenties, and few men over the age of thirty continue to ejaculate during sleep. Kinsey's group pointed out, however, that several cases of nocturnal ejaculation in older males up to age eighty had been verified.

Nocturnal ejaculation provides a physiologic "safety-valve" for accumulated sexual tension that has not been released in another fashion. Men who have reached high and sustained levels of sexual arousal without ejaculating, no matter how the arousal came about, are thus able to discharge this physiologic tension in a completely natural reflex.

Kinsey's group also found that women can experience orgasm during sleep. They noted: "As with the male, the female is often awakened by the muscular spasms . . . which follow her orgasms." Thirty-seven percent of their sample reported orgasm during sleep by age forty-five, but only about 10 percent of females had such an experience in any given year. Eight percent of their sample had sleep-associated orgasms more than five times per year and only 3 percent averaged more than twice a month.

We have found that almost all women who report orgasms during sleep have previously been orgasmic by other means. A small number of women are distressed by sleep-associated orgasms because they fear they either may have been unknowingly masturbating or are "oversexed." One married woman told us:

> One month I was awakened by orgasms four or five different times. I have a good sex life with my husband, and I hardly ever masturbate, so I couldn't figure out why this was happening to me. I started to think that perhaps I was becoming a nymphomaniac, a person who could never get enough sexual satisfaction. Fortunately, I was able to discuss this with a woman who's a psychiatrist, and she set my mind at ease.

Until more people become aware of the natural occurrence of female orgasms and periodic vaginal lubrication during sleep, it is likely that similar reactions will occur.

Explicitly sexual dreams, like wakeful sexual fantasies, are quite common. Seventy percent of females and nearly 100 percent of males have erotic dreams according to Kinsey and his colleagues. The content of sexual dreams may sometimes be alarming because behavior is depicted that might be objectionable as an actual event. While most people realize that dreams are not equivalent to action, others are distressed because they fear the impulse that the dream represents. Persistently disturbing sexual dreams may in some cases be a sign of an underlying sexual conflict that might benefit from professional counseling.

OTHER FORMS OF SOLITARY SEX

Interest in reading sexually explicit materials or in looking at pictures of sexual acts or organs is hardly new. The erotic art of ancient Greece, India, Africa, and Japan is one indication of the cultural universality of such interest. Throughout the centuries, books about sex have been widely and eagerly read. While the legal aspects of pornography and obscenity are beyond the scope of this discussion, we will look briefly here at the behavioral side of using erotic materials for solitary sexual arousal.

Today, it is practically impossible to find a high-school student in America who has not come across some form of sexually explicit material (erotica). Although there are bookstores, movie theatres, and videotape clubs catering to the "twenty-one or older" crowd, there are ample supplies of so-called soft-core erotica in men's and women's magazines, bestselling novels, advertising campaigns, comic books, and general-release movies to guarantee that anyone remotely interested in viewing such materials can have the opportunity.

There are many reasons why people show an interest in the use of erotica. Erotica provides a source of knowledge and comparative information about sexual behavior. These materials often produce sexual arousal which can be prolonged or abbreviated depending on a person's appetite at a particular moment. Like sexual fantasies, erotica triggers the imagination and so helps people deal with forbidden or frightening areas in a controlled way. Erotica gives people an opportunity to imaginatively rehearse acts that they hope to try or are curious about. Finally, just like Westerns or spy thrillers, erotica can provide a kind of pleasurable recreation or entertainment separate and apart from its sexual turn-on effect.

There seem to be few differences in the sexual arousal induced by words, photographs, or movies. Some people prefer the more vivid, real-life action of cinema, whereas others prefer to let their imaginations expand on a drawing or photograph or find that the printed word offers a greater ease in erotic interest. Such differences are matters of style and preference in just the same way that one person prefers a concert to a movie while someone else likes seeing a play, no matter what the subject matter. In contrast, the content of erotica, rather than its style of presentation, does have a specific effect. People are more likely to be sexually aroused by content to which they relate, rather than by portrayals of sexual acts which they find uncomfortable or offensive.

The sexual arousal that occurs with the use of erotic materials is not simply psychological. Many investigators have noted specific physiologic changes in people who watch erotic pictures or movies, read erotic passages, or listen to tape recordings of erotic stories. Men often experience penile erection while women undergo changes in vaginal blood flow or lubrication. Although one study suggested that viewing a sexually explicit movie elevated circulating testosterone in men, other studies have not found any evidence of hormone changes.

In the past, it was generally assumed that men responded more frequently and powerfully to erotic readings, pictures and films than women did. Research evidence, however, indicates that this is not the case: both sexes respond to erotica in similar ways. To be certain, some females — having been taught that it is not "ladylike" to allow oneself to be intrigued or excited by such materials — avidly avoid any exposure to erotica or do their best to block their own spontaneous responses by an act of will. The same reaction may occur among women who object to pornography on political grounds as exploitive of females. Other females may be more open to the opportunity but have difficulty noticing mild sexual arousal even when physiological changes such as increased vaginal blood flow can be detected. The male, in contrast, usually has more obvious external evidence of his arousal.

Another consideration is that while both males and females have a similar capacity to respond to erotica, the *type* of erotica (content, style, plot) may also be important in determining the response pattern. In the past, it was thought that males tend to be more "object" oriented and respond to stark close-ups of sexual action, while females pay more attention to the style, setting, and mood. Recent studies, however, show that males and females are actually quite similar in what they find erotically arousing.

Some of the time — exactly how often no one knows — people use erotic readings, pictures, or movies to accompany masturbation. This is certainly what springs to mind when we think of a teenager and his collection of *Playboy* centerfolds or a businessman who has slipped out of his office in midafternoon to watch a skin flick at the local theatre. At other times, erotica is used primarily to stimulate sexual desire — heightening the appetite but not providing the main course. Erotic materials are also used to try to turn on one partner, turn someone into a partner, or to otherwise enrich an interpersonal sexual experience. We will not dwell on this aspect here since our discussion is about solitary sex.

How the use of erotica affects behavior is a complicated question that provokes much controversy at present, with no single answer readily apparent. In the United States, President Lyndon Johnson established a special Commission on Obscenity and Pornography in 1968, which reviewed a large body of research over the next two years. The Commission (1970, pp. 28–29) noted:

> When people are exposed to erotic materials, some persons increase masturbatory or coital behavior, a smaller proportion decrease it, but the majority of people report no change in these behaviors. Increases in either of these behaviors are short lived and generally disappear within 48 hours. . . .
>
> In general, established patterns of sexual behavior were found to be very stable and not altered substantially by exposure to erotica.

In addition, in Denmark after hard-core pornography became widely available (and legal) in 1965, the rates of many sex crimes decreased substantially, while studies in America showed that rapists, child molesters, and other sex offenders actually had less exposure to sexually explicit materials during adolescence than other adults. Furthermore, repeated heavy exposure to erotica seems to lead to satiation and boredom rather than changes in sexual behavior. Thus, a number of authorities have concluded that reading pornographic materials or viewing sexually explicit pictures or films doesn't turn people into sexual maniacs or incite men to rape or act in sexually impulsive ways.

There is another aspect to this question, however, since a

number of observers believe that in recent years there has been a considerable increase in the appearance of violence in pornography. Since both a 1960s presidential commission and a more recent task force of the National Institute of Mental Health have found clear evidence linking pictorial portrayals of violence in the media to increased aggressive behavior by observers, there is now much concern about whether this fusion of violence and pornography may have specific negative effects on behavior. In addition to other uncertainties about violent pornography, many people are also concerned with the ways in which pornography debases and "objectifies" women, portrayals that they feel may contribute to sex discrimination by showing women as "mindless" sex objects. Thus, even if such materials do not directly affect behavior, by reinforcing existing stereotypes and prejudices about men and women they may strengthen or even create types of attitudes that are ultimately expressed in behavior.

Violent pornography can be defined as depictions of sex in which force or coercion is used against women. In the past decade, violent pornography has become prominent in films such as *Maniac, Texas Chainsaw Massacre,* and *Tool Box Murders,* and has been shown graphically in issues of *Hustler* and similar men's magazines. Now evidence linking the viewing of such violent pornography to aggression against women is beginning to emerge.

Neil Malamuth and Edward Donnerstein have been in the forefront of researchers studying this relationship. Their work has shown that it is the violence, rather than the sexual content, of such materials that produces negative effects. For instance. Donnerstein recently conducted a set of experiments that showed that exposure to X-rated depictions of sexual violence against women often increases the acceptance of myths about rape (such as the notions that women secretly want to be raped and enjoy the experience) and lead many men to say that they would commit rape if they were certain they wouldn't be caught. In addition, viewing X-rated violent films increases men's aggressive behavior against women in laboratory settings and decreases male sympathy and sensitivity toward rape victims when the subjects are viewing videotapes of simulated rape trials. Specifically, Donnerstein (1983) observed:

Most startling, the men by the last day of viewing graphic violence against women were rating the material as significantly less debasing and degrading to women, more humorous, more enjoyable, and claimed a greater willingness to see this type of film again.

Malamuth's studies have previously shown that hostility toward women predicts rape-related attitudes, motivations, and behaviors. In addition, he has found that men who score high on the likelihood-to-rape scale have more arousal fantasies after exposure to slides and tapes depicting rape than after exposure to mutually consenting coitus. However, some of the research findings are confusing, since "even men who score low on the likelihood-to-rape scale are sometimes highly sexually aroused by portrayals of rape" (Cunningham, 1983), and women themselves are also sometimes highly aroused by eroticized depictions of rape. On the contrary, women who listened to a description of rape that was realistic, emphasizing the victim's fright and pain without any attempt at eroticizing the scenario, generally registered lower genital responses and lower levels of subjective sexual arousal.

Although these studies are thought-provoking, they must be interpreted cautiously at present for several different reasons. Thus far, for example, they have involved relatively small samples and have been conducted primarily in college student populations. Furthermore, these studies have involved experimental methods that lead to somewhat artificial judgments of attitudes and (potential) behaviors; for ethical reasons, a true field-study of the effects of violent pornography has not yet been conducted. Despite these limitations, however, the findings described above are particularly noteworthy because they have been based on the effects of viewing relatively brief amounts of violent pornography. Since loss of sensitivity to violence and even a proclivity toward violent behavior may well be a cumulative effect, studies are now under way examining aggressive behavior and attitudes toward violence after repeated, prolonged exposure.

There are several other sides to the question of the long-range effects of erotica. There is evidence that sexually explicit materials can sometimes help people to overcome sexual problems or can lessen their sexual inhibitions. Sometimes, however, these materials provoke anxiety — particularly when people compare their physical attributes or sexual response patterns to the stars of erotica. These heroes and heroines are not only highly attractive but engage in instantaneous, endless passion (the hallmark of erotica). Some people understandably respond to these images with guilt, embarrassment, or self-doubt.

Given the trend of the last fifteen years toward a greater acceptability and accessibility of sexually explicit items — and the recent boom in uncensored cable TV and home video systems — it seems important to gather more complete data on the effects of erotica. We should not overlook the possibility that use of erotica is sometimes accompanied by problems; neither should we be frightened by old negative attitudes reborn in the guise of modern morality.

CHAPTER THIRTEEN

Heterosexuality

Thirty-six-year-old twice-married, twice-divorced woman in group therapy session: You know, heterosexuality isn't as simple as I'd like it to be.

MANY PEOPLE would probably feel a certain twinge of sympathy for this woman's plight. Although a great majority of our population is heterosexual, being heterosexual is no guarantee that you've found sexual happiness. There are, for example, those nagging questions of sexual technique. There are also certain matters of sexual conduct that have as much to do with personal and cultural values as with anything else. In this chapter, after beginning with a discussion of sexual techniques we will shift our focus to examining certain aspects of heterosexual behavior: premarital, marital, extramarital, and postmarital.

TECHNIQUES OF
HETEROSEXUAL ACTIVITY

There is always a danger that any discussion of sexual technique will sound like a mechanical checklist that implies that good sex is simply a matter of pushing the right buttons at the right time. Fortunately, sex usually involves more than mechanical coupling. It draws upon feelings, moods, desires, and attitudes that are expressed in the physical interaction and that contribute significantly to the quality of the shared experience. At the risk of saying the obvious, there is not just one way of having good sex — sexual technique is, as much as anything else, a matter of communication between partners in which each person conveys to the other a sense of what feels good and what doesn't. As we discuss sexual techniques with some attention to their physical (and practical) details, keep the preceding thoughts firmly in mind.

Noncoital Sex Play

Many people describe all sexual activity between partners other than intercourse by the term "foreplay," which implies that these acts are (or should be) preliminary to intercourse, making intercourse the "main event." However, foreplay is a misleading term because intercourse is not always the focal point of sex; some people prefer other forms of sexual activity instead of coitus. Furthermore, if coitus is first and other sexual acts follow, should these then be called "afterplay"? To avoid such problems, we prefer to discard the term "foreplay" entirely and talk instead about noncoital sex play.

Touching and Being Touched

Touching can be many different things. At one level, it is primarily a wordless way to communicate a willingness, a wish, or a demand to make love. At another level, while touching serves the same communicative purpose, it is valued and enjoyed for its sensual pleasures almost as much as intercourse or orgasm. At still another level, touching is a source of comfort and security — an affirmation of togetherness, commitment, and trust. Touching can also be a mechanical, unemotional way of manipulating another body. In this approach, the essence of sexual interaction seems to be in knowing how to move a hand, where to place a mouth, or when to use a tongue in a joining of separate, almost disembodied anatomical parts. This mechanical kind of touching turns persons into objects, regardless of gender.

Touching another person satisfies the human need not to feel alone, while being touched satisfies the need to be desired as a physical presence. In touching and being touched by a trusted and trusting person, one experiences not only the pleasure of being alive but also the joy of being a sensual creature.

Touching need not involve the hands only. Many varieties of skin-to-skin contact lead to feelings of warmth, tenderness, and closeness. Kissing is a fine example of a touch that can be immensely sensual or more important as a symbol of affection and intimacy. Some people enjoy passionate, almost continuous kissing during sex, while others prefer only an occasional kiss or no mouth-to-mouth kissing at all. Psychiatrist Marc Hollender theorizes that women have a greater need for being held and cuddled than men do, although he emphasizes that this does not mean that sex is less important to women. He does speculate, however, that sometimes a woman's need to be held leads her to participate sexually in exchange for cuddling and affection from her partner.

The act of touching can be unstructured and exploratory, or it can be focused in a more stimulative fashion. While touch as a vehicle for sexual arousal will be discussed in some detail in just a moment, it is also relevant to point out that many people find that touch in the form of a massage — with or without sexual stimulation — permits them to relax and to develop an awareness of their bodies that enhances the quality of a sexual experience.

Touching the Genitals

Many forms of genital stimulation can result in sexual pleasure and arousal. The genital regions in both sexes are highly sensitive to touch, and this sensitivity tends to increase as erotic excitation mounts. A touch that might be unarousing or even uncomfortable to a person who is not sexually excited can be pleasurable or electrifying as physical passion rises; conversely, a touch that is "just right" in the beginning moments of sexual play may be "too little" or "too slow" or otherwise out of sync at a later moment.

During genital touching, many people presume that their partner would like just the same type of stimulation they enjoy. As a result, men often stimulate the clitoris vigorously, mimicking the rapid, forceful stroking typical of male masturbation. In contrast, women are often worried about stroking the penis too vigorously, or touching or squeezing the scrotum too roughly, not wishing to hurt their partner. In addition, men and women often rely on erroneous assumptions about what would or would not turn their partner on; probably the most common example is that many men routinely insert a finger or fingers deeply into the vagina early in genital play although relatively few women find this arousing and some find it distracting or uncomfortable.

These observations underscore the importance of clear communication between sexual partners not only to enhance sexual pleasure but to protect your partner from making you uncomfortable physically or psychologically. One person can't know with any real accuracy what another is feeling or wants at a given moment without some form of communication. Since none of us is an infallible mind-reader, it is helpful to develop open lines of information exchange. But since words may disturb a beautiful mood, nonverbal messages — conveyed by a touch, a move, a look — are often best suited to the occasion unless they don't succeed in getting the message across, in which case words become necessary.

Not only does the type of genital play that a woman prefers vary from one woman to another, the same woman may have different preferences at different times. Many women enjoy

The preferred pattern of genital touching varies considerably from person to person and from time to time.

firm, sustained rubbing of the shaft of the clitoris (as we mentioned in chapter 3, direct stimulation of the tip of the clitoris is frequently uncomfortable), while others prefer clitoral stimulation alternated with caresses of the vaginal lips, the mons, or the perineum. Some women enjoy insertion of a finger into the mouth of the vagina, or gentle, teasing stroking just outside the vaginal opening. Deep vaginal penetration is usually not pleasurable unless a woman is highly aroused, and even then she may get little out of this form of stimulation, permitting it to occur primarily because she feels that it excites her partner. There are wide individual differences in this matter, as the following comment from a twenty-four-year-old woman shows:

> The thing that I like best of all about sex is when I'm really turned on and Tim is finger-fucking me. If he can get three or four fingers crammed inside, I have my biggest and best orgasms.

The tissues of the vulva and vagina may be irritated by too much touching or too much pressure if there is not enough lubrication present. Since the clitoris and the vaginal lips have no lubrication of their own, bringing some lubrication from the vagina to these areas is often helpful. Saliva or artificial lubricants such as K-Y Jelly, hypoallergenic lotions, or baby oil can also be used to reduce friction and to provide another dimension to genital touching.

In the last chapter, we mentioned that many women enjoy using a vibrator during masturbation. A vibrator can be incorporated into partner sex as well, but it is important to talk this through together. Some men feel that the vibrator is a kind of mechanical intruder; others worry that the vibrator is desired only because they can't do the job properly; yet many men are perfectly happy to share in a variety of sexual stimulation that increases their partner's pleasure. Some couples have made their vibrator an integral part of their sex lives, even taking it with them on vacations, while others use it only from time to time.

There are many similarities between male and female sexuality, as we have noted before, and genital touching preferences are no exception. Men do not only want one kind of touch, and there is variability between men and in the same man at different times regarding the genital touches that create pleasure or arousal.

When the penis is not erect, most men prefer a light, playful stroking or caressing of the penis, the inner thighs, and the scrotum. If touching is restricted to the penis (or if penile stimulation is too vigorous) while the penis is flaccid, it is unlikely to be very arousing and may actually be somewhat threatening since the man may begin to worry that he isn't responding swiftly enough. Once erection begins to occur, the firmness of touch applied to the penis in stroking or squeezing motions can be comfortably (and arousingly) increased. One of the most common complaints we have heard from men is that their female partners don't grasp the penis firmly enough once it is erect. Men usually prefer an up-and-down stroking of the penis with the fingers encircling the shaft; direct manual stimulation of the head of the penis may be uncomfortable or irritating.

Some men enjoy having the scrotum gently squeezed or lightly stroked or cupped in their partner's hand, while other men prefer not receiving any direct scrotal stimulation. If the testes are "rubbed the wrong way," it may be quite uncomfortable and a real dampener to sexual feelings. Many men enjoy some form of tactile stimulation focused at the frenulum (the small fold of skin just below the coronal ridge on the underside of the penis), although relatively few women seem to be aware of this fact. Men may also enjoy having saliva, lotion, or oil applied to the penis to enhance their arousal. Care should be taken (in either sex) that the lotion or oil applied to the genitals is not too cold or too hot and that it contains no alcohol, since alcohol tends to irritate the male urethra and female genital tissues.

It may be surprising, but relatively few heterosexual men and women have taken the time to show their partners how they like to have their genitals touched. Of course, this can be accomplished partly in conversation alone ("I really liked how you did that tonight — would you try that again another time?") But conversation sometimes leaves a few doubts ("What does he mean by a firmer touch?"), and a "hands-on" demonstration is often the simplest way of conveying an accurate message. One person can place a hand on top of his or her partner's hand, showing them what they mean by "firm" and "light" or just *where* to stroke, since small differences in positioning may make all the difference in the world.* You can also show your partner exactly what you like by doing it yourself in his or her view. For a variation on this theme, ask them to put their hand on top of yours so they can actually feel the rhythm of the movement.

Oral-Genital Sex

Stimulation of the male genitals by the use of the tongue, lips, and mouth is called "fellatio," and oral stimulation of the female genitals is called "cunnilingus." Fellatio or cunnilingus can be used to induce or heighten sexual arousal or to produce orgasm. Either form of oral-genital sex can be done with one partner stimulating the other individually or with simultaneous reciprocal stimulation (the simultaneous version is sometimes called "69" because the inverted, side-by-side position of the numbers is similar to the position commonly used for this form of sex play).

There are a vast number of techniques and combinations of techniques for oral-genital stimulation that can be pleasurable and arousing. No one way is the "right" way to do it. Licking, sucking, kissing, and nibbling can feel good anywhere on the genitals; the pressure (light, firm, or in between), speed (fast, slow, or changing), and type of motion employed can be varied considerably to attain different effects. The moistness and warmth of oral-genital contact is highly erotic for many people. Some enjoy a teasing, stop-start approach; others prefer a more direct, sustained type of stimulation. Here too, finding out what your partner likes is a matter of open communication.

In cunnilingus, many women are highly aroused by oral stimulation of the clitoris. This can take the form of gentle tongue movements over the shaft and tip of the clitoris, more rapid, focused licking, or sucking the clitoris either gently or in a rougher fashion. During high levels of arousal, a few women enjoy having the clitoris bitten gently. Other techniques some women enjoy are: oral stimulation of the clitoris combined with

---

* This point is readily apparent to anyone who has had an itch on their back that they couldn't reach themselves. Trying to direct someone else to just the right spot — "A little higher . . . to the left . . . now up a little . . . no, back down a little lower and toward the middle . . ." — can be terribly frustrating. Just the same is sometimes true of telling someone in words what feels good sexually.

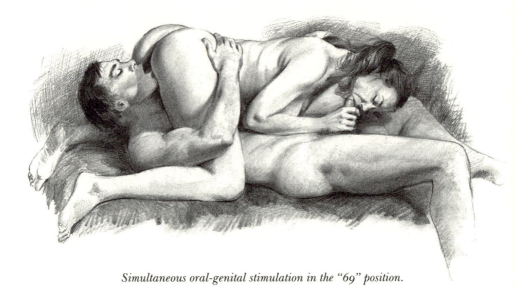

Simultaneous oral-genital stimulation in the "69" position.

manual stimulation of the vagina; oral stimulation of the minor lips (the area just outside the vagina); having the tongue thrust in and out of the vaginal opening; having the clitoris stimulated manually (either by their partner or by themselves) while oral stimulation is directed at other parts of their genitals; and having their partner blow into the vagina or on the clitoris.

In fellatio, methods of stimulation include sucking the glans or shaft of the penis by engulfing it in the mouth, licking various parts of the penis or scrotum, nibbling or kissing anywhere along the genitals. Although fellatio is often referred to as a "blow job," most men don't enjoy a real blowing motion (it's not like playing a saxophone). The frenulum is often particularly sensitive to oral stimulation and most men find that the glans of the penis is also exquisitely sensitive to warm, moist caresses. Many men enjoy having the scrotum lightly stroked during fellatio and the area just beneath the scrotum is often quite sensitive to manual or oral massage.

Some women are uncomfortable with fellatio because they have a sensation of gagging if they take the erect penis into their mouth. This sensation is often due to a reflex response called the "gag reflex," which can be triggered by pressure at the back of the tongue or in the throat; it is a real physiologic event, not an imaginary happening. Even when a woman can comfortably accommodate part of the penis in her mouth, if her partner thrusts in the throes of his own excitation, it may push the penis so far in that the gag reflex takes over. There are two solutions to this problem. First, the woman can grasp the shaft of the penis so that she has full control over the depth of penile penetration into her mouth, preventing sudden thrusts or jabs. Second, the gag reflex can be fairly easily reconditioned in most people by gradually inserting the penis a bit more deeply over a number of occasions until the reflex is minimized, or even practicing by inserting a cylindrical object (or some fingers) into the mouth.

Another difficulty a woman may have with fellatio is not wanting the man to ejaculate in her mouth. A couple can agree in advance that the man will withdraw before ejaculation; alternately, many women have found that with a little experience they can overcome this concern. Some women prefer to rinse the ejaculate out of their mouth promptly because they don't like the taste of semen; others don't mind it much; and still others swallow the ejaculate. While there are no health risks to swallowing semen, it is unlikely that this has beneficial health effects (preventing acne or preserving youthfulness) either.

While many people are enthusiastic about the pleasures of oral-genital sex, others consider it "dirty," perverted, sinful, embarrassing, or simply unappealing. Among those with reservations about this type of sexual activity, many find that with a little effort (and practice) they can easily develop a personal comfort level for oral-genital sex. This comment from a twenty-eight-year-old man illustrates some of the dilemmas:

> At first, when I thought about oral sex on a woman I was scared — scared that I would be turned off by the smell and the flavor, and scared that I wouldn't know how to do it right. I was also into this trip where I didn't think it was a very "manly" thing to do, although I can't really remember where I got that idea. But then I got involved with a beautiful woman who sort of eased me into it, helping me take my time and all. After just a few tries, my fears disappeared and I sort of threw myself into the action.

Although many people have been taught to think of the genitals as unclean, routine bathing or showering that includes carefully washing the genitals with soap and water will ensure cleanliness. From a scientific viewpoint, oral-genital contact is no less hygienic than mouth-to-mouth kissing. The natural secretions of the genitals are relatively clean and each person's genital odors partly reflect the type of foods they eat. Many couples like to shower or bathe together before engaging in oral sex, and some people who consider oral sex the most intimate form of sex will only engage in this activity with a partner to whom they are particularly close. For others, oral sex is a stopping point to prevent the intimacy (and reproductive risk) of coitus, as shown by this comment from a twenty-two-year-old woman:

> While I was in high school I learned that if I gave a guy a good blow job he wouldn't pressure me into screwing. So I stayed a virgin — except for my mouth — for five years of a very active sex life. Everyone was happy!

One last word about oral-genital sex: some people incorrectly think that fellatio or cunnilingus are homosexual acts, even if experienced by heterosexual couples. While many homosexuals engage in oral-genital sex, so do a majority of heterosexual couples. The activity itself is neither homosexual nor heterosexual.

Anal Sex

Stimulation of the anus during sexual activity can be done in several different ways: manually, orally, or by anal intercourse. Although anal sex is sometimes thought of as a strictly "homosexual" activity, a large number of heterosexual couples occasionally incorporate some variety of anal stimulation into their noncoital sexual play.

Anal stimulation can be the primary focus of sexual activity or an accompaniment to other types of stimulation. For instance, many couples sometimes include manual stimulation of the anus (either lightly rubbing the rim or inserting a finger into the anus) during coitus, and others use this technique during oral-genital sex. Anal sex in any of its forms can be highly arousing and lead to male or female orgasm. But many people have strongly negative attitudes toward anal sex, an act which they may regard as being unclean, unnatural, perverted, disgusting, or simply unappealing.

Although anal intercourse can be pleasurable, it can also be a source of discomfort in both a physical and emotional sense. The anal sphincter tightens ordinarily if stimulated, and attempts at penile insertion may be distressing even if done slowly and gently. If the penis is forced into the anus, injury is possible. To minimize risk, it is wisest to use an artificial lubricant liberally and to dilate the anus gently by manual stimulation before attempting insertion.

One final note about anal stimulation. Anything that has been inserted into the anus should not be subsequently put into the vagina unless it has been thoroughly washed. Bacteria that are naturally present in the anus can cause vaginal infections, so moving from anal intercourse (or finger insertion) to vaginal intercourse (or finger insertion) is unwise.

Coital Sex

For many people, the hallmark of heterosexuality is penile-vaginal intercourse, or coitus. Hundreds of marriage manuals have offered instruction in the "how-to's" of coital connection; although many of these books speak of "making love," they usually wind up conveying the message that the best sex is attained by following their mechanical blueprints. As Germaine Greer put it in *The Female Eunuch* (1972, p. 37):

> The implication that there is a statistically ideal fuck which will always result in satisfaction if the right procedures are followed is depressing and misleading. . . . Real satisfaction is not enshrined in a tiny cluster of nerves but in the sexual involvement of the whole person.

We have talked about the individuality of sexual responsivity in many places in this book. Each person is a unique sexual being with personal preferences and idiosyncracies molded by past experiences, current needs, mood, personality, and a host of other variables. Each of us may also find that our body responds in different ways at different times, even if external circumstances are almost the same. When we add a partner into this equation, it's no wonder there is no way of describing the ideal type of coitus. We will discuss selected aspects of techniques of coital sex first in terms of coital positions and then in terms of styles.

Man-on-Top, Face-to-Face

The most common coital position in the United States is with the woman lying on her back, legs spread somewhat apart, with the man lying on top of her. This position, which is sometimes called the "missionary position," offers a relative degree of ease of penile insertion and also permits as much eye contact and kissing as desired. If the woman wishes, she may raise her legs in the air or wrap them around the man's back or shoulders, which causes deeper penetration of the penis in the vagina. This position also gives the best chance of conception, since the semen pools in a position in the vagina closest to the mouth of the cervix.

Despite its popularity, the man-on-top position has some disadvantages. Many women feel "pinned" underneath the weight of their partner and find it difficult to do much pelvic movement. The woman also has little control over the depth of penetration, and if her partner is lying against her body it may be difficult for either person to stimulate the clitoris manually. While the man has maximum freedom of movement, he may find it tiring to support his weight on his elbows and knees and, as his muscles fatigue, he may "tense up" physically. In addition, men tend to have less control over ejaculation in this position than in many others. This position is also apt to be uncomfort-

Man-on-top, face-to-face intercourse position.

able if the man is considerably heavier then the woman or if the woman is in the later stages of pregnancy.

Woman-on-Top, Face-to-Face

Another popular coital position is for the woman to be on top. In this position, the woman either can be sitting up to a degree or can lie down against her partner. In contrast to the man-on-top position, in this position the woman has considerable control over coital movements, thrusting, and tempo. The woman is free to caress her partner's body with her hands, and the man has his hands available for stroking her breasts, genitals, or other body parts. The visual stimulation of this position and the man's freedom from supporting his weight may encourage his stroking. The woman-on-top position is generally the best one to use for a man who wants to gain greater control over ejaculation and is also the position used most often in sex therapy when a woman has difficulty reaching orgasm, since either partner can stimulate the clitoris manually in this position. It is also well-suited to the later stages of pregnancy.

A few drawbacks to the woman-on-top position should be

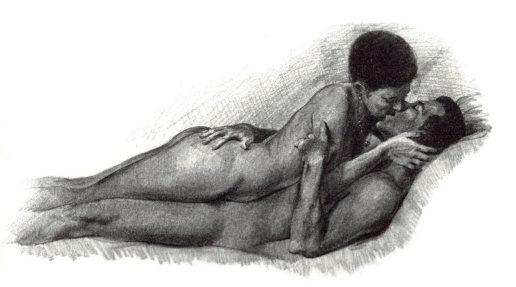

One variation of the woman-on-top, face-to-face intercourse position.

Another variation of the woman-on-top, face-to-face intercourse position.

mentioned. Some couples feel uncomfortable with this position because they believe the man should "always" be on top. Since the woman may seem to be the "aggressor" by being above the man, and since she has greater control over pelvic movements, a few men feel that their masculinity is threatened in this position. At a more practical level, some men find it difficult to engage in active pelvic thrusting in this position.

When the woman-on-top position is used, it is important to insert the penis properly. The woman should never hold the erect penis at a 90-degree angle to the man's body and try to sit down on it, as this can be uncomfortable to either partner. Instead, the penis should be held at a 45- to 60-degree angle (pointed in the direction of the man's head) since this matches the angle of the vagina when the woman is in a forward-leaning position; the woman can then slide back onto the penis. With a little bit of practice, this maneuver becomes simple.

Rear-Entry

In rear-entry positions, the man faces the woman's back and the penis is placed into the vagina from behind. Coitus can be ac-

Rear-entry intercourse, "doggy style" (above) and "spoon position" (below).

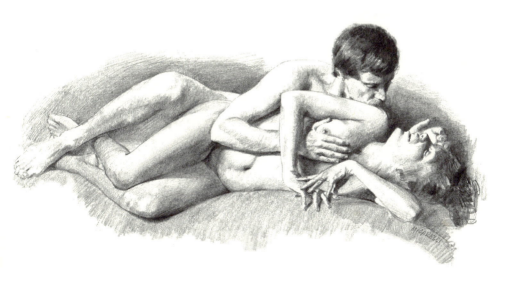

complished with the woman on her hands and knees in the "doggy style"; the woman may lie face down with her hips propped up by a pillow; or the couple may lie on their sides, with the man's front to the woman's back in the "spoon position." Rear-entry can also be done in a sitting or standing position. In most of these positions, the man can usually reach around his partner's body to stimulate her clitoris or breasts, but mouth-to-mouth kissing is difficult and eye contact is lost. Many couples feel that rear-entry coitus is less intimate for these latter reasons, and some couples object to this position because it seems too much like anal intercourse. However, the sensations of thrusting against the buttocks can be pleasurable, and when the woman's legs are close together more stimulation of the penis is possible.

Side-to-Side, Face-to-Face

In this position, the partners are facing each other but lying on their sides. Since neither person is burdened by the other's weight, this is often a relaxed position where a lot of leisurely caressing and cuddling can be included. Both partners have at least one free hand. The primary drawback of this position is

Side-to-side, face-to-face intercourse position.

that inserting the penis in the vagina can be tricky. Thus, many couples begin in a different face-to-face position and then roll onto their sides. There are two other disadvantages in the side-by-side position: first, the penis is a bit more likely to slip out of the vagina than in other positions, and second, there is less mechanical leverage to achieve vigorous pelvic thrusting.

Timing, Tempo, and Other Themes

Above and beyond the countless embellishments in coital positioning that are possible, there are many stylistic variations that affect coital sex. The surroundings chosen for a sexual encounter can range from a bedroom to an automobile to an outdoor setting. People can make love in the dark, with dim lighting, or in broad daylight. Some people like to have background music for sex, others like to use a water bed or lots of pillows, and still others prefer sex in a hot tub or in front of a crackling fire.

Timing and tempo are also important ingredients. Just as a person's appetite for food can be met in a variety of ways — a quick snack, an elegant gourmet meal, or a simple, satisfying steak — sexual appetite can also be served by a "quickie," by a long, leisurely episode of lovemaking, or by a technically straightforward sexual encounter. One version is not always better than another — its quality and enjoyment depend on the needs and responses of the people involved.

Many heterosexual couples approach coitus in a businesslike fashion — sex is always at the same time (usually late at night), with the same "routines," the same position, and sometimes identical dialogue. Although this can lead to boredom for some people, for others it is a perfectly comfortable and satisfying situation — predictability is not always a liability. On the other hand, many couples enjoy a more creative approach to their sex lives, varying not only the positions and stimulatory techniques they use but also the time when sex occurs, the setting, and who plays the most active role.

Although it was the traditional view of our culture that women should be relatively passive sexually and let men initiate, set the tempo, and end the activity (by ejaculating), this pattern has undergone considerable change in the last decade. Our impression from thousands of research and clinical interviews is that many women are now playing a far more active role in sexual interactions in general and in coital sex in particular. Being less bound by old gender role stereotypes means that many women now feel freer to show or tell their partners what they like and do not like sexually. This openness benefits both sexes by improving the chance that the woman will enjoy sex and by relieving the man of the responsibility of having to be the expert who knows just what to do to satisfy his partner.

A few specific examples can illustrate how this works. It is often easier for the woman to insert the penis in the vagina than for the man because: (1) the woman knows when she's "ready"; (2) the woman knows precisely where the opening to the vagina is; and (3) the woman can adjust the angle of insertion to whatever feels most comfortable. If the man tries to control insertion, he is far more likely to have to hunt for the vagina (in contrast, women have little difficulty locating the position of an erect penis!);* he also frequently proceeds to intercourse before his partner is ready. Another example of an active role a woman may take during lovemaking is to communicate what tempo of penile thrusting she prefers. Many men automatically start deep, vigorous thrusting during intercourse, but most women become more stimulated by slower, shallower thrusting, at least in the early stages of coitus.

We cannot conclude this section about sexual techniques without stressing again that sexual enjoyment usually relates less to mechanical proficiency than to how two people relate to one another. This does not mean that good sex requires a meaningful long-term relationship, but it most often depends on effective communications between partners. Working at sex — trying to develop erotic artistry through diligence, practice, and having a single-minded purpose — is more likely to interfere with the spontaneous enjoyment of sexual experiences than to produce a memorable sexual encounter.

---

* A simple experiment can document this point fairly well. In a dark room compare the difference in feeding your partner and feeding yourself. You should have no problem placing food in your own mouth — you know instinctively just where it is — but when you try to guess where your partner's mouth is, you are liable to deliver the food to the chin, the nose, or some other slightly-off-target area. A similar problem can occur when the man tries to always insert the penis — it can lead to some stumbling and fumbling, which can be an annoying distraction.

PERSPECTIVES ON
HETEROSEXUAL BEHAVIOR

Now we will shift our attention from matters of sexual technique to an examination of heterosexual behavior and attitudes toward such behavior. Unfortunately, most of the descriptive information about sexual behaviors and attitudes available from various survey studies has methodological shortcomings. Nevertheless, it is useful to become acquainted with some of the findings of these surveys as long as we recognize that they are all approximations of the actual patterns of behavior they seek to measure.

Premarital Sex

Premarital sex is often talked about as synonymous with premarital intercourse, but people can be sexually active prior to marriage without having coital experience. Another problem with the term "premarital sex" is its implication that marriage is the goal of each and every person in our society.

Statistics about age at first sexual intercourse and experience with other types of sexual behavior during adolescence were given in chapter 8, but a few other facts about premarital intercourse are intriguing. Kinsey and his colleagues shocked many people in 1953 when they reported that half of the women they studied had sexual intercourse before marriage. Among men studied, 68 percent of men with a college education, 85 percent of men with a high-school education, and 98 percent of men with only an elementary school education had premarital coitus.

Two of the more recent sex surveys indicate that, at least among younger women, there has been a sizable increase in premarital coital experience. Hunt's data (1975) showed that 81 percent of married women between the ages of eighteen and twenty-four had premarital intercourse. The *Redbook* survey done in 1975 reported that 90 percent of women under twenty-five said they had premarital intercourse, and among all women in their survey who were married between 1970 and 1973, 89 percent had such experience. While these major differences from the Kinsey findings may reflect some sampling biases (for instance, the 100,000 women who answered the *Redbook* survey were younger, better educated, and more financially secure than the population as a whole), there is no question that premarital intercourse among women is more widespread today that it once was and that the age at first intercourse is declining.

These changes seem to reflect a shift between 1963 and 1975 in attitudes toward premarital sex in America: the percentage of adults who thought that premarital intercourse was always wrong dropped from 80 to 30 percent during this time, as shown by three national surveys. Data from the *Playboy* study, while supporting this finding, also put it in a slightly different perspective — even among eighteen- to twenty-four-year-olds, premarital coitus for a man in circumstances other than a strongly affectionate relationship was thought to be wrong by 29 percent of the men and 53 percent of the women queried. Premarital coitus for a woman in the same circumstances was thought by 44 percent of the men and 71 percent of the women to be wrong.

Thus, it appears that approval of premarital coitus for many teenagers and young adults is still restricted to love relationships or relationships with strong caring and affection. Recreational or casual sex, as opposed to relationship sex, is far less common, although it is written about so extensively that many young adults are convinced that most of their contemporaries have "freer" sexual attitudes than they do. The fact that most young adults see premarital coitus as justified when it occurs in a legitimate, committed relationship but less so when it is purely casual and sensuous shows that older cultural values have not been discarded but have changed.

Not all adolescents or young adults are comfortable with the idea of nonvirginity, although the reasons for their concerns vary considerably. One useful framework for examining the diversity of sexual philosophies among the unmarried has been described by D'Augelli and D'Augelli in 1977. According to these authors, *inexperienced virgins* are individuals who have had little dating experience until college and usually have not thought much about sex; *adamant virgins* are people who firmly believe that intercourse before marriage is improper (and often have a strong religious basis for this belief); *potential nonvirgins* are individuals who have not yet found the right partner or the right situation for coital sex and often seem to have a high fear of pregnancy; *engaged nonvirgins* are those whose coital experience has usually been only with one partner (typically someone they love or care deeply about) and only in the context of a committed relationship; *liberated nonvirgins* are people who have more permissive attitudes toward premarital intercourse and value the physical pleasures of coitus without demanding love as a justification; and *confused nonvirgins* are those who "engage in sex without real understanding for the motivation, the place of sex in their lives, or its effects on them." People in the latter category often use sex as a way of attempting to establish a relationship and after the relationship terminates usually feel ambivalent about having had intercourse.

A number of other notable changes have occurred in premarital sexual behavior in the last three decades. First, relatively few young men today are sexually initiated by prostitutes or have premarital intercourse with prostitutes, although Kinsey and his colleagues found in 1948 that more than one-quarter of college-educated men who had not married by age twenty-five and 54 percent of high-school-educated men had premarital intercourse with a prostitute. Second, use of oral-genital sex among young unmarried men and women has increased dramatically compared to Kinsey's day, with the percentage of people using fellatio more than doubling and the percent trying cunnilingus rising from 14 to 69 percent. Third, there seems to be a greater degree of premarital sexual experimentation in recent years compared to Kinsey's time. People are more willing to try a wider range of coital positions, drugs (especially cocaine and marihuana) to enhance sexual and sensual feelings, and anal intercourse. The trend toward experimentation is also shown by recent evidence that women are having premarital sex with more partners than in the past. Finally, the average frequency of premarital coitus has increased, particularly among women. In 1953, Kinsey and co-workers reported that one-third of twenty-one- to twenty-five-year-old women had premarital intercourse, averaging a frequency of once every three weeks, while Hunt found in 1975 that two-thirds of his eighteen- to twenty-four-year-old sample had premarital coital experience with an average frequency of just over once a week. The percentage of women who are orgasmic with premarital coitus has increased substantially too.

Another change in premarital sex behavior has been the greater acceptance of *cohabitation* — unmarried heterosexual couples living together. The implications of this development for sexual behavior are discussed in chapter 7.

Although the research is not sophisticated enough to draw firm conclusions about the effects of premarital intercourse on marital relationships, some studies have indicated potential problems. Many studies in the 1960s concluded that marriages tend to be rated as more successful when premarital chastity has been maintained. This view, however, was challenged by some who suggested that the findings might only show that the same people who are ashamed or afraid of having premarital intercourse might also be ashamed or afraid of admitting that their marriages were unhappy. One large-scale survey in 1969 found that people with extensive premarital sexual experience tend to have numerous extramarital affairs and also noted that the more premarital coital partners a person had, the greater tendency to have less happy marriages.

We suspect that better designed studies might show more positive effects of premarital sex — for example, better sexual communication and less sexual inhibition. We also want to point out that couples who break up premaritally because of lack of sexual compatibility (a group that has not yet been studied at all) may be actually doing each other a favor.

Marital Sex

More than 90 percent of Americans have married by their early thirties. By ages forty-five to fifty-four, only about 4 percent of women and 6 percent of men in the United States have never been married. In this section, we will examine patterns of marital sex behavior in terms of coital frequency, orgasm, techniques, and satisfaction in traditional marriages and then look briefly at the sexual implications of alternative marriage styles.

Frequency of Marital Coitus

The average American couple has intercourse two or three times per week in their twenties and thirties, after which the frequency slowly declines. Past age fifty, coital frequency averages once a week or less. "Average" frequencies do not tell the whole story, however. In each study on the frequency of marital coitus, a broad range of individual variation was found. Some young married couples have no coital activity, while other couples have intercourse several times a day. While, in general, coital frequency rates decline with the length of marriage, some couples clearly develop better sexual relationships as time goes by and may be more coitally active after fifteen or twenty years of marriage then they were early in their marital lives.

There does not seem to be any correlation between the frequency of coitus in marriage and educational or occupational status. The effects of strong religious feelings on marital sexuality are uncertain. Kinsey and his co-workers found that less religious husbands had higher marital coital rates but such an effect was not found for women. According to the *Playboy* survey, churchgoing wives had lower rates of marital coitus than less religious women. In the *Redbook* survey, the frequency of marital coitus was virtually identical for strongly religious and nonreligious women, although strongly religious wives were more likely to report that they always play an active role in lovemaking.

Orgasm

Although Kinsey and his colleagues reported that men reached orgasm in essentially all of their marital coitus, Hunt found that 8 percent of husbands in the forty-five and older age group did not have orgasm anywhere from occasionally to most of the time, and 7 percent of the twenty-four- to forty-four-year-old married men did not have orgasm in at least one-quarter of their coital experiences. These statistics do not mean that a man who doesn't ejaculate during intercourse necessarily has a problem: in some cases, the man may have already ejaculated in another type of sexual play, and in other cases, the man simply may not feel the need.

Kinsey and his co-workers reported that marital coital orgasm had not been experienced by one-quarter of women after one year of marriage, but by the end of twenty years of marriage, this figure had fallen to 11 percent. In the same study, it was found that 45 percent of wives reported orgasms in 90 to 100 percent of coital experiences in their fifteenth year of marriage. Hunt found that 53 percent of wives reported coital orgasms "all

or almost all the time" in marriages of fifteen years median duration. In the *Redbook* survey, 63 percent of wives reported coital
orgasms all or almost all of the time, and only 7 percent had
never experienced coital orgasm.

For some married couples, lack of orgasmic responsiveness in
coitus can be a major problem, as shown by this comment from
a thirty-nine-year-old woman:

> Although I feel like I am pretty open sexually, willing to try
> things and fairly easily turned on, after eighteen years of mar-
> riage I still haven't had an orgasm with intercourse. Years ago, it
> didn't bother me all that much, but for the last five years I've felt
> worse and worse about it. My husband and I have tried every-
> thing possible, all to no avail. After awhile, I started to lose inter-
> est in sex and to get very depressed. Our marriage has been badly
> hurt by this problem and I don't know what to do. [This woman
> and her husband subsequently entered sex therapy with us and
> were able to correct the lack of coital orgasm quite easily. Some-
> times with long-standing problems, it is easier to get objective
> help from others than to try to "do-it-yourself."]

For other couples, lack of female coital orgasm poses no problem or threat, as a thirty-one-year-old woman explained:

> I've never had an orgasm during intercourse, but that's never
> really bothered me. In fact, I think my husband was more con-
> cerned about it than I was. I enjoy the closeness and touching of
> sex, and I get aroused, too — it's just the orgasm that's missing.
> If I need the release of an orgasm I can masturbate, but orgasm
> just isn't the most important part of sex for me.

Sexual Techniques

Marital sexual techniques have undergone some remarkable
changes in the last forty years if data from available surveys are
to be believed. For example, while oral-genital sex was avoided
by large numbers of the married men and women who participated in the Kinsey studies, a majority of married people today
include fellatio and cunnilingus in their sexual repertoires.
Eighty-seven percent of wives in the *Redbook* survey reported
using cunnilingus often or occasionally, and 85 percent reported
fellatio with a similar frequency. Similarly, although Kinsey and

his colleagues did not report statistics on anal intercourse in marriage in their initial volumes, an updated report on Kinsey's unpublished data showed that less than 9 percent of married respondents had experience with this type of sexual act. In contrast, nearly one-quarter of the married women in Hunt's study and 43 percent of the wives in the *Redbook* survey had tried anal intercourse. (Most women indicated, however, that this was the least liked type of sexual activity.)

The other major changes in marital sexual technique include an increased amount of time in sexual play and the use of a wider variety of coital positions. While Kinsey and his coworkers found that precoital play was often limited to a few kisses among those with a grade-school education and averaged about twelve minutes among the college-educated, Hunt found that at both educational levels the time spent in sexual play before intercourse averaged fifteen minutes. Similarly, while Kinsey believed that three-fourths of married men ejaculated within two minutes after inserting the penis in the vagina, Hunt's data indicate that marital coitus now lasts an average of ten minutes. This change may reflect a greater awareness by married men and women today that women are more likely to enjoy sex more and to be orgasmic if intercourse is unhurried. Greater use of a diversity of coital positions in marriages today may be due to an increased awareness that the "missionary position" often limits the sexual options available to the woman.

Sexual Satisfaction

How does the quality of a couple's marriage contribute to their sexual satisfaction? And how does the nature of a married couple's sex life relate to their overall marital satisfaction? Since relatively little research has been done on these subjects, these questions can only be partially answered.

The original Kinsey reports did not evaluate the relationship between sexual adjustment and marital happiness, but a later analysis of the data led Gebhard to conclude that women were much more likely to be orgasmic in "very happy" marriages (rated by self-report) than in other marriages. It is not clear whether the very happy marriages led to a better sexual climate or vice versa.

Data from the *Playboy* survey indicate that a large majority of married men and women who described their marital sex as very pleasurable rated their marriages as very close. Among those who rated their marital sex as displeasing or lacking in pleasure, almost no one rated their marriage as very close and just a few thought their marriages were fairly close. Interestingly, in people rating their marriages as not close or fairly distant, 59 percent of the men still reported that marital sex was "mostly" or "very" pleasurable, as compared to 38 percent of the women. Apparently more men than women are able to enjoy sex when their marriage is strained.

The *Redbook* survey found a strong correlation between the frequency of intercourse and satisfaction with marital sex for women. It also noted that 81 percent of the women who were orgasmic all or most of the time in marital coitus rated the sexual side of their marriage as good or very good, while only 52 percent of women who were occasionally orgasmic and 29 percent of women who were never orgasmic (or who did not know if they were) felt that their sexual relationship was good. A strong correlation was also found between a wife's ability to communicate her sexual desires and feelings to her husband and the quality of marital sex.

In 1983, Blumstein and Schwartz also found a correlation between the frequency of marital sex and sexual satisfaction. Among those having sex three times a week or more, 89 percent of husbands and wives were satisfied with the quality of their sex lives, while only 53 percent of those having sex between once a week and once a month were satisfied. In marriages with a sexual frequency of once a month or less, the satisfaction rate dropped to only 32 percent. Another important factor that was linked to sexual satisfaction was equality in initiating or refusing sex. Eighty percent of husbands and wives who reported that sexual initiation in their relationship was equal were satisfied with the quality of their sex lives, versus 66 percent of those who said that sexual initiation was one-sided. Similarly, 80 percent of married men and women who reported being able to refuse sex on an equal basis were satisfied with the quality of their sex lives, compared to 58 percent of husbands and 61 percent of wives who reported that sexual refusal was not equal.

Intriguingly, Blumstein and Schwartz's data indicate that heterosexual men who receive and give oral sex are happier with their sex lives and with their relationships than those who do not. Performing and receiving oral sex, however, does not seem to be linked to sexual satisfaction for heterosexual women, possibly because many of them see both fellatio and cunnilingus as a form of submissiveness or degradation. Instead, it appears that intercourse is more essential to sexual satisfaction for heterosexual women than for heterosexual men.

Research surveys are not the only way of elucidating the relationships between sexual and marital satisfaction, and attempts to categorize just what satisfaction is may be misleading. One person may be satisfied with a marriage that provides economic security and freedom from major conflict, while another person considers that kind of marriage as tolerable but unsatisfactory. Likewise, one person may judge how satisfying a sexual relationship is primarily in terms of coital frequency, while to someone else a variety of forms of sex play and the quality of both partners' sexual responses may be the basis for making a judgment. If both spouses largely agree on what they require for sexual and marital happiness, the chances seem to be higher that they will be able to attain it. On the other hand, many marriage counselors and sex therapists can attest to the fact that many marriages are troubled by some major form of sexual distress. It has also been suggested that the high prevalence of extramarital sex may also relate to a lack of marital sexual satisfaction since, as the sociologist Ira Reiss says, if other things are equal, "the more marital sexual satisfaction, the less the desire for extramarital relationships." We will return to this point shortly.

Alternative Marriage Styles

Although marriage may be defined by legal criteria that strictly regulate who the spouses can be, there are many types of marriage that differ in various ways from the traditional one man—one woman (monogamous) marriage. In the following paragraphs, we will briefly describe certain aspects of alternative (nonlegal) marriage styles as they relate to heterosexual behavior.

Triads, or three-partner marriages, can consist of one man and two women or two men with one woman, with the latter form being less frequent. Triads tend to form more naturally than larger-group marriages, which usually require some design and planning, and seem more likely to be motivated by existing friendship or love. Triads are also likely to be formed on the basis of pre-existing sexual relationships, although sex is often not the primary reason behind this choice of alternative marriage style. Here's how one twenty-six-year-old woman involved in a triadic marriage with another man and woman explained it:

> Most people think our marriage is just an excuse for freewheeling sex, but that's not true at all. If sex was what we wanted, why go to all the trouble of marrying in a threesome, which makes almost everyone think we're weird? The real reason we married is that we all love each other, and we want to stay together, sharing our lives. Sex is just one part of that sharing, although it's a part we usually enjoy.

Group marriages consist of four or more partners who genuinely regard themselves as married to each other although they do not all have a legally recognized relationship. To date, there has only been a limited amount of research on group marriage, with the work of Constantine and Constantine (1973) providing the best overview of the subject. The participants in group marriages tend to be in their twenties or early thirties and many of them have previously been legally married and have one or two children. The most common form of group marriage in America consists of two men and two women; in some widely publicized cases one man has had five to ten wives. Individuals in group marriages tend to have liberal nonreligious backgrounds, an unconventional outlook, and a high need for change and autonomy, but they enter group marriages for just the same reasons that two people might get married. The potential advantages of group marriage are both economic and emotional: it can satisfy a need for variety (including sexual variety) and it can present opportunities for interpersonal growth. On the other hand, most group marriages do not survive very long: sexual jealousy is fairly common and arranging sexual attention and combinations can be difficult. As Murstein (1978, p. 132) notes, "In a six-person group there are nine possible heterosexual pairings for the night and fifty-seven different relationships of from two to six people." In general, men appear to have greater difficulty adapting to group marriages than women do.

A few other types of nontraditional marriage styles will be discussed in the next section, while celibacy as a choice for heterosexual marriage will be considered in chapter 15.

Extramarital Sex

Extramarital sex can be defined as any form of sexual activity between a married person and someone other than his or her spouse. Although considered sinful, criminal, or immoral through most of the history of the Western world — and sometimes punished by whippings, fines, brandings, and even death — the practice has both persisted and become fairly commonplace.

In 1948 Kinsey and his co-workers estimated that half of all married males in their sample had had extramarital coitus (this figure was approximate because many men were reluctant to discuss this area openly). Kinsey's group also noted in 1953 that by age forty, 26 percent of married women had had extramarital sex. In 1975 Hunt believed that these estimates were still accurate, but the *Redbook* survey found that among thirty-five- to thirty-nine-year-old wives, 38 percent had extramarital sexual experience. In a survey done by *Cosmopolitan* in 1980, half of married women eighteen to thirty-four years old and 69.2 percent of married women thirty-five or older had had extramarital sexual activity. While some other recent surveys have also reported higher figures for extramarital sex — for example, Hite found that two-thirds of the married men in her sample had extramarital experience, and a 1983 *Playgirl* survey found a 43 percent incidence in married women — these numbers contrast sharply with the data of Blumstein and Schwartz which show that only 26 percent of husbands and 21 percent of the wives they studied had any form of extramarital sex involvement. Unfortunately, much of the research on extramarital sex is methodologically limited and the reasons that motivate people to engage in extramarital sex are only poorly understood at the present time.

The Extramarital Experience

While most people believe that extramarital sex is always wrong, there has been a traditional double standard that rationalizes to a certain degree extramarital sex for men while more strongly condemning it for women. In some European countries, for instance, having a mistress is regarded as a privilege of wealthy married men. Similarly, many societies permit female heterosexual prostitution (a sort of temporary "rent-a-mistress") as a means of providing for a presumed male need for sexual variety while protecting against destruction of the bonds of matrimonial relationships. Since prostitution is seen by most people as extramarital sex at a purely physical level, it is not as threatening as other types of extramarital sex that carry the risk of emotional involvement that might eventually lead to the breakup of a marriage.

Perhaps for this same reason, some married people feel most comfortable with extramarital sex that is purely and directly aimed at physical pleasure: the "one-night stand." The background circumstances that lead to this brief encounter of extramarital sex vary tremendously — a lonely businessman on an overnight trip, a bored housewife who feels the walls are closing in on her at home and makes the rounds of the city bars, men or women trying to prove to themselves that they're not really getting old or that they still have sex appeal — the list could go on indefinitely. The extramarital one-night stand is often so impersonal that the participants don't know each other's names. No commitment is made and none is intended; it's really sex with "no strings attached."

How do people react to having such an experience? Some people find exactly what they're looking for: a release of pent-up tension, a means of getting even with their spouse for something, a way of satisfying their curiosity, a change of pace from their ordinary sexual diet, or a temporary form of escape. Others find the experience to be empty, guilt-provoking, awkward, or frightening. These comments illustrate the types of reactions we've frequently heard:

> *A thirty-one-year-old woman:* I'd been married for almost ten years and had always been faithful, but I kept wondering what it would be like to have sex with someone else. One night I was out with some friends, and we met a few guys who bought us drinks and talked with us awhile. One of them was real good-looking and flirting with me, and I sort of flirted back. We went off to a motel for three or four hours, and it was beautiful sex, fantastic sex, just like in a novel. But that was the end of it, and it just felt good to know that I'd had the experience. I never told my husband and I don't plan to.

> *A thirty-six-year-old man:* My wife and I have very old-fashioned values and we both took our marital vows seriously, meaning no screwing around with anyone else. I never worried about it too much, since I wasn't the type to be running around anyway. But one night when I was working late a secretary asked me for a ride home, and then invited me in for coffee. Well, she was just divorced a few months, and she wanted more than coffee, and I was perfectly happy to oblige. But it was a stupid thing to do — not much fun, and lots of guilt about it afterwards — and I don't think I'd do it again.

The extramarital affair contrasts sharply with the one-night stand in that there is a continued sexual relationship over time. The affair may be relatively short-lived (lasting just a few weeks) or may go on for years. A popular Bernard Slade play, *Same Time Next Year,* tells the story of an extramarital affair that occurred once a year on the same weekend for decades. An extramarital affair can be mainly for sex or it can blossom into a relationship on its own with sex playing a relatively minor part and companionship and conversation being more important.

Affairs are probably less frequent than one-night stands for both practical and personal reasons. Unless a person's spouse knows about, and approves of, a continued extramarital liaison (a statistically unlikely possibility at present), the partners in the affair must create time to be together, find a place to meet (and preserve their anonymity), and explain their absence to their spouses.* For these reasons many affairs involve secretive (and hurried) get-togethers during the day, since it may be easier to get an hour or two away from work than a similar time away from home at night.

At the personal level, many people feel that having an affair is like "playing with dynamite." Although it may be sexually satisfying and emotionally fulfilling, they are cautious about letting the affair take over their lives or pose a threat to their marriage. Many people who have had affairs state that they loved their spouse during the time of their extramarital involvement and did not want to jeopardize that relationship.

Since thorough research is lacking, it is difficult to say how frequently extramarital sex contributes to divorce. Clearly, some men and women are so upset at discovery of their spouse's infidelity that this leads to a broken marriage. On the other hand, it is not likely that many solid marriages break up for this reason alone. It is also impossible to know how often an affair occurred because the marriage was already weak or troubled or how often an affair was simply "the straw that broke the camel's back." The experience of many marriage counselors indicates that it isn't so much the notion of a spouse having sex outside the marriage that causes the rift (we've never encountered a marriage that broke up because a man visited a prostitute, for instance) as the idea of an emotional involvement with someone else along with the sex. In other words, for some couples, extramarital sex poses

---

* Extramarital sex between a married man and a single woman or a married woman and a single man is likely to offer greater convenience than an affair between two married persons. They have a place to go (the single partner's residence) and less of a scheduling problem, since the single partner is not as likely to be tied down by home and family responsibilities. But these types of affairs can create their own brand of problems: the single person may press the married one to divorce and remarry, for example.

a threat to the intimacy of marriage in both a physical and psychological sense. For other couples, extramarital involvements are no source of concern — they judge their marriage primarily in terms of how they interact together and how adequately their relationship meets their needs.

The *Redbook* survey uncovered another interesting aspect of extramarital sex. Married women who work outside the home seem to have a higher likelihood of extramarital sex: 27 percent of full-time housewives have had extramarital sex, but 47 percent of wives who work part-time or full-time outside the home have had extramarital intercourse. It is not clear whether these statistics mean that women who work outside the home have more liberal attitudes to begin with, or have extramarital sex more often because of greater opportunity, or if some other explanation applies.

Although extramarital sex offers excitement, variety, and the thrill of the forbidden, the available data show that the overall pleasure of extramarital sex is somewhat lower for men and women in general than their overall sexual pleasure in their marriages. While extramarital sex may release sexual inhibitions for some people and help them to become fully responsive for the first time in their life, it may lead other people to sexual problems.

Consensual Extramarital Sex

When two spouses agree that one or both of them is free to engage in extramarital sexual activity, there may be many reasons behind the decision. They may be looking for a way to preserve their personal freedom, hoping to improve the quality of their marriage, seeking to live by a particular personal philosophy, or simply trying to add variety and excitement to their lives.

Both partners may be interested in extramarital sex and one person may agree to let the other do as he or she pleases without any intention of participating themselves. For one couple, talking about the extramarital experience in vivid detail may be a source of turn-on, while another couple decides that extramarital sex is okay for them only if they don't discuss it together.

Consensual (i.e., with consent) extramarital sex can take many forms. A decade ago, Nena and George O'Neill wrote a bestselling book called *Open Marriage,* in which they suggested that traditional marriages often presented few options for choice or change. Their "open marriage" concept — as opposed to the "closed," or traditional, marriage — envisioned a flexible relationship in which both spouses were committed to their own and their partner's fulfillment and growth. The O'Neills stressed that an open marriage involved a willingness to negotiate change and to discard the expectations of a closed marrriage, particularly the idea that one partner is able to meet all of the other's needs (emotional, social, economic, intellectual, and sexual). The O'Neills emphasized role equality and flexibility between spouses, but many people presumed that their book was primarily an endorsement of consensual extramarital sex as a growth experience. However, the O'Neills later wrote:

> While some benefits were noted, it was observed that by and large these [extramarital sex] experiences did not occur in a context where the marital partners were developing their primary marriage relationship sufficiently for this activity to count as a growth experience. Frequently it obscured relationship problems, became an avenue of escape, and intensified conflicts. (O'Neill and O'Neill, 1977, p. 293)

Nevertheless, the open marriage approach to discarding the notion of sexual exclusivity and restrictiveness seems to work well for some couples. It should be noted, however, that an open marriage does not need to involve outside sexual relationships.

Swinging is another form of consensual extramarital sex. Here, married couples exchange partners with other couples, with all parties agreeing to the arrangement. Most surveys indicate that only 2 to 4 percent of married couples have ever engaged in swinging, and less than half of these couples have done it on a regular basis.

The husband usually first brings up the idea of swinging, and the wife's reaction is typically one of shock. Only a few couples who discuss the idea ever attempt to put it into reality, and even here many get cold feet and don't go through with their plans at the last minute.

Although a couple will occasionally be introduced to swinging by people they already know, this is generally an "underground" activity. The interested couple usually must turn to swingers' magazines or sexually explicit newspapers to either answer an ad or place one themselves. Here are several examples of ads from swingers' publications:

> *Attractive Couples.* He, mid-30s, muscular, virile. She, mid-20s, shapely, sexy. Desire open-minded couples under 35 for relaxed get-togethers.

> *Sensual Black Couple,* male 32 and hung, female 28 and 36-24-36. Into partying and fun. Photos required for reply.

> *Super Attractive Couple* — she bi, 28 beautiful, natural redhead; he, 30, 6'1", 175 lbs., handsome and good build. Seeking extremely attractive couples for friendship and fun. Both educated, sincere, gentle, with sense of humor. Photo, phone a must.

Swingers may either get together for a two-couple "party" or may meet in groups with many couples. There are usually drinks served and pornographic movies may be shown to get people in the mood. The sexual activity may take place entirely in heterosexual twosomes, with each couple retreating behind a closed bedroom door ("closed swinging") or, if the door is left open, it's a signal for anyone who wishes to come in and join the fun. While "open swinging" commonly involves two women having sex together, male homosexual contact is less frequent and in some groups of swingers it is entirely barred. Several factors account for female homosexual relationships among swingers: most men are hardly matches for sexually aroused multiorgasmic women. While the men rest, they often experience sexual restimulation if they watch women make love to each other. Presumably the sight of other men making love to their spouses (while they are not similarly occupied) can be too ego-threatening, while the sight of two women making love usually is not.

Variations on the swinging theme include group sex in private clubs such as the now defunct Sandstone in California or in clubs that charge admission, such as Plato's Retreat in New York City. Swinging can also be an activity with a prominent social side: groups of swingers occasionally vacation together, organize picnics, or go for a day at the beach.

Different studies have defined the backgrounds and personalities of swingers in different ways. In some studies, swingers appeared to be conservative, traditional, and religious. In others they were portrayed as liberal, nonreligious (or even antireligious), and antiestablishment. All studies agree that swingers as a group tend to have more premarital experience, more premarital partners, and more (and earlier) interest in sex than "nonswingers." Although there is no evidence that swingers have abnormal personalities, two studies found that swingers are more likely to have had counseling or psychotherapy than nonswingers.

The positive side of swinging includes having a shared activity that truly eliminates the double standard. Couples may feel pleased by having a variety of sexual partners without deceiving each other; some couples find that their own sex lives are improved. But the other side of the coin bears examining too, especially since most couples withdraw from swinging after brief experimentation. They may find that swinging has led to jealousy, feelings of inadequacy, guilt, rejection, and even to sexual dysfunction and divorce.

Nonmarital Sex

Despite the popularity of marriage in our society, there are still many people who remain single by choice or by lack of marital opportunity. Current statistics indicate, for example, that almost 2 million Americans are now living in cohabitation relationships. Others become single, after once having been married, by divorce or death of a spouse. Although largely neglected by researchers, these individuals also have sexual needs on which they act in various ways.

The Never-Married Single

Although in the past it was often presumed that a man or woman who reached age thirty without marrying was flawed in one way or another, today it is clear that many people choose to be single as a creative option in their lives. The choice may be a rejection of the restrictions and responsibilities of marriage or it may be based on other factors, such as economic independence. Increasing numbers of women are placing career objectives ahead of marriage as a goal today, and men are also reconsidering whether being married is really the way they want to live.

Some of these people decide to marry after a long period of singlehood, while others remain single throughout their lives.

Relatively few people choose to be single to preserve a particular sexual life-style, but the image of the "swinging singles scene" — freely accessible sex with little or no interpersonal responsibility beyond the requirements of the moment — has been burned into the public mind by the media. While singles bars continue to flourish in many locales, many people dislike the impersonal nature of sexual shopping around, as these two comments reveal:

> *A twenty-eight-year-old woman:* I spent much of last year making the rounds at the singles bars looking for fun. After a while, the "lines" and faces all blurred together — every guy claimed he was a doctor, a lawyer, or a corporate vice president — and the deceptiveness got to be too much. Even the sex wasn't very good . . . it certainly wasn't worth the agonies of trying to find someone to make it with.

> *A thirty-four-year-old man:* I tried the dating bars for a month or so, but I couldn't really bring myself to continue. Sure there was lots of sex available — by 11:00 at night a lot of gals got desperate — but it was all a phony scene, one that had no real humanity to it.

On the other hand, some people find the freedom and variety of the singles bars exciting and fulfilling.

> *A twenty-nine-year-old woman:* I'm a lawyer, and all day long I've got to think, to use my brain. Sometimes at night I just like to throw away that identity and go out for a good time. If I meet an attractive guy, I invite myself to his place — that way I don't have to worry about throwing *him* out, I can get up and leave when I'm ready. And sometimes it just feels good to make contact with someone else, with no commitments or obligations.

The singles bars are populated by several different groups of people. The under-twenty-five singles are generally in a "premarital" stage. Most of the over-thirties are divorced people and married men or women on the prowl for extramarital sex. The twenty-five-to-thirty-year-olds include both the not-interested-in-marriage-now group, the younger divorced crowd, and some people looking for the perfect mate.

Other nonmarried singles choose different approaches to their sexual needs, meeting prospective partners at work, through their families, or in other social settings. In recent years, computer dating services have become a popular means of trying to meet someone with similar interests where a sexual opportunity may develop. Some nonmarried people form relatively exclusive long-term sexual relationships in which neither partner has marriage as an objective. Others maintain nonexclusive long-term sexual relationships with a small number of partners whom they see on a rotating basis. Others avoid lasting relationships in favor of a series of one-night stands, and others choose celibacy as best for themselves.

One infrequent but interesting aspect of nonmarital singlehood is *communal living,* where a group of people pool their economic resources and personal energies into a cooperative venture of sharing that sometimes includes sexual sharing but always includes some degree of emotional sharing and support. Communal living can be seen as a form of group cohabitation, and communes are certainly not new — in the mid-1800s hundreds of communes existed, some of which included rather free and open sex as a central theme. Some communes are organized on the basis of political or religious beliefs, while others are far less structured and seem to exist mainly on the philosophy of "do your own thing." Most communes have relatively short life spans, perhaps because sexual jealousy and possessiveness crop up fairly frequently.

The Previously Married Single

People who have been married and then become single through divorce or death of a spouse remarry at extraordinarily high rates, as we mentioned in chapter 7. Most divorced men and women become sexually active within a year following their divorce, although older divorced people are somewhat slower in this regard than those under age forty.

Since almost half of all marriages end in divorce, there is less stigma toward divorce now than in the past. The divorced woman is no longer viewed as "used goods" by men, and people are not regarded as being "failures" if their marriages don't last forever. The person who becomes divorced, however, may find it somewhat difficult at first to adjust to the idea of nonmarital sex, to work out the specific details of how to meet people, and to handle the logistics of sexual activity (where to go, what to tell the kids, and so on).

Widowed men and women sometimes choose to abstain from sexual activity after their spouse's death but a large majority of widowers and 43 percent of widows engage in postmarital coitus. The widow over sixty is sometimes handicapped by a lack of available male partners. Interestingly, the widower at any age is likely to have a wider selection of partners, as women outlive men and men are relatively free to choose from women of all ages. Widows, conversely, are expected to choose partners close to their own age or older. As previously stated, many people think that the elderly have no sexual needs or feelings, but this is far from true: sexual interest is often maintained and sometimes improves with age. Among divorced or widowed persons, it is common to hear that postmarital sex is more pleasurable and fulfilling than in their prior marital experience.

CHAPTER FOURTEEN

Homosexuality and Bisexuality

IN all known societies heterosexual behavior is the preferred pattern of most people most of the time; however, homosexual behavior has existed throughout history, and in many societies homosexuality in certain forms is accepted or even expected. In Western society, the subject provokes strong reactions from many people. Homosexuality has been described, for example, as "loathsome and disgusting," "grossly repugnant," "degenerate," "foul," and "immoral, indecent, lewd, and obscene." These statements were not made in a public opinion poll: they can each be found in judicial decisions from important court cases in the 1970s. Similarly, Charles Socarides, a psychiatrist writing in the *Journal of the American Medical Association* in 1970, called homosexuality "a dread dysfunction, malignant in character, which has risen to epidemic proportions."

Despite such strongly negative attitudes toward homosexuality, the last decade also provided a remarkable set of counterbalancing events. In 1974 the American Psychiatric Association officially decided that homosexuality was not an illness. Increasing numbers of courts began to uphold the civil rights of homosexuals on the basic premise that discriminating against people on the basis of their sexual preferences was illegal. Homosexual men and women in many walks of life — sports, science, government, the arts — publicly announced their sexual orientation and began using slogans such as "gay is beautiful" to raise the consciousness of the country and combat stereotypes. Homosexuals in the United States have also recently emerged with substantial political power. In early 1984, for example, a top aide to

Democratic presidential-hopeful Walter Mondale said that gays "may be the most important new force in American politics."

In this chapter, after first offering some definitions of homosexuality and bisexuality and briefly examining their history, we will discuss a number of complex and often controversial issues relating to these sexual orientations.

DEFINING TERMS

The word "homosexual" comes from the Greek root "homo," meaning "same," although the word itself was not coined until the late nineteenth century. It can be used either as an adjective (as in a homosexual act, a homosexual bar) or as a noun that describes men or women who have a preferential sexual attraction to people of their same sex over a significant period of time. While most homosexuals engage in overt sexual activity with members of the same sex and generally do not find themselves particularly attracted sexually to people of the opposite sex, neither of these two conditions is required to fit the definition we have offered. It is clear that a person with no sexual experience whatever may still consider himself or herself homosexual; also, many homosexuals are able to be aroused by heterosexual partners or heterosexual fantasies.

Bisexuals, in contrast, are men or women who are sexually attracted to people of either sex. Usually, but not always, the bisexual has had overt sexual activity with partners of both sexes.

The best statistics available on the numbers of homosexuals in the U.S. are from the Kinsey reports, which found that 10 percent of white American males were more or less exculsively homosexual for at least three years of their lives between ages sixteen and fifty-five; 4 percent were exclusively homosexual on a lifelong basis. Thirty-seven percent of the white male population had in adolescence or adulthood at least one homosexual experience which led to orgasm. In their female sample, Kinsey and his associates found that by age forty, 19 percent had experienced some same-sex erotic contact, but only 2 or 3 percent of women were mostly or exclusively homosexual on a lifelong basis.

Many homosexuals today prefer the term "gay" as a synonym for homosexual, believing either that it sounds less harsh or judgmental or that it makes a social-political statement of pride in this sexual orientation. Other homosexuals feel uncomfortable with the word "gay" and reject its use completely. To provide appropriate background for a contemporary look at homosexuality and bisexuality, we will first look back in history.

HISTORICAL PERSPECTIVES

Homosexuality was clearly condemned in the earliest Jewish tradition. In the Bible we are told: "And if a man lie with mankind, as with womankind, both of them have committed abomination: they shall surely be put to death; their blood shall be upon them" (Leviticus 20:13).

Yet in ancient Greece, homosexuality and bisexuality in certain forms were widely accepted as natural in all segments of society. Plato's *Symposium* praised the virtues of male homosexuality and suggested that pairs of homosexual lovers would make the best soldiers. Many of the Greek mythological gods and heroes, such as Zeus, Hercules, Poseidon, and Achilles, were linked with homosexual behavior. Although some Greek literature and art portrayed sexual relations between two women or two adult men, most of the homosexual relations seemed to occur between grown men and young adolescent boys. Clearly, most Greek men married; yet homosexual activity was not seen as shameful or sinful.

In the early days of the Roman Empire, homosexuality was apparently unregulated by law, and homosexual behavior was common. Marriages between two men or between two women were legal and accepted among the upper classes, and several emperors, including Nero, reportedly were married to men.

Although most historians who have written on the subject suggest that Christianity more or less from its beginnings strongly condemned and persecuted homosexuality, a new study states that this was not the case at all. In a book called *Christianity, Social Tolerance, and Homosexuality,* John Boswell, a historian, argues that for many centuries Catholic Europe showed no hostility to homosexuality.

The primary ammunition for the Church's position against homosexuality came from the writings of Saints Augustine and Thomas Aquinas, who both suggested that any sexual acts that could not lead to conception were unnatural and therefore sinful. Using this line of reasoning, the Church became a potent force in the regulation (and punishment) of sexual behavior. While some homosexuals were mildly rebuked and given prayer as penitence, others were tortured or burned at the stake.

In the Middle Ages, accusations of homosexuality became one of the weapons of the Inquisition, whose dedicated investigators rarely failed to extract an appropriate "confession" from their suspect, whether guilty or not. These "confessions" were used to portray purported homosexuals not only as sexually deviant but as heretic and treasonous.

The negative attitudes toward homosexuality that stemmed from religious beliefs dominated Western thought until the medical view of sexuality began to emerge in the eighteenth and nineteenth centuries. This was hardly a sign of progress, however, since the medical view simply substituted the word "illness" for "sin." For example, Krafft-Ebing's *Psychopathia Sexualis* linked homosexuality to genetic flaws and a predisposing weakness of the nervous system. By the start of the twentieth century, it was generally agreed that homosexuality was an illness with which a person was born.

Although there is still a lively scientific debate on the origins of homosexuality, it appears that tolerance for homosexuality is once again on the upswing. In 1957 the *Wolfenden Report* in England recommended that laws against any form of private sexual behavior between consenting adults be repealed. In 1969 a few nights of summertime demonstrations on Christopher Street in Manhattan's Greenwich Village protesting police raids on a homosexual bar marked the beginning of an era of gay political activism that expanded into a full-fledged gay rights movement. By 1979 thirty-nine American cities, towns, and counties had enacted ordinances that banned discrimination against homosexuals in housing and jobs. The new visibility of the gay community has activated political and social opposition from many quarters, including the so-called Moral Majority, and tensions exist not only between heterosexuals and homosexuals, but within the homosexual community itself.

In his book, *The Homosexualization of America*, Dennis Altman pointed out:

> No longer sinners, criminals, perverts, neurotics, or deviants, homosexuals are being slowly redefined in less value-laden terms as practitioners of an alternative lifestyle, members of a new community. In a self-proclaimed pluralistic society like the United States, this is probably the most effective way to win tolerance, if not acceptance.

However, Altman's view, which was undoubtedly correct at the beginning of the 1980s, is rapidly being undermined by a new development that may ultimately disrupt much of the social progress made by gays in the last two decades. This new development comes in the form of a deadly disease known as AIDS (for "acquired immune deficiency syndrome," discussed in detail in chapter 20), which predominantly affects homosexual males and which has alarmed much of the country because of its mysterious nature, apparent sexual transmission, and high fatality rate. If the number of AIDS cases continues to grow at the alarming pace seen from 1980 to 1985, it may give people an excuse once again to view homosexuality as a form of disease and provide ammunition to those who want to discriminate against gays.

THEORIES ON THE ORIGINS OF HOMOSEXUALITY

Why do some people become homosexual? Is it a lifelong condition over which a person has no control? Is it an entirely voluntary choice, consciously and deliberately made at a certain phase in life? Is it mainly a response to the role models a child is exposed to at home or at school?

Each of these questions has important implications for political, legal, and religious interpretations of homosexuality, but unfortunately, the basic problem is that no one really knows what "causes" heterosexuality either. It may simply be that so far the wrong research questions have been asked. Nevertheless, it is useful to understand some of the viewpoints on the origins of homosexuality.

Many homosexuals claim that their sexual orientation is the result of biological forces over which they have no control or choice. Several types of evidence have been examined to see if this might be so. In 1952, an investigator named Kallman reported findings that supported the earlier viewpoint that homosexuality was a genetic condition. This study examined the sexual orientations of sets of identical and fraternal male twins where one twin was homosexual. Its underlying assumption was that since both twins were exposed to the same prenatal and postnatal environments, a genetic cause for homosexuality would show up as a high *concordance rate* among identical twins, since they have identical genes; that is, both twins would be homosexual rather than one being homosexual and one heterosexual. A lower concordance rate would be expected among fraternal twins, since their genetic makeup is different. Astonishingly, Kallman found 100 percent concordance in identical twins for male homosexuality and only 12 percent concordance in fraternal twins. This finding is astonishing because very few phenomena in biological research show 100 percent outcomes or matching. Subsequent studies have failed to replicate these results, and the genetic theory of homosexuality has been generally discarded today.

Instead, recent research has led many to speculate on the possibility of hormonal factors causing or predisposing to homosexuality. First, it has been well documented that prenatal hormone treatments of various types can lead to male or female homosexual behavior patterns in several different animal species. Second, some scattered findings show that prenatal sex hormone excess or deficiency in humans may be associated with homosexuality. For example, some preliminary studies of human females with the adrenogenital syndrome — a prenatal androgen excess — indicate that these individuals may be more likely to develop a lesbian orientation. Similarly, there are a few reports of homosexuality in men with Klinefelter's syndrome, which is usually marked by a prenatal androgen deficiency, although the statistics are far from comprehensive. Third, a great deal of attention has focused on a comparison of hormone levels in adult homosexuals and heterosexuals. While several studies have found either lower testosterone or higher estrogen in homosexual men, and one study found higher blood testosterone in lesbians

than in heterosexual women, other studies have failed to replicate these findings.

This body of research has major limitations. For example,
treating adult homosexuals with sex hormones does not alter
their sexual orientation in any way. The experimental animal
models of homosexuality do not appear to be a good parallel to
homosexuality in humans. The relatively rare instances of prenatal hormone excess or deficiency linked to homosexuality in
humans may be special cases without much relevance to sexual
development in general. And the conflicting reports on the sex
hormone status of adult homosexuals leave many questions un-
answered. Most notably, there may possibly be many "types" of
homosexuality (and heterosexuality) which — until discovered
— will confound attempts at pinpointing the biological influences on sexual orientation.

Despite the interest in possible hormone mechanisms in the
origin of homosexuality, no serious scientist today suggests that
a simple cause-effect relationship applies. Instead, the possibility
that prenatal hormones may influence brain development in
ways that could predispose individuals to certain adult patterns
of sexual behavior is being considered.

Psychological theories about the origins of homosexuality can
be traced back to Freud, who believed that homosexuality was
an outgrowth of an innate bisexual predisposition in all people.
Under ordinary circumstances, the psychosexual development
of the child would proceed smoothly along a heterosexual
course. Under certain circumstances, however, such as improper
resolution of the Oedipal complex, normal development might
be arrested in an "immature" stage, resulting in adult homosexuality. Furthermore, since Freud thought that all people have
latent homosexual tendencies, he believed that under certain
conditions — such as continuing castration anxiety in males —
overt homosexual behavior might occur for the first time in
adulthood.

Freud's views on homosexuality are difficult to pin down, because he wrote relatively little on the subject. Although psychoanalysis seems to have reinforced the notion of homosexuality
as a form of mental illness, Freud took a fairly neutral stance on
the subject in a letter to the mother of a homosexual son:

Homosexuality is assuredly no advantage, but it is nothing to be ashamed of, no vice, no degradation, it cannot be classified as an illness; we consider it to be a variation of the sexual development. Many highly respected individuals of ancient and modern times have been homosexuals, several of the greatest men among them (Plato, Michelangelo, Leonardo da Vinci, etc.). It is a great injustice to persecute homosexuality as a crime and cruelty, too. (Historical Notes . . . , *American Journal of Psychiatry,* 1951)

While it is not clear whether Freud was simply trying to reassure this distraught mother or genuinely held to the sentiments shown in this letter, it is clear that many later psychoanalysts took positions strongly opposed to homosexuality.

Since Freud had suggested that disordered parent-child relations might lead to homosexuality, in 1962 the psychoanalyst Irving Bieber and his colleagues evaluated the family backgrounds of 106 homosexual and 100 heterosexual men seen as patients. They found that many of the homosexual men they saw had overprotective, dominant mothers and weak or passive fathers, whereas this family constellation was infrequently seen in their heterosexual subjects. Bieber discarded the Freudian notion of psychic bisexuality and suggested that homosexuality results from fears of heterosexual interactions.

Further research on this apparent "cause" of male homosexuality has had mixed results. In 1965, Bene found that homosexual men had relatively poorer relationships with their fathers than did heterosexual men, and described their fathers as "ineffective," but there was no indication of maternal overprotection. In 1966, however, Greenblatt found that fathers of homosexual men were good, generous, dominant, and underprotective, while mothers were free of excessive protectiveness or dominance. More recently, Siegelman reported that for groups of heterosexuals and homosexuals who were well adjusted psychologically, there were no apparent differences in family relationships. Likewise, Bell, Weinberg, and Hammersmith found no support in 1981 for Bieber's theory in their important study of homosexuality, which is probably the most extensive one done to date.

Reviewing such findings, a former American Psychiatric Association president, Judd Marmor, suggests that although there seems to be "a reasonable amount of evidence that boys exposed to this kind of family background have a greater than average likelihood of becoming homosexual," not all people who have this background become homosexual. As Marmor also notes:

> Homosexuals can also come from families with distant or hostile mothers and overly close fathers, from families with ambivalent relationships with older brothers, from homes with absent mothers, absent fathers, idealized fathers, and from a variety of broken homes.

Pursuing this same line of research, Wolff found that among one hundred lesbians compared to heterosexual women, the most prominent parental characteristics were a rejecting or indifferent mother and a distant or absent father. Thus, she believed that female homosexuality arises from a girl's receiving inadequate love from her mother — leading her to continually seek such love from other women — combined with her poor relationship with her father that prevented her from learning to relate to men. To these observations we must add that many homosexuals come from perfectly well-adjusted family backgrounds. Unfortunately, many parents assume the blame for a child turning out gay and agonize over "What did we do wrong?" The present evidence simply does not show that homosexuality only or usually results from improper parenting.

A relatively newer direction of research into the causation of homosexuality has been largely based on principles of behavioral psychology. In this viewpoint, sexuality is considered to be primarily a learned phenomenon. Proponents of this view believe that the psychological conditioning associated with the reinforcement or punishment of early sexual behavior (and sexual thoughts and feelings) largely controls the process of sexual orientation. Thus, people's early sexual experiences may steer them toward homosexual behavior by pleasurable, gratifying same-sex encounters, or by unpleasant, dissatisfying, or frightening heterosexual experiences.

Sexual fantasies can also be conditioned. A positive sexual encounter with a homosexual partner can become the raw material for fantasy during masturbation, which is positively reinforced when it is followed by orgasm. In addition, a variety of other factors may influence a person's early sexual conditioning.

Currently, attention has focused on children who show atypical gender-role behavior ("sissy" boys and "tomboyish" girls) who are thought by some researchers to have a greater likelihood of becoming homosexual.

The behavioral view also suggests why some heterosexuals change their sexual orientation to homosexuality in adulthood. According to Feldman and MacCulloch, if a person has unpleasant heterosexual experiences combined with rewarding homosexual encounters, there may be a gradual shift in the homosexual direction. Although some homosexuals who have "switched" after an earlier period of heterosexual life do not fit this picture exactly, it is common to find many who do. The observation that some female rape victims shift to lesbianism also supports this viewpoint.

In summary, we should point out again that there is no firm agreement about what "causes" homosexuality or heterosexuality. When discussing a few of the best-known theories, we cautioned in almost every instance that the data are not complete and that current thinking must be left open, subject to better studies and research questions. Several of these theories may be correct and may account for a certain percentage of homosexuals in our society. Yet some years from now, all these ideas may also appear terribly foolish and outdated. Although we now believe that homosexuality is primarily a by-product of postnatal events, we are open to the possibility that prenatal programming may yet be proven to play an important role.

THE PSYCHOLOGICAL ADJUSTMENT OF HOMOSEXUALS

Through most of the last hundred years, the prevailing notion was that homosexuality was an illness. Recalling Krafft-Ebing's belief that homosexuality resulted from hereditary defects and the psychoanalytic notion that homosexuality resulted from incapacitating fears of castration, it is not difficult to see how this viewpoint was so logically accepted.

Further "proof" of the psychological maladjustment of homosexuals was provided by a fair number of enterprising scientists. Some who conducted studies of homosexuals in prisons concluded, not surprisingly, that these individuals were less emotionally healthy than heterosexuals chosen from everyday life. By the 1950s, the trend was to select homosexuals being seen by psychiatrists, ignoring the basic sampling error this strategy introduces: since people usually go to psychiatrists because of emotional difficulties, it was almost preordained that this type of sampling would lead to the (unwarranted) conclusion that homosexuals are mentally ill.

Fortunately, a more sophisticated line of research was undertaken by the psychologist Evelyn Hooker in the late 1950s. Hooker selected a group of sixty men — thirty homosexuals and thirty heterosexuals — who were neither psychiatric patients nor prisoners and who were matched for age, education, and IQ. She gave them all personality tests and obtained detailed information about their life histories, and then had a group of expert psychologists evaluate the tests without knowing which belonged to the homosexual and which to the heterosexual men. The results showed that the "raters" could not distinguish between the two groups, providing the first objective indication that homosexuality is not necessarily a form of psychological maladjustment.

In an important series of studies, the psychiatrists Marcel Saghir and Eli Robins extended the work begun by Hooker. They not only chose nonpatient populations but also decided to compare male and female homosexuals to unmarried heterosexuals, since the rate of certain psychiatric illnesses is higher in single persons. Their overall conclusion was straightforward: the majority of homosexuals studied were well-adjusted, productive people with no signs of psychiatric illness. Relatively few differences were observed between the homosexual and heterosexual groups, although an increased rate of alcoholism was found in lesbian subjects.

For the last fifteen years, many research studies have evaluated the performance of homosexuals and heterosexuals on a variety of psychological tests. A recent review of data from dozens of these studies concluded that there are no psychological tests that can distinguish between homosexuals and heterosexuals and there is no evidence of higher rates of emotional instability or psychiatric illness among homosexuals than among heterosexuals.

These results do not imply that homosexuals are *always* emotionally healthy, any more than similar results could prove that heterosexuals never get depressed or become anxious. But the underlying fact is that homosexuality, by itself, is not a form of mental illness nor is it typically associated with other signs of mental illness.

HOMOPHOBIA

The hostility and fear that many people have toward homosexuality is called *homophobia*. The origins of homophobia are just as uncertain as the origins of homosexuality, but some psychologists believe that it is partly a defense that people use to insulate themselves from something that strikes too close to home. Thus, in cases of brutal beatings or murders of homosexuals — perhaps the ultimate expression of homophobia — the motivation may be partly to stamp out any inherent homosexual impulses that may lurk in the attacker's heart.

Homophobia is expressed in many ways in our society. Homosexuals are ridiculed by jokes or by derogatory terms like "fairy," "faggot," or "queer." Parents are fearful that any "feminine" interest shown by a boy may lead to a homosexual life, so they are quick to provide footballs, toy guns, and model airplanes even if their son is not very interested in those items. Police use a variety of techniques to apprehend men engaging in homosexual acts, including entrapment (having a policeman pose as a willing homosexual partner) and use of hidden cameras in public restrooms; if these forms of law enforcement were to be directed at heterosexual activity, the hue and cry would be deafening!

Homosexuals have been banned from the military and the ministry (although not in all church denominations). Homosexuals have been denied housing, jobs, and bank loans. To many people, the homosexual is seen as a sick or even "contagious" individual who may "infect" others and so propel them to a life of homosexual depravity. This view was suggested by an attorney who stated that most male homosexuals find sex partners by walking up to a man standing at a bathroom urinal and grabbing hold of the man's penis (*Gay Liberation* v. *University of Missouri,* 1977).

In light of such homophobia, it is not surprising that much of the straight world thinks homosexuals should be treated to convert them to heterosexuality. While such treatment is often possible with homosexuals who are highly motivated and desire such a change, as described by Hatterer, Masters and Johnson, and Marmor, it is unwarranted and unethical for homosexuals who have no wish to *be* converted.

TECHNIQUES
OF HOMOSEXUAL AROUSAL

The sexual techniques used by homosexual men and women generally mirror those of heterosexual partners, but homosexuals seem to be more willing to experiment and to be more attentive to style. This is partly because heterosexuality is more subject to convention and partly because heterosexual variations are regarded as "abnormal" by many people; since *no* techniques of homosexual stimulation are socially approved, there are fewer automatic restrictions that limit the behavior of homosexual partners.

The physical approaches to erotic arousal used by homosexual men and women must be viewed in context, since there are wide differences in the styles and techniques employed depending on time, place, and circumstance. Although individuality is primary, a few general observations are useful: sexual openness is likely to be greater in a committed relationship rather than in anonymous sex; a leisurely, relaxed approach to sex depends on privacy and environmental comfort; and much homosexual interaction is a matter of mutual agreement and doing what feels good at the moment rather than following a preordained script.

One other fact is pertinent to this section. We have previously shown that the physiological responses of homosexual men and women are no different from those of their heterosexual counterparts. Laboratory observation of the sexual responses of 94 homosexual men and 82 homosexual women in more than 1,200 sexual response cycles involving masturbation, partner manipulation, or oral-genital sex showed that more than 99 percent of these cycles resulted in orgasm. This percentage almost exactly matches the earlier observations of heterosexual men and women studied under identical conditions. Of course, equivalent physiology does not suggest equivalence in all other ways, but it is helpful to realize that the ways in which our sexual responses work are not controlled by our individual sexual orientation, whether it is homosexual or heterosexual.

Lesbian Sex

Kinsey and his co-workers found that two-thirds of homosexual women had orgasms in 90 to 100 percent of their lesbian contacts, while only two-fifths of women in their fifth year of marital intercourse had this high a rate of orgasmic frequency. They also found that among women with extensive lesbian experience, 98 percent had used genital touching, 97 percent had manually stimulated the breasts, 85 percent had orally stimulated the breasts, 78 percent had experience with cunnilingus, and 56 percent had used a genital apposition technique (rubbing the genitals together).

Manual stimulation of the genitals is the most widespread and frequent form of lesbian sex. Saghir and Robins found that all of the fifty-seven lesbians in their study had used this technique, while in 1978 Bell and Weinberg noted that approximately 80 percent of their sample had used it in the past year, with more than 40 percent reporting a frequency of once a week or more.

In contrast to married heterosexual couples — where the "action-oriented" man seems to hurriedly reach for the breasts or move directly to genital stimulation — committed lesbian partners usually share full body contact, with holding, kissing, and general caressing for some while before they make a specific approach to breast or genital touching. Furthermore, when a committed lesbian couple begins breast play, the two women usually give it lengthier and more detailed attention than heterosexual couples. Masters and Johnson noted that for committed lesbian couples:

The full breast always was stimulated manually and orally with particular concentration focused on the nipples. Interestingly, almost scrupulous care was taken by the stimulator to spend an equal amount of time with each breast. As much as 10 minutes were sometimes spent in intermittent breast stimulation before genital play was introduced.

The lesbian caressing her partner's breasts seems to do so with more attention to her partner's responses, while men often approach heterosexual breast stimulation more for their own arousal than for their partner's pleasure. This concern is also shown by the fact that lesbian lovers realize that breast touching can be painful just before a period, while many men seem oblivious to this fact.

During genital stimulation in lesbian couples, the clitoris is rarely approached first, in contrast to the pattern shown in marital sex, where direct clitoral stimulation was the first form of genital contact in about half of the observed episodes. Besides starting with more relaxed genital play, lesbians do not usually insert a finger deeply into the vagina. When vaginal stimulation occurred, it was usually in the form of play around or just inside the mouth of the vagina. In spite of the common belief that lesbians usually use a dildo or object inserted in the vagina to simulate heterosexual intercourse, a distinct minority of lesbians employ this technique.

Two patterns of genital play are most common in lesbian encounters: (1) a prolonged, nondemanding approach of repeatedly bringing the partner to a high level of arousal which is then allowed to recede, in a "teasing" pattern; and (2) stimulation involving more continuity and rapidly increasing intensity until orgasm is reached. Although these two approaches are sometimes combined, most lesbian couples seem to prefer and consistently use one or the other.

Cunnilingus is the preferred sexual technique among lesbians for reaching orgasm. Lesbians generally are more effective in stimulating their partners via oral-genital sex than heterosexual men and usually involve themselves with more inventiveness and less restraint than heterosexual couples. This is probably because a woman is more apt to know what feels good to another woman on the basis of her own personal experiences; lesbian lovers may also be less embarrassed about genital tastes and odors than their heterosexual counterparts. In keeping with these points, it was noted that committed lesbian couples characteristically used a leisurely, less demanding approach to cunnilingus than married heterosexuals. While lesbians who have oral sex most frequently seem happiest with their sex lives and their relationships, about one-quarter of lesbians say they rarely or never use this form of stimulation.

Body rubbing techniques involving total body contact and specific genital-to-genital rubbings are also enjoyed by some lesbians but seem to be a less important source of attaining orgasm. Relatively few lesbians use techniques of anal stimulation.

A recent survey that included 772 lesbian couples and 3,547 married heterosexual couples found that lesbians had genital sexual activity considerably less often than married heterosexuals. However, another study that involved individuals rather than couples found that lesbians had sex more often than heterosexual women and also had more frequent orgasms, a greater number of partners, and a higher degree of sexual satisfaction.

Gay Male Sex

Like their lesbian counterparts, male homosexuals in committed couples tend to take their time in whatever form of sexual interaction they are involved in instead of hurrying along in a goal-oriented effort as many married heterosexual couples do. Male homosexuals also tend to deliberately move more slowly through excitement and to linger at the plateau stage of arousal, using more freeflowing, inventive styles of sexual play than married heterosexuals. This general description has its exceptions. Some committed homosexual couples are completely goal-oriented and push ahead with the sexual action at a frenzied pace, while some married heterosexuals enjoy a much more leisurely, unstructured pattern of sex. But the overall contrast between the two groups is striking.

In the initial stages of sexual interaction, most committed homosexual male couples begin with a generalized approach of hugging, caressing, or kissing. Nipple stimulation — either manually or orally — is frequently incorporated into the early touching, almost invariably leading to erection for the man being stimulated. (Interestingly, few wives stimulate their husbands' nipples as part of sexual play.)

A "teasing" pattern of genital play is frequently used in gay male sex. This often involves selective attention to the frenulum of the penis and the use of a variety of touches or caresses to enhance erotic arousal. Many of the men in committed homosexual relationships studied by Masters and Johnson said that they stimulated their partners the way they liked to be stimulated. Others said that they had discussed genital stimulation techniques directly with their partners and had learned in this fashion what was most pleasing. In contrast, very few heterosexual married couples have specifically discussed the man's preferences in techniques of penile stimulation, and the "teasing" approach is infrequently employed.

There are no major differences in the techniques used for fellatio between homosexual or heterosexual couples, presuming that there are equivalent amounts of experience with this type of sexual play. Fellatio seems to be the most common form of gay male sexual activity, with more than 90 percent having experience in giving and receiving such stimulation.

Anal intercourse is another common male homosexual practice. Saghir and Robins found that 93 percent of the homosexual men they studied had experienced anal intercourse with a male partner. However, Bell and Weinberg found that 22 percent of the gay white males in their study had not performed anal intercourse in the preceding year, and the frequency of anal intercourse was considerably less than for fellatio.

Some authorities have suggested that male homosexuals be classified as "active" or "passive" depending on whether they prefer to be the "insertor" or the "insertee" in anal intercourse. The fact is, most gay men who participate in anal sex enjoy both roles; other gay men find the idea of anal sex discomforting or repulsive.

Recently, another form of anal sex has become popular in certain gay communities. This practice, known as "fisting" or "handballing," involves the insertion of the hand into the rectum (usually after prior cleansing with an enema) followed by movement geared at producing sexual stimulation. While this type of sexual activity is not unique to homosexual males, having been reported in both heterosexuals and lesbians, it appears to be predominantly a practice of the gay male community. Devotees of this form of sexual stimulation, which is often combined with the use of illicit drugs, point out that it requires considerable trust, slowness, and gentleness, and sometimes describe it as an ecstatic, transcendent type of experience. There are substantial health risks associated with "fisting," including the risk of damaging the anus or rectum and the risk of contracting hepatitis B.

PERSPECTIVES ON HOMOSEXUAL BEHAVIOR

Although it is just as foolish to suppose that all homosexuals are alike as it would be to imagine that all vegetarians behave the same way, dress similarly, and fit a general personality profile, many people in our society have made just such an assumption. The resulting stereotypes tell us that we can identify a homosexual by appearance (e.g., the limp-wristed, lisping, mincing male or the short-haired, "butch" female), by profession (the male hairdresser or interior decorator), by personality (maladjusted, overly emotional and impulsive), and by life-style (unmarried men or women over thirty are "suspect"; if they live with another person of the same sex, they are *doubly* suspect).

These stereotypes are largely inaccurate. While it is true that a small number of homosexual men carry themselves in an effeminate way, this group amounts to no more than 15 percent of the homosexual population. There are also many men who are totally *heterosexual* who speak in an effeminate voice or whose other mannerisms appear effeminate. The number of gay women who have a "masculine" appearance is also very small, and a woman who looks "masculine" is not necessarily homosexual. Similarly, there are homosexual doctors, lawyers, truck drivers, professional athletes, and politicians as well as gay hairdressers or designers. No occupational group is purely heterosexual or homosexual.

There is no evidence that most homosexuals are emotionally maladjusted — a fact that is particularly remarkable when one considers the antihomosexual prejudices of our society. Finally, there is no such thing as a "homosexual life-style" that would accurately describe how most gays live. This is not surprising, since there are people who are exclusively homosexual on a life-long basis while others are exclusively homosexual for just a few years. There are also "closet" homosexuals who try to pass as straight in the everyday world (including many homosexuals who are in heterosexual marriages) and others who have openly announced their homosexuality. There are militant homosexuals and more conservative ones; there are homosexuals who remain in long-term, committed relationships while others prefer independence and a more casual approach to sex — there are endless examples of homosexual diversity.

Recognizing these facts, it is nevertheless possible to discuss certain common aspects of homosexual behavior as they apply to the lives of many gay people just as we examined heterosexual behavior patterns.

Discovering Homosexuality

Some homosexuals say that they were aware of being gay as early as age five or six, while others don't make the discovery until sometime in adulthood. However, it is not very likely that the young child has a real sense of homosexual orientation. The sense of being "different" during childhood that some homosexuals recall as adults is not always an accurate barometer of later sexual orientation, since many "straight" adults also felt "different" as children. Furthermore, adult recollections of childhood feelings and behaviors may possibly be influenced by social expectations of what homosexuals "should" have felt (Ross, 1980).

Most of the available research on self-discovery of a homosexual identity suggests that this process is most likely to occur during adolescence for males and at a somewhat later time for females. At earlier ages, the child is exposed to role models that are exclusively heterosexual (at least in a visible sense) whether at home, at school, on television, or in children's books. This fact, combined with the automatic assumption in our society that everyone is heterosexual unless "proven" otherwise, means that children are almost invariably conditioned to think of themselves as heterosexuals destined to live a heterosexual life.

How do people "discover" that they are homosexual? No single pattern fits everyone. Many gay males report having first had some type of sexual contact with another boy as young teenagers which led them initially to suspect that they were homosexual. Only after a period of identity confusion did they begin to actually think of themselves as homosexual, to seek out other homosexuals, and to devise ways of dealing with their homosexuality in a heterosexual world.* Others suspect that they may be homosexual *before* having any same-sex sexual activity and confirm their suspicion by positive responses to sexual experiences with male partners or by feeling more comfortable with homosexual friends than heterosexual ones. For some homosexuals, the process of self-discovery occurs only after much effort is spent trying to fit the expected heterosexual mold but finding that it just isn't comfortable.

Although some lesbians come to a firm discovery of their sexual identity in adolescence, more typically whatever homosexual feelings they have during this time are pushed aside as "a passing phase." There may be close emotional attachments formed with other females that never progress to the stage of physical contact, or specifically sexual experiences may not be labeled as homosexual. A large number of lesbians do not adopt this sexual orientation until after a heterosexual marriage.

There is, of course, a big difference between *discovering* homosexuality and *accepting* it. Some gay men and women have no difficulty at all in this sphere, but much more often there is conflict and uncertainty over the implications of being homosexual in a heterosexual society. Some who have labeled themselves as homosexual seek therapy to "cure" themselves of this self-perceived problem; others feel enthusiastic and even energized by their homosexuality. Although only one out of twenty homosexual men and women studied by Bell and Weinberg expressed a great deal of regret over being homosexual, approximately one in three had considered giving up their homosexual activity. These authors also noted that gay men are more likely than lesbians to have difficulty accepting their homosexuality, which they speculated might be because homosexuality is more often seen by males "as a failure to achieve a

---

* It is important to remember that homosexual and heterosexual experimentation is very common in childhood or adolescence. The fact that someone finds pleasure in a sexual act with another person of the same sex does *not* necessarily mean that he or she is homosexual.

'masculine' sexual adjustment," whereas lesbians "more often experience their homosexuality as a freely chosen rejection of heterosexual relationships" (Bell and Weinberg, 1978, p. 128).

A Typology of Homosexuals

Bell and Weinberg (1978) studied 979 homosexual men and women who were recruited by personal contacts, public advertising, use of mailing lists, and special recruitment cards distributed in gay bars, gay baths, and by gay organizations. Although their sample is neither truly cross-sectional nor representative of all homosexuals in America, it provided a broad-based opportunity for analyzing important information about homosexual feelings and behavior. One of the more interesting findings to emerge from this study was the existence of a typology of sexual experiences that allowed for comparisons between groups. Approximately three-fourths of their sample could be assigned to one of these types on the basis of statistical criteria.

Close-coupled homosexuals lived in one-to-one same-sex relationships that were very similar to heterosexual marriages. They had few sexual problems, few sexual partners, and infrequently engaged in cruising (deliberately searching for a sexual partner). Ten percent of the male homosexuals and 28 percent of the lesbians in Bell and Weinberg's sample were categorized in this group.

Open-coupled homosexuals lived in one-to-one same-sex relationships but typically had many outside sexual partners and spent a relatively large amount of time cruising. They were more likely to have sexual problems and to regret their homosexuality than close-coupled homosexuals. Eighteen percent of the gay males and 17 percent of the lesbians in Bell and Weinberg's study were classified as open-coupled.

Functional homosexuals were those who were not "coupled," who had a high number of sexual partners and few sexual problems. These individuals — 15 percent of gay males and 10 percent of lesbians — tended to be younger, to have few regrets over their homosexuality, and to have high levels of sexual interest.

Dysfunctional homosexuals (12 percent of gay males and 5 percent of lesbians) were not "coupled" and, while scoring high in number of partners or amount of sexual activity, had substantial numbers of sexual problems.

Asexual homosexuals were low in sexual interest and activity and were not "coupled." They tended to be less exclusively homosexual and more secretive about their homosexuality than others. Sixteen percent of the male homosexuals and 11 percent of the lesbians were grouped in this category.

These types also proved to have important connections to a person's social and psychological adjustment. In general, close-coupled homosexuals tended to be at least as happy and well adjusted as heterosexual men and women. The dysfunctionals and asexuals, on the other hand, tended to be worse off psychologically than heterosexuals and had considerable difficulty coping with life. Male dysfunctionals were "more lonely, worrisome, paranoid, depressed, tense, and unhappy than any of the other men," and dysfunctional lesbians were more likely than other women "to have needed long-term professional help for an emotional problem." The asexuals were generally loners who, despite being lonely, had little interest in becoming involved with friends or socializing in the gay community. Asexual homosexual men had the highest incidence of suicidal thoughts.

The existence of these homosexual types *"proves"* nothing — there are probably corresponding "types" of heterosexuals that could be identified, with some showing better social and psychological adjustment and others being more troubled. The point we want to reinforce is that all homosexuals are not alike; there is just as much diversity among homosexuals as among heterosexuals.

Coming Out

"Coming out" is a process in which homosexuals inform others of their sexual orientation. It can be a long process of self-disclosure that begins cautiously with telling a best friend (and waiting to see the reaction that is provoked), then progresses to include a group of close friends, and finally — in its most complete form — lets family members, colleagues at work, and more casual acquaintances know as well. Coming out can also be accomplished in a remarkably brief time, although a great deal of thought and planning may have gone into the decision.

Many homosexuals find that it is considerably easier to come out in the gay world than to their heterosexual friends and family. They may choose to live a life of "passing" as heterosexual to avoid social disapproval, economic repercussions, or other potential problems while they are still known to other gays as "out of the closet" in a limited sense. Here is what one twenty-six-year-old homosexual man said about this issue:

> Philosophically, I'd love to announce my gayness to the world. But it'd probably cost me my job, and it would cause so many problems for my family (especially my father, who's a minister) that I just don't see the point. I lead one life at night, and another during working hours or family get-togethers, and it's really no big deal.

Others consider such a solution unsatisfying, dishonest, or lacking in trust or conviction. However, coming out to a hostile world can create tremendous agony in a person's life. On the other hand, finding support from family and friends can be reassuring and gratifying, as this explanation from a twenty-four-year-old woman makes clear:

> For my whole college career I was afraid to let my parents know I was gay. I talked about coming out, and worried about it tremendously, but I couldn't really bring myself to do it. Then, just before I graduated, something clicked in my head that said I had to come out *NOW*. I just couldn't believe how accepting my parents were. It certainly helped me feel better about myself.

"Coming out" doesn't usually work out this easily, however. Most parents are terribly upset at finding out that their son or daughter is homosexual, and many of them urge the child to seek therapy to "correct the problem." In some cases, parents refuse to see or speak to a gay child; in other cases, while taking a less severe attitude of reproach, parents (or other family members) clearly remain very uncomfortable with the idea of having a homosexual relative.

Despite a current push for homosexuals to "come out of the closet" to indicate pride in their sexual orientation and to work for political gains, most homosexual men and women remain secretive. It appears that homosexuals with lower social status are somewhat more likely to be open about their sexual orientation, while those who are better educated or who have more income are more likely to keep their homosexuality hidden.

Partners and Relationships

Not all homosexuals engage in sexual activity on a frequent basis, but in general homosexual men tend to be more sexually active than homosexual women. Most research has also shown that homosexual men tend to have many more sexual partners than lesbians or heterosexual men and women.

While many homosexual males engage in quick, impersonal sex with strangers, others prefer to enter into long-term affectionate homosexual relationships. Tripp suggests that these ongoing homosexual relationships seem to be rare because they are far less visible than either long-term heterosexual relationships or short-lived homosexual liaisons. Bell and Weinberg note that the relative instability of long-term homosexual relationships may exist partly because they are not encouraged socially or sanctioned legally. In addition, some observers believe that one effect of the recent AIDS epidemic has been to encourage monogamy and close-coupled relationships among gay men.

Coleman believes that male homosexuals are at a disadvantage in learning intimacy and relationship skills because they have few role models to follow and because there has traditionally been "a lack of public support for these relationships." He also points out that "lingering negative attitudes about homosexuality can sabotage efforts to establish or maintain a relationship." Extending this view, McCandlish states that "society's homophobia and resulting social isolation" and lack of family support often produce such stresses on stable relationships that "what might have been minor and even growth-producing difficulties, instead overwhelm the couple and force a premature end to the relationship."

Another perspective on this topic can be gained by recalling the differences in socialization that typically affect males and females in Western society. As Bell and Weinberg suggest, socialization tends to orient males (straight or gay) to sexual variety, whereas females (straight or gay) are more oriented to monogamy. As a result, many young males want a number of sexual partners, while most females want the intimacy that they are more likely to find in a one-to-one relationship. When heterosexual males form relationships with females, they are socialized by the female to become more monogamous, but this is less likely to happen for males in homosexual relations. Thus, many gay males are promiscuous because they've had few social learning experiences to help them to develop intimacy skills.

We have worked with homosexual couples who have been together for decades in relationships that were every bit as close and committed as heterosexual marriages. We have also seen homosexuals who passed through a series of medium-term relationships lasting a few years each. In general, these patterns mirror the descriptions of love discussed in chapter 9, suggesting that there may not be that many major differences between heterosexual and homosexual committed relationships. This view is now supported by a good deal of research that shows that homosexual couples are generally well adjusted and indistinguishable from heterosexual couples.

THE GAY WORLD

The gay world, like the heterosexual world, is not simply or easily described in a few paragraphs. In large urban areas, the gay community may exist as a full-fledged entity, complete with places for social and sexual contacts (bars and baths), gay merchants, gay churches, gay clinics, and gay recreational groups. In other areas, there is no organized homosexual community, and sexual contacts may be hurriedly made in public restrooms, parks, and pickup bars.

Although in the past homosexuals were often forced to go to gay bars or to "cruise" in certain designated locations, today there are mushrooming numbers of homosexual organizations that provide new meeting grounds without the stigma of "being on the prowl." Cities like Washington, San Francisco, and New York have dozens of support organizations for gay physicians, lawyers, teachers, and parents (a sizable number of homosexuals have had children in heterosexual marriages). Other organizations aim at providing various forms of counseling services for homosexuals, ranging from religious advice to help in dieting to finding nonjudgmental medical care. On many college and university campuses, gay students have formed groups to provide peer support and acceptance. Homosexually oriented newspapers and magazines, both national and regional, provide additional information about the homosexual subculture in America and also carry personal advertisements that allow interested parties to get together for sexual purposes.

However, the emerging pride of the homosexual community is a complex phenomenon — all is not happiness and tranquility. Karlen described the other side of the coin in 1978, in the pre-AIDS era, as follows:

> No one knows better than homosexuals that gay is a euphemism. There is a squalid side of the life — lavatory gropings, prostitution, rampant venereal disease, play-acting, promiscuity, mercurial and crisis-ridden romances, abuse of alcohol and drugs, guilt, suicide. Almost all homosexuals except gay militants have said to me that the causes are as much inherent in homosexuality as in the antihomosexuality of the rest of society. The gay world has a bruising, predatory quality that gives many in it a far grimmer view than their heterosexual sympathizers hold.

The shadow of the AIDS epidemic is accentuating a view of the sordid, secretive side of homosexuality, and leading to considerable disagreement within the gay community. For instance, some homosexual men are saying that monogamy is a sign of sanity and gays who are still having sex with numerous partners are really "homophobic"; while other gays are insisting on keeping the baths and steamhouses open as a sign of their freedom of sexual choice. Where such schisms will lead in the next decade is anyone's guess.

HOMOSEXUALITY AND THE LAW

Laws in our society have been strongly antihomosexual for the last few centuries. Underlying this legal posture was the notion that if homosexuals were allowed to function freely as heterosexuals do, they would undermine the moral fabric of our society and possibly recruit large numbers of people into homosexual acts. In the last ten years, however, the law has been turning somewhat from foe to friend of homosexuals. Although homosexual acts are still illegal in most places, ordinances banning "crimes against nature" — which usually apply equally to heterosexual and homosexual noncoital acts — are now rarely enforced if the acts involve consenting adults in private. Even when a conviction is obtained, the usual sentence is to seek psychiatric counseling.

Homosexuals are still banned from being employed by the FBI, the CIA, or the military. A representative of the Defense Department's Office of Manpower told a group of Harvard students that the military believed homosexuals "harmed military discipline, morale, and effectiveness" (*The New York Times*, April 19, 1983), and in fiscal 1983, the U.S. Navy alone discharged 1,167 homosexuals from its ranks (*The New York Times*, December 2, 1983). Gay organizations have generally been denied tax-exempt status by the Internal Revenue Service. Even in San Francisco, which probably has the largest homosexual population in America, when county supervisors approved a law in 1983 giving homosexuals partnership rights similar to those of married heterosexuals (for instance, coverage of partners by city health and pension plans), it was vetoed by the mayor. Legal advances, however, have been made as well.

Gay parents — particularly lesbian mothers — have been slowly but increasingly successful in winning child custody cases in the courts. Employment discrimination against homosexuals has been banned in most federal agencies, and gay people may now leave the military with honorable discharges. While in the past the U.S. Immigration and Naturalization Service barred homosexuals from visiting this country or becoming citizens, a recent ruling by the Ninth Circuit U.S. Court of Appeals declared that since homosexuality was no longer considered a mental disorder, this practice could no longer continue. Discrimination in housing and against homosexual prisoners has also been banned in many states.

These advances, which have been part of a general movement to protect the civil rights of minority groups, still have a long way to go to eradicate legal inequities against homosexuals in a society that is still "antigay." But the loosening of harsher practices is being viewed by some as promising a more equitable treatment under the law in future times.

BISEXUALITY

Although homosexuality has been extensively studied and written about in the last fifteen years, bisexuality has received far less attention. It is difficult to estimate the incidence of bisexuality in our society today. Kinsey found that 9 percent of single thirty-year-old women could be classified as bisexual, while about 16 percent of single thirty-year-old men fit the same description. However, these numbers may exaggerate the active incidence of bisexuality, which we suspect is actually less than 5 percent in our society if it is defined in terms of sexual activity with male and female partners in the past year.

Bisexuals are sometimes called "AC/DC" (based on terminology used to describe two types of electric current), "switchhitters" (borrowed from baseball lingo for a person who bats from either the right or left side of home plate depending on who's pitching), or people who "swing both ways" (this too is a baseball phrase that might also relate to "swinging" as a type of sexual behavior). In the late 1970s, bisexuality became stylish in certain circles where it was regarded as a sign of sexual sophistication and being "open-minded," but since bisexuality has been identified as a high-risk factor for AIDS, its popular appeal has decreased considerably.

People move into bisexuality in a number of different ways. For many, it is a form of experimentation that adds spice to their sex lives but doesn't become the main course. For others, it represents a deliberate choice that permits participation in whatever feels best at the moment. Some men and women seem to alternate their choice of sex partners randomly, depending on availability and circumstances to dictate which gender is involved at a particular time. Most often, whichever of these patterns applies, people with bisexual experience have a decided preference for one gender, but this is not always true.

Masters and Johnson recently described a subgroup of bisexuals they called *ambisexuals*, who were men or women who had no preference whatsoever over the gender of their sex partners, had never become involved in a committed sexual relationship, and had frequent sexual interaction with both men and women. The ambisexuals accepted or rejected any sexual opportunity primarily on the basis of their physical need, with the personality or physical attractiveness of a potential partner having much less to do with their choice.

In some instances, the bisexual has had a long-term heterosexual relationship, which is then followed by a long-term homosexual relationship (or vice versa). After such a sequence, the person involved may develop new views on traditional notions that limit sex-object choice. Another pattern sometimes seen is concurrent involvement in heterosexual and homosexual relationships, as described in this comment from a twenty-three-year old woman:

> I had been dating a guy I was very friendly with for about a year with a good sexual relationship. Then I suddenly found myself making it with my roommate, who slowly but expertly introduced me to how two women can make love. I really enjoyed both kinds of sex and both personal relationships, so I continued them for some while until my graduate school career was over and I moved to a new town.

Research on female bisexuality has shown that some women who identify themselves as bisexual say that they have different emotional needs, some of which are best (or exclusively) met by men, and others by women. We have occasionally come across this same explanation from bisexual men, but much more often the male bisexual explains his sexual life-style in terms of a need for variety and creativity. Some bisexuals of either gender say that their sexual openness is a sign that they aren't biased against homosexuality and that they aren't sexist.

Three different sets of circumstances seem to be particularly conducive to bisexuality. Sexual experimentation in a relationship with a close friend is quite common among women and can also occur with two male friends or with a male homosexual who develops a casual but friendly relationship with a woman. Group sex also is another avenue for bisexual experimentation; while males usually initiate the group activity, females often feel more comfortable engaging in same-sex contact. Finally, some people come to adopt a bisexual philosophy as an outgrowth of their personal belief systems. For instance, some women who have been active in the "women's movement" find that they are drawn closer to other women by the experience and translate this closeness into sexual expression. This same process can be a form of subtle intellectual coercion, however:

> I was an ardent feminist but also as straight as I could be. As I worked extensively with women's groups, I began to feel more and more pressure to "try" a sexual experience with another woman, with the implication being that if I didn't, I really wasn't into sisterhood and was enslaved by male cultural propaganda. I finally gave in to this pressure and had an awful time. Shortly after that, I began drifting away from the movement because it hit too raw a nerve in me.

A few other circumstances of bisexual behavior are notable because they are usually not labeled as bisexual by the participants. Under conditions of prolonged sexual segregation, heterosexual people often turn temporarily to same-sex experiences. This is true of both male and female prisoners and members of the military. Similarly, many men who participate in brief homosexual encounters in public restrooms (euphemistically called the "tearoom trade") are heterosexually married and do not think of themselves as bisexual. Young male prostitutes who cater to a homosexual clientele generally see this as a detached, depersonalized act done "for the money," thus their self-perception remains strongly heterosexual.

The nature of bisexuality remains very much a puzzle at the present time. There are no good leads on what "causes" bisexuality, and the varied pattern of bisexual biographies wreaks havoc with many theories about the origins of sexual orientation. It is very possible that as more is learned about this subject our understanding of the complexities of human sexuality will be improved.

CHAPTER FIFTEEN

The Varieties of Sexual Behavior

SEXUAL BEHAVIOR, like human behavior, is varied and complex, and defies simple schemes of classification. In this chapter, we will shift our focus from the more common patterns of sexual behavior considered in the last three chapters and look at the diversity shown in less typical sexual variations.

It isn't very difficult to decide that a person who is sexually aroused only when riding a camel is abnormal or that a couple whose sex life consists primarily of intercourse several times a week at night in the privacy of their bedroom seems almost "too normal." Most people believe they know intuitively how to rate sexual behavior as normal or abnormal. Nevertheless, trying to define what is sexually normal and what is not is one of the more perplexing problems in sexology today. Let's see where the difficulties arise.

Most dictionary definitions of "normal" say that it is primarily a matter of conforming to a usual or typical pattern. What is unusual or atypical not only varies from culture to culture but also varies over time, as we have noted throughout this book. But there is still more complexity in establishing what is normal. From a sociological perspective, behavior that falls outside the accepted customs and rules of a particular society is considered deviant. From a biological viewpoint, normality implies natural and healthy. A psychological view of abnormality stresses that it produces a personal, subjective sense of distress — such as excessive nervousness, depression, or guilt — or that it interferes with a person's ability to function adequately in ordinary social and occupational roles. Statistically, normality becomes a matter of numbers: what is rare is abnormal, what is common is not.

The point we are making is twofold: first, defining normality is not as simple as it seems; second, the distinction between normal and abnormal is somewhat arbitrary because it generally involves value judgments of one type or another. Thus, it is important to note that in many instances there is no clear-cut separation between normal and abnormal. While it is easy to say that a person who masturbates twice a week is not exhibiting abnormal sexual behavior and that a person who compulsively masturbates a dozen times a day *is,* where do you draw the line? Once a day? Three times a day? Six times a day? Is the behavior abnormal only if it is compulsive? Is it abnormal only if it continues persistently over time?

Given these potential difficulties in determining just what normality means, we suggest that you be aware of your own feelings about the topics we'll be discussing and try to determine how those feelings color your reaction to understanding these different forms of sexual behavior.

Using words like "normal" or "abnormal" to describe people or behavior, or using other word combinations that may sound scientific or "official" (such as healthy, well-adjusted, and law-abiding versus diseased, pathological, deviant, sick, or criminal) is called "labeling." Labels affect how we view other people and how they view us in return; labels may also affect how we feel about ourselves. In general, labels that identify a person as "different" are likely to lead people to cautiously distance themselves from or perhaps reject that person; labels that indicate sameness or familiarity (in association with something viewed as "normal" or "good") tend to foster acceptance.

Stigmatization refers to the negative effects labeling can produce, such as branding people as undesirable or discrediting them in various ways with social, legal, and economic consequences. For instance, people labeled as forgers may be picked up by the police as suspects if bad checks are being passed in their community even if they have reformed completely. A veteran with a dishonorable discharge may have a hard time getting a job. Stigmatization operates very strongly in terms of sexual labels, too — how would you react to being told that someone was a child molester?

In the past, some of the forms of sexual behavior discussed in this chapter were called sexual deviations, perversions, or aberrations. These labels inevitably led to stigmatization and were also applied in a somewhat arbitrary fashion, since their underlying concept was based on a notion of cultural conformity. To avoid these problems as much as possible, we prefer to speak about sexual variations and to use the relatively neutral term "paraphilia" — derived from Greek roots meaning "alongside of" and "love" — to describe what used to be called sexual deviations. While this strategy serves our purpose fairly well at this time, it is possible that in the future this term will also become a source of stigmatization and a new, unstigmatized alternate will be needed.

THE PARAPHILIAS

A *paraphilia* is a condition in which a person's sexual arousal and gratification depend on a fantasy theme of an unusual sexual experience that becomes the principal focus of sexual behavior. A paraphilia can revolve around a particular sexual object (e.g., children, animals, underwear) or a particular sexual act (e.g., inflicting pain, making obscene telephone calls). The nature of a paraphilia is generally specific and unchanging, and most of the paraphilias are far more common in men than in women.

A paraphilia is distinguished from sporadic sexual experimentation just as drug dependence is different from episodic, recreational drug use. The person with a full-blown paraphilia typically becomes preoccupied with thoughts of reaching sexual fulfillment to the point of being seriously distracted from other responsibilities. In addition, types of sexual activity outside the boundaries of the paraphilia generally lose their turn-on potential unless the person supplements them with the paraphilic fantasy.

While some of the paraphilias may seem so foreign to you that it's hard to see how they could be arousing to anyone, paraphilic acts, often in watered-down versions, are commonly used by sex partners wishing to add a little variety to their ordinary techniques. For example, some people get turned on by very explicit sexual language, others want to be bitten, scratched, or slapped during sex, and others find that watching their partner undress is highly arousing. Each of these innocuous acts if magnified to the point of psychological dependence could potentially be transformed into a paraphilia.

With this general background, we will now discuss some of the major types of paraphilias.

Fetishism

In *fetishism,* sexual arousal occurs principally in response to an inanimate object or body part that is not primarily sexual in nature. The fetish object is almost invariably used during masturbation and is also incorporated into sexual activity with a partner in order to produce sexual excitation. Fetishists usually collect such objects and may go to great lengths, including theft, to add just the "right" type of item to their collection. One man we encountered who had a fetish for women's high-heeled shoes had gradually accumulated a hoard of more than a thousand pairs, which he catalogued and concealed from his wife in his attic.

Among the long list of objects that have served as fetishes, the most common are items of women's clothing such as panties, brassieres, slips, stockings or panty-hose, negligees, shoes, boots, and gloves. Other common fetish objects include specific materials such as leather, rubber, silk, or fur or body parts such as hair, feet, legs, or buttocks. While a few fetishists are aroused by drawings or photographs of the fetish object, more commonly the fetishist prefers or requires an object that has already been worn. This object, however, does not function as a symbolic substitute for a person (the former owner). Rather it is *preferred* to the owner because it is "safe, silent, cooperative, tranquil and can be harmed or destroyed without consequence," as the psychiatrist Robert Stoller notes. In the great majority of cases, the person with a fetish poses no danger to others and pursues the use of the fetish object in private.

In some cases, the fetishist can become sexually aroused and orgasmic only when the fetish is being used. In other instances, sexual responsiveness is diminished without the fetish but not completely wiped out. In the second set of circumstances, the fetishist often engineers sexual arousal by fantasizing about the fetish. For a small number of fetishists, the fetish object must be used by a partner in a particular way for it to be effective: for instance, the genitals must be rubbed by silk, or a partner must wear black garters and high-heeled shoes.

There is a thin line of distinction between fetishism and certain types of sexual preferences. A person is not described as a fetishist if sexual arousal is dependent on having an attractive partner; a man who is turned on by a woman in black, lacy lingerie is also not labeled as a fetishist as long as this is not the primary focus of his arousal. Deciding exactly when individual preference blends into too much dependency is not always easy.

Transvestism

A transvestite is a heterosexual male who repeatedly and persistently becomes sexually aroused by wearing feminine clothing (cross-dressing).* Although many transvestites are married and are unremarkably masculine in everyday life, their cross-dressing may be accompanied by elaborate use of makeup, wigs, and feminine mannerisms in a masquerade that sometimes can fool even the most skilled onlooker.

Transvestism must be distinguished from male transsexualism, where the person wants an anatomical gender change and wants to live as a woman. Transsexuals usually are not sexually aroused by cross-dressing, but in a few cases, transvestism may evolve into transsexualism. Transvestites can also be distinguished both from female impersonators (who are entertainers) and from male homosexuals who occasionally "go in drag" (cross-dress). These types of cross-dressing are not associated with sexual arousal, and there is no psychological dependence on wearing feminine clothing as a form of tension release.

Cross-dressing usually begins in childhood or early adolescence, with the histories of many transvestites indicating that as children they were punished by being dressed in girl's clothes. In one of the cases we have seen, a six-year-old boy was deliberately "taught" to cross-dress by his transvestite father, who in

---

* It is interesting to note that women who cross-dress (wear male clothing) are often regarded as fashionable and are not diagnosed as transvestites. While cross-dressing among women is not rare, it has not generally been associated with sexual excitement.

turn had been guided into this behavior by his father when he was a child. In adulthood, most transvestites confine their cross-dressing to the privacy of their own homes, although a few may wear women's panties under their usual masculine clothing throughout the day. In many cases, the transvestite's wife is fully aware of her husband's cross-dressing and may actually help him perfect his use of makeup or select attractive clothing styles. In other cases, the wife may be confused and upset by discovering her husband's passion for cross-dressing and may insist that he seek treatment or may simply file for divorce. Other wives reluctantly tolerate the cross-dressing but don't give it their approval, as this forty-two-year-old woman explains:

> I can't really understand why John enjoys this bizarre business, and I live in constant fear that he'll be discovered by our children. But he doesn't hurt anyone with his dressing up, and it doesn't take anything away from our love, so I really can't complain too much — I just have to accept things as they are.

While most transvestites are exclusively heterosexual (the majority of married transvestites have children), a small percentage cross-dress while cruising heterosexual bars or social clubs. Their prospective male partners may be completely unaware that a masquerade is going on and may become involved in limited varieties of sexual activity. The transvestite, for example, may perform fellatio on them or masturbate them manually, claiming to be genitally "indisposed" because of menstruation or some other reason. Obviously, the "truth-in-advertising" laws are not being met here, or to paraphrase a Flip Wilson line, "What you get is not exactly what you see."

Voyeurism

In our society, looking at nude or scantily clad women is an acceptable male pastime, as shown by attendance at topless bars, Las Vegas revues, and the binoculars that are trained on the Dallas Cowboy Cheerleaders instead of the action on the football field. Recently, there has been a corresponding phenomenon among women who can view the nude or seminude male in "women-only" clubs that feature attractive male go-go dancers or in women's magazines that show male frontal nudity. The social acceptability of such interests illustrates our earlier point about the continuum between normality and abnormality because the pleasures of "looking" can be transformed into another form of paraphilia.

The *voyeur*, or Peeping Tom, is a person who obtains sexual gratification by watching others engaging in sexual activity or by spying on them when they are undressing or nude. Voyeur comes from the French verb meaning "to see." Peeping Tom comes from the legendary nude ride of Lady Godiva; Tom the Tailor was the only townsman who violated Lady Godiva's request for privacy by "peeping." In *voyeurism*, "peeping" (or fantasizing about peeping) is the repeatedly preferred or exclusive means of becoming sexually aroused. While women may be "voyeuristic" in the sense of becoming sexually excited by seeing others nude or watching others in sexual acts, cases of female dependency on voyeurism for sexual response are very rare.

Voyeurism is mainly found among young men and often seems to burn out by the middle-age years. Voyeurs frequently have a great deal of trouble forming heterosexual relationships, and many have only limited amounts of heterosexual experience. In fact, being a voyeur allows such a man to avoid social and sexual interaction with women; many voyeurs confine their sexual activity to masturbation while peeping or while fantasizing about previous peeping escapades.

The voyeur prefers to peep at women who are strangers, since this confers a novelty and forbidden quality on the act. The voyeur is often most sexually excited by situations in which the risk of discovery is high, which may also explain why most voyeurs are not particularly attracted to nudist camps, burlesque shows, nude beaches, or other places where observing nudity is accepted.

While it might be thought that peepers are harmless individuals because they avoid personal contact, this is not always the case. Some voyeurs have committed rape, burglary, arson, or other crimes.

Exhibitionism

Exhibitionism is a condition in which a person repeatedly and preferentially exposes the sex organs to unsuspecting strangers to obtain sexual arousal. While exhibitionism is found almost exclusively in males, a few cases of female exhibitionism have been reported. Many exhibitionists are impotent in other forms of heterosexual activity and seem to be pushed by an "uncontrollable urge" which leads to their impulsive behavior.

The peak occurrence of exhibitionism is in the twenties, with relatively few cases after age forty. (This probably indicates that most exhibitionists gradually stop exposing themselves as they become middle-aged.) According to one study, the typical exhibitionist is married, above average in intelligence, satisfactorily employed, and without evidence of serious emotional problems. Most exhibitionists also tend to be passive, shy, and sexually inhibited men. In many instances, a particular episode of exhibitionistic behavior is triggered by a family conflict or a run-in with an authority figure.

Although an act of exhibition usually produces sexual excitation in the performer, it is not always accompanied by erection or ejaculation even if the man masturbates while exposing himself. For some men, the primary intent of exhibitionism is to evoke shock or fear in their victims; without such a visible reaction, they derive little pleasure from the act. Apparently such men are trying to "prove" their masculinity by an unmistakable anatomical display.

More exhibitionists are caught by the police than people in any other category of paraphilia. The need to risk being caught may be an important element of the turn-on, leading some exhibitionists into behavior almost guaranteed to result in arrest. The exhibitionist may repeatedly "perform" at the same street corner or use a parked car (which can be easily identified) as the theater for his "act."

It is generally agreed that the exhibitionist is unlikely to rape or assault his victims, but there are apparently a few exceptions to this finding. In one case, an exhibitionist who was unsatisfied with his victim's response slapped her in the face; in another instance, an exhibitionist became so enraged at being ignored by his victim that he ran after her, dragged her into an alley, and forced her to perform fellatio.

Just as observing nudity or sexual activity is relatively acceptable in our society under certain conditions, displaying one's body in sexually provocative garb (low-cut dresses, open-necked shirts, "see-through" blouses, tight-fitting pants) is acceptable too.

Obscene Telephone Calling

Twentieth-century technology has contributed at least one new type of paraphilia through the widespread availability of the telephone in our society: some people repeatedly make obscene telephone calls as a means of obtaining sexual excitement.

The obscene telephone caller is almost always male and typically has major difficulties in interpersonal relationships. The relative safety and one-sided anonymity of the telephone — the caller usually knows the name and phone number of the person to whom he is speaking — allows an idealized masturbatory experience with no need to worry about a face-to-face confrontation.

There are three different types of obscene telephone calls. In the first (and probably most common), the caller boasts about himself and describes his masturbatory action in explicit detail. In the second type, the caller directly threatens his victim ("I've been watching you," "I'm going to find you"). In the third type, the caller tries to get the victim to reveal intimate details about her life. This is often done by the caller claiming to be conducting a "telephone research survey" about a subject such as women's lingerie, menstruation, or contraception. More than a few obscene telephone callers announce themselves to their victims as sex researchers.*

Sometimes the obscene telephone caller repeatedly calls the same victim; more often, unless the "victim" shows a willingness to stay on the phone and play his game (and a surprising number of women do), the caller moves on to other victims. The compulsive obscene phone caller must be distinguished from the adolescent who occasionally indulges in the same activity as a prank without giving much thought to the distress caused to others.

* We have had many complaints over the years from women who received phone calls inquiring about their sex lives from men who claimed to be representatives of the Masters & Johnson Institute. Incredibly, many of these women talked at great length to the bogus "interviewer" and had second thoughts only after the call was completed. Readers should be aware of the fact that *no* legitimate sex research is conducted by telephone surveys.

The victim can report obscene phone calls to the telephone company and the police, but the chances of catching the caller are fairly low unless the call can be traced by keeping the caller on the line or the caller can be trapped into "meeting" the victim under police supervision. Obtaining an unlisted telephone number may help stop obscene telephone calls, and some women prefer to list their names in the phone book with last name and first initial only in order to make themselves a less obvious target.

Sadism and Masochism

Sadism is the intentional, repeated infliction of pain on another person to achieve sexual excitement. It is named after the Marquis de Sade (1774–1814), a French author who wrote extensively about cruelty as a means of obtaining sexual gratification. *Masochism* is a condition in which a person derives sexual arousal from being hurt or humiliated. It is named after an Austrian novelist, Leopold Baron von Sacher-Masoch (1836–1905), whose *Venus in Furs* (1888) gave a detailed description of the pleasure of pain.

The exact incidence of sadism and masochism is not known, but several surveys indicate that 5 to 10 percent of men and women describe such activities as sexually pleasurable on an occasional basis. Many of these people have probably engaged in mild or even symbolic sadistic or masochistic behavior, with no real physical pain or violence involved. Judging from our research, sadomasochism is only infrequently a full-fledged paraphilia. Giving or receiving physical suffering is thus infrequently the preferred or exclusive means of attaining sexual excitement. And, contrary to the mistaken notion that most women are masochists, both sadism and masochism occur as paraphilias predominantly in men.

Forms of sadism run the entire gamut from "gentle," carefully controlled play-acting with a willing partner to assaultive behavior that may include torture, rape, or even lust-murder. Some sadists require an unconsenting victim to derive pleasure; others become sexually aroused with a consenting partner only if the suffering is obvious.

Similarly, masochism can range from mild versions to extremes. In the mild forms of masochism, activities like bondage (being tied up for the purpose of sexual arousal), being spanked, or being "overpowered" by physical force are mainly symbolic enactments under carefully controlled conditions with a trusted partner. At the opposite end of the spectrum are genuinely painful activities such as whippings, semistrangulation, being trampled, and self-mutilation. The masochist who desires "heavy" pain or bondage may have great difficulty in finding a cooperative partner. For this reason, some masochists resort to inflicting pain on themselves in bizarre ways, including burning themselves, hanging themselves (which causes several dozen deaths annually), or searching out the services of a prostitute who will provide the necessary stimulation.

Although sadomasochistic activities in their extreme forms can be physically dangerous, most people who try these varieties of sex do so with a commonsense understanding of the risks involved and stay within carefully predetermined limits. The allure of sadomasochistic sex, for many people, seems to lie in its erotic nature and its sense of "breaking the rules" of ordinary sexual conduct. In sharp contrast is the relative handful of people whose sexual arousal is dependent on sadomasochism, whose preoccupation with pleasure derived from giving or receiving pain becomes almost all-consuming. As one man told us, "In the heat of my sexual passions, I would stop thinking about the real world and its consequences."

The psychological meaning of sadomasochism is unclear at present. Noting that many masochists are men who occupy positions of high status and authority (such as executives, politicians, judges, and bankers), some experts theorize that private acts of submissiveness and degradation provide the masochists with an escape valve from their rigidly controlled public lives. Seeking sexual pain or humiliation may also be a way of atoning for sexual pleasure for a person who was raised to believe that sex is sinful and evil. Conversely, sadists either may be seeking a means to bolster their self-esteem (by "proving" how powerful and dominant they are) or may be venting an internal hostility that they cannot discharge in other ways.

An entire industry has evolved in support of sadomasochistic sex (often abbreviated as S-M). There are equipment supply catalogues that advertise shackles, whips, nail-studded chains, mouth gags, and other devices of torment. There are magazines with picture spreads on S-M activities and detailed "how-to" articles. In some large cities, S-M bars have opened, and "private" clubs featuring dungeons and regular "social" hours exist. Sexually explicit newspapers usually carry ads like these:

"Mistress Alexandra" — I was born to dominate men. My perfect, young, sensuous body brings men to their knees. . . . Slaves desire so much to please me that they will submit to penis torture, nipple discipline, rectum stretching, enemas, whichever entertains me. I will, if necessary, enforce obedience with my cat-of-nine-tails, or other dungeon devices. Mon.–Sat. Noon 'til 10 p.m. Call _____ .

MISTRESS INGA'S WORLD — Come visit her dungeon room built out of 7" solid stone, equipped with everything from a whipping post to a suspension system. Call _____ .

"Submissive Cherry" — Most girls grow up needing to be touched gently "like China dolls." Ever since I was a little girl I wanted to be dominated and put in my place. Now I am a big girl with a luscious bottom that needs to be spanked. I would like to be your total body slave. Reasonable Rates. Call _____ .

As we pointed out in chapter 11, sadomasochistic fantasies are very common, but most people who find such fantasies arousing have no desire to have the real-life experience.

Zoophilia

Engaging in sexual contact with animals is known as *bestiality;* when the act or fantasy of sexual activity with animals is a repeatedly preferred or exclusive means of obtaining sexual excitement, it is called *zoophilia.*

Kinsey and his colleagues found that 8 percent of the adult males and 3.6 percent of the adult females they studied reported sexual contact with animals. For the females, this usually involved sexual contact with household pets, while for males, this often involved farm animals such as sheep, calves, or burros (animal sex contacts were two to three times as common in rural males as in city dwellers, according to Kinsey's findings). Male bestiality generally included vaginal intercourse, while female bestiality was more likely to be limited to having the animal perform cunnilingus or masturbating a male animal. A few adult women have trained dogs to mount them and regularly engage in intercourse with their pet.

Bestiality usually involves curiosity, a desire for novelty, or a desire for sexual release when another partner is unavailable. Zoophilia sometimes involves sadistic acts that may harm the animal.

Pedophilia

Pedophilia (literally, "love of children") describes adults whose preferred or exclusive method of achieving sexual excitement is by fantasizing or engaging in sexual activity with prepubertal children. While some authorities state that pedophilia occurs only in males, there are specific cases of women having repeated sexual contact with children. About two-thirds of the victims of pedophiles are girls (most commonly, between ages eight and eleven).

The pedophile, or child molester, is a complete stranger to the child in only 10.3 percent of cases, showing that the popular stereotype of the child molester as a stranger who lurks around schools and playgrounds with a bag of candy is generally incorrect. In about 15 percent of reported cases, the pedophile is a relative, making the sexual contact a form of incest. The actual percentage of cases involving relatives may be higher than this, since fewer cases of this type may be reported to the police out of concern for "protecting" the family member. (Incest will be discussed in more detail in chapter 16.) Most pedophiles are heterosexual and many are married fathers; a substantial number have marital or sexual difficulties and alcoholism figures prominently in these cases, too.

MacNamara and Sagarin caution that it is hard to know how often pedophiles who have been caught say they were drunk as an excuse to reduce the stigma and lessen the chances of punishment: "By claiming drunkenness, a man is saying in effect that he is not the sort of person who in a sober state would become involved in an act of this type." Thus, he may convince others that instead of needing punishment, psychiatric care, or rehabilitation, he simply needs to stop getting drunk.

There are three distinct age groups where pedophilia is common: over age fifty, in the mid-to-late thirties, and in adolescence. Strictly speaking, the person who has only isolated sexual contacts with children is not a pedophile and may be expressing sexual frustration, loneliness, or personal conflict.

Several different types of pedophiles have been distinguished. According to Cohen, Seghorn, and Calmas, the most common is the *personally immature pedophile* — a person who has never succeeded in developing interpersonal skills and is drawn to children because he feels in control of the situation. His victims are usually not strangers and the sexual contact is not impulsive, often beginning with a drawn-out "courtship" in which he befriends the child with stories, games, and disarming companionship. In contrast, the *regressed pedophile* usually has developed strong heterosexual relationships without much difficulty; during some point in adulthood, however, he begins to develop a sense of sexual inadequacy, has problems dealing with everyday stresses, and often becomes alcoholic. His sexual contact with children is apt to be impulsive and with strangers, sometimes reflecting a sudden, uncontrollable urge that comes over him. The *aggressive pedophile* (the least common version) often has a history of antisocial behavior and may feel strong hostility toward women. They are most likely to assault their victims and may cause severe physical harm.

According to the psychologist Nicholas Groth, who works extensively with sex offenders, 80 percent of pedophiles have a history of being sexually abused when they were children. Exactly why these men who were themselves sexually victimized should in turn victimize others is uncertain at present, but the tendency may relate to defects in personality development partly caused by the trauma of sexual victimization.

There is no single pattern of sexual activity that fits all pedophiles. While fondling the child's genitals or having the child touch his own genitals may be the most common pedophilic act, there are many instances of intercourse, fellatio, and other varieties of sexual stimulation.

Most societies take a dim view of pedophilia, and laws against sexual contacts between adults and children are strongly enforced. But the sexual abuse of children has become another "big business" today, with numerous instances of children being pushed into prostitution or being used for the production of pornography. The *Chicago Tribune* (May 16, 1977) reported that a nationwise homosexual ring had been discovered which shipped young boys around the country for prostitution. In Los Angeles, the René Gunyon Society is working to decriminalize sex between children and adults, using the slogan "Sex by age eight, or else it's too late!" The North American Man-Boy Love Association, which argues for the right of adults to have sex with underage boys, recently accused the FBI of conducting a "witch hunt" when arrests involving two of their group's members were made in connection with kidnapping charges. Even more recently, charges of child sexual abuse have been leveled at school and day-care personnel (including some women as well as men) in various parts of the country. A strong public outcry against such practices has occurred in the past few years as the dimensions of this problem have become more widely publicized.

Other Paraphilias

There are a number of other paraphilias that are relatively rare and about which fairly little is known. *Apotemnophilia* refers to persons with a sexual attraction to amputations, who may sometimes try to convince a surgeon to perform a medically unnecessary amputation on them to increase their erotic satisfaction. Persons with this type of paraphilia seek out sexual partners who are amputees.

Coprophilia and *urophilia* refer, respectively, to sexual excitement deriving from contact with feces and urine. *Klismaphilia* is sexual excitement preferentially or exclusively resulting from the use of enemas. *Frotteurism* is sexual arousal that results from rubbing the genitals against the body of a fully clothed person in crowded situations such as subways, buses, or elevators.

Necrophilia is sexual arousal from viewing or having sexual contact with a corpse. This bizarre paraphilia has sometimes led people to remove corpses from cemeteries, or seek jobs in morgues or funeral homes.

Causes and Treatment of the Paraphilias

There is very little certainty about what causes a paraphilia. Psychoanalysts generally theorize that these conditions represent "a regression to or a fixation at an earlier level of psychosexual development resulting in a repetitive pattern of . . . sexual behavior that is not mature in its application and expression" (Sadoff, 1975, p. 1539). In other words, an individual repeats or reverts to a sexual habit arising early in life. Castration anxiety and Oedipal problems are seen as central issues. According to Robert Stoller, these conditions are all expressions of hostility in which sexual fantasies or unusual sexual acts become a means of obtaining revenge for a childhood trauma usually related to parents inhibiting their child's budding sexuality by threats or punishment. The persistent, repetitive nature of the paraphilia is caused by an inability to completely erase the underlying trauma.

Behaviorists, instead, suggest that the paraphilias begin via a process of conditioning. Nonsexual objects can become sexually arousing if they are frequently and repeatedly associated with pleasurable sexual activity (most typically, masturbation). Particular sexual acts (such as peeping, exhibiting, bestiality) that provide an especially intense erotic response (often heightened because of their "forbidden" nature) can, under certain circumstances, lead the person to prefer this sexual behavior. However, this is not usually a matter of classical conditioning alone: there must usually be some predisposing factor, such as difficulty forming person-to-person sexual relationships or poor self-esteem.

Whatever the cause, it is apparent that paraphiliacs rarely seek treatment unless they are trapped into it by an arrest or discovery by a family member. Most of the time, the paraphilia produces such immense pleasure that giving it up is unthinkable. In this sense, the paraphilia is very much like an addiction, which has led many workers in the field to attempt to understand such sexual behavior in terms of other types of addictions. Among paraphiliacs in therapy, there may be deliberate attempts to lull the therapist into believing the behavior has been eradicated when it continues in full force.

The literature describing treatment approaches is fragmentary and incomplete. Traditional psychoanalysis has not appeared to be particularly effective with the paraphilias and generally requires several years in treatment. Therapy with hypnosis has also had mixed results. Current interest is focused primarily on a number of behavioral techniques that include (1) *aversion therapy,* which attempts to reduce or extinguish behavior by conditioning; for example, electric shocks may be given to a person while he views photographs of the undesired behavior; (2) *desensitization procedures,* which neutralize the anxiety-provoking aspects of non-paraphiliac sexual situations and behavior by a process of gradual exposure; (3) *social skills training,* generally used in conjunction with either of the other approaches, and aimed at improving a person's ability to form interpersonal relationships; for example, a man may be coached in how to talk with women, how to overcome his fear of rejection, and how to express affection; and (4) *orgasmic reconditioning,* wherein a person might be instructed to masturbate using his paraphilia fantasy and to switch to a more appropriate fantasy (for example, intercourse with his wife) just at the moment of orgasm. With this reconditioning process, the person is gradually taught to become aroused by more acceptable mental imagery, and other appropriate techniques, such as fading or satiation, are used to reduce substantially the arousal of the undesirable fantasy.

Recently, drugs called antiandrogens that drastically lower testosterone on a temporary basis have been used in conjunction with these forms of treatment. The antiandrogens lower sex drive in males and also reduce the frequency of mental imagery of sexually arousing scenes. Thus, these drugs usually lessen the compulsiveness of the paraphilia, allowing concentration on counseling without as strong a distraction from paraphiliac urges. MPA (medroxyprogesterone acetate, also known by its trade name, Depo-Provera) has been the main antiandrogen used in this country.

HYPERSEXUALITY

People with extraordinarily high sex drives, which are insistent and persistent but rarely lead to more than fleeting gratification or release despite numerous sex acts with numerous partners, are considered to be *hypersexual,* or "oversexed." In women, this condition has been called "nymphomania"; in men, it is called "satyriasis" or "Don Juanism."

There has been little scientific study of these conditions, which often seem to be considered more of a joke than anything else. There are no absolute criteria for defining hypersexuality. The central features of most studied cases are that (1) sexual activity is an insatiable need, often interfering with other areas of everyday functioning; (2) sex is impersonal, with no emotional intimacy; and (3) despite frequent orgasms, sexual activity is generally not satisfying.

To many men, the idea of a woman with a greater sex drive than their own is somewhat threatening, so they may use the label of nymphomania to preserve their own egos: the label "proves" that the woman is abnormal. Similarly, men with sexual dysfunction sometimes accuse their wives or partners of being "oversexed" in an effort to hide their own fears and sense of inadequacy, just as some women who do not enjoy sex or who object to the frequency of their husband or partner's amorous desires accuse him of being oversexed. In our society, a man who is highly sexed and who has many sexual partners is generally (often enviously) called a "stud," while a woman with the same characteristics is often called a "nympho," which carries a negative connotation.

CELIBACY

A very different form of sexual behavior is *celibacy,* or abstention from sexual activity. Celibacy can be a conscious and deliberate choice, or it can be a condition dictated by circumstance (poor health, unavailable partner, etc.).

In some religions, a vow of lifetime celibacy is expected of those who join the clergy. Various religions also exalt the purity and holiness of celibacy for their lay members. Celibacy can also be practiced on a temporary or periodic basis, where it can allow some people to have more of a sense of control over their lives, to devote more attention to nonsexual aspects of their relationships, or to take a "rest" from the pressures of sexual interaction.

In the last few years, celibacy has become more talked about as a sexual alternative. An article in *The Village Voice* (January 23, 1979) reported that "coming out" as celibate is the latest sexual vogue. In 1980 a book appeared titled *The New Celibacy: Why More Men and Women Are Abstaining from Sex — and Enjoying It*, suggesting:

> Celibacy is a way of breaking boundaries, old patterns of behavior that exist between mind and body, between the self and others. It enables one to be free of sexuality in order to evaluate and experience the joys of life without sex. If the results of being celibate for some time lead to becoming sexual once again, fine; it will be bringing about an even more sexually alive state than before. If one chooses to remain celibate because other nonsexual experiences turn out to be very fascinating, then too there will be clear benefits resulting from the celibate exploration.

For people who get no pleasure out of sex, celibacy may be a welcome relief, like being released from imprisonment. Others may choose celibacy even though they enjoy sex because they find that abstinence rejuvenates their lives or creativity. Of course, some people find celibacy to be a frustrating and unfulfilling choice and may quickly reject it.

There are no health risks known to result from celibacy. As mentioned in chapter 12, if physical sexual tensions mount to a critical level, they are discharged by orgasms during sleep. But it is clear that while celibacy is a positive sexual alternative for some people, it is not right for everyone.

PROSTITUTION

Prostitution is difficult to define since humans have always used sex to obtain desirables such as food, money, valuables, promotions, and power. For practical purposes it is best to define a prostitute as a person who for immediate payment in money or valuables will engage in sexual activity with any other person, known or unknown, who meets minimal requirements as to gender, age, cleanliness, sobriety, ethnic group, and health. Some societies lack prostitution while in others, particularly urban societies, prostitution is tolerated or exists in spite of efforts to eliminate it.

The major reason for the existence and extraordinary persistence of female prostitution is that it is an easy solution to the problem faced by economically disadvantaged women. Virtually all the prostitutes in Europe and America today entered their occupation for economic reasons; they were not captured by "white slavers" nor motivated by pathological sexual needs. Some needed money for an emergency or a drug habit, others drifted into it through accepting gifts and finally money from boyfriends who gradually became more numerous and less known; some were beguiled by a pimp's promises; and still others simply realized — as one said — "I was sitting on a fortune."

Here are two different stories that show how a woman finds her way into "the life."

> *A nineteen-year-old call girl:* I was one of six children in a poor family with nowhere to go. I was having sex regularly by the time I was 13, and at age 17 I realized I might as well get paid for it. Everyone knew who the pimps were, and I just connected with a guy who set me up in his stable. Now I make plenty of money and I help my family out.

> *A twenty-four-year-old massage parlor attendant:* I was divorced at twenty-one with a kid and no talent. The only legitimate work I could find was at the minimum wage and that was the pits. A friend told me about this place, and now I get about $500 a week for locals [masturbating the male customer] and blow jobs. If a guy wants to get laid, that'll cost him $50 extra. In another couple of years I'll quit this job, but I'll have some money saved.

Female prostitutes can be classified as house girls who work in brothels, street girls who solicit in public, B-girls who meet their clients in bars, and call girls who accept appointments, usually via telephone and often only with recommended clients. To this list, one must now add massage parlors and other not so subtly disguised cover-ups such as "escort services." Many, but by no means all, have pimps with whom they share their income. The pimp provides affection and protection, arranges for bail, is available for an occasional loan, and often helps obtain customers. A pimp generally (depending on one's viewpoint) is supported by, or manages, several prostitutes.

Female prostitutes seldom have orgasm in their business contacts but are normally orgasmic in private life. Some will engage in certain sexual techniques with a customer but never with a husband or boyfriend and vice versa. Prostitution may be a regular occupation or an occasional source of extra income, but age ultimately reduces attractiveness and forces retirement.

Although Xaviera Hollander's famous book, *The Happy Hooker*, presents a generally rosy picture of prostitution, there is another side to it too. The detrimental side of female prostitution is not the sexual activity itself but the evils that often accompany prostitution: exploitation by organized crime and/or pimps, sexually transmitted disease, drug addiction, the physical risks of "kinky" sex or assault by a customer, and the inability to save money for future needs. These evils can be eliminated or at least minimized, as in Denmark, where organized prostitution and pimping are strictly prohibited and where the prostitutes are required to have regular employment in addition to their prostitution. Most nations, however, content themselves with futile attempts to wholly suppress prostitution or to confine it to specific areas.

While female prostitution is almost exclusively heterosexual (men paying women), male prostitution is almost exclusively homosexual (men paying men). In male prostitution, there are no pimps; large-scale organization is absent; the price is much less; male brothels are extremely rare (although call boys exist); and the relationship with the customer is not the same. In the United States the male prostitute often presents himself as a "straight" (heterosexual) male who has orgasm as a result of the customer's activities and who often does nothing to the customer. Elsewhere in the world, the male prostitute is active rather than passive and seeks to provide the customer with an orgasm.

Male prostitutes are often known as "hustlers" since "hustling" (by male or female prostitutes) is the act of soliciting a prospective customer. As with female prostitutes, customers are referred to as "Johns," "scores," or "tricks." Male prostitutes sometimes assault and rob their customers, knowing that there is almost no likelihood their crime will be reported to the police.

A recent study found that about seven out of ten young male prostitutes are only "part-timers" who continue to pursue conventional educational, vocational, and social paths while earning money by selling sex. Male prostitutes who sell their services to women are generally known as gigolos. However, gigolos primarily cater to wealthy, older women.

Prostitution exists primarily because men are willing to pay for sex. Men seek out prostitutes for a variety of reasons. Some men are temporarily without sexual partners because they are traveling or in military service; others with a physical or personality handicap cannot easily obtain partners. In some societies sex with nonprostitutes is very difficult to arrange. Some males seek special techniques that their usual partner will not provide; others do not want to invest the time, emotion, and money in an affectional relationship and simply prefer to buy physical sex. While the increase in nonmarital sexual intercourse in the United States has diminished the prevalence of prostitution, there will always be some customers such as the men just described.

LEGAL ASPECTS
OF SEXUAL BEHAVIOR

Most laws pertaining to sexual behavior in the U.S. and in most of the Western world can be traced back to sexual prohibitions from the Judaeo-Christian tradition. The original intent of these prohibitions was to preserve moral order as defined by particular sets of religious values. Today, our Constitution demands that our legal system function apart from any religious influence, but this does not totally alter the body of law that has been passed down through the centuries.

Most laws about sexual behavior in America are found at the state or local level. Since these laws vary incredibly from one state to the next and undergo considerable change from year to year (based on both new legislation and new judicial decisions), it is not possible to provide a comprehensive, up-to-date listing of what is legal and what is not on a state-by-state basis. Instead, we will provide an overview of some controversies related to laws about sexual behavior, illustrating our discussion with several specific areas of interest. Legal aspects of rape and incest will be considered separately in chapter 16.

Private Sex with
Consenting Adults

It's safe to say that in any twenty-four-hour period, millions of ordinary Americans unwittingly engage in sexual acts that are defined as criminal and could lead to imprisonment. In most states, oral-genital sex is illegal — even between husband and wife — and extramarital sex, premarital sex, homosexual acts, and prostitution all fall within the long arm of the law.

Laws that regulate private sexual behavior between consenting adults are a source of dismay to many who argue that these are "victimless" crimes: no one is hurt by the activity, and the government should not be poking its nose in the bedrooms of the nation. Others note that laws pertaining to sexual behavior are no more arbitrary than laws pertaining to business, sports, taxes, or education and point out that many laws depend on first making a moral judgment that gets transformed into legislation.

While most people agree that there are indeed "victimless" crimes, not everyone agrees on just what "victimless" means. The prostitute is regarded as a victim by many; women may be victimized by an unintended pregnancy; anyone can be victimized by sexually transmitted disease. Defining victimization broadly, some people believe that unrestricted sexual permissiveness may lead to the downfall of our civilization.

Despite these philosophical and political disputes, if current laws regulating sexual behavior *were* enforced in a strict and uniform manner, our prisons would have to accommodate the great majority of our population. Recognizing this dilemma and recognizing that the greater common good could be better served if our criminal justice system turned its attention to crimes of violence and of a serious nature, the American Law Institute drafted a Model Penal Code that recommends abolishing laws that regulate the private sexual behavior of consenting adults. The major provisions of this code have now been adopted by a number of states (including Illinois, Connecticut, Colorado, Oregon, and Hawaii), although political pressures have prevented it from becoming more widespread. The basic problem is that most politicians need to run for reelection and do not want to be accused by ultra-conservative groups or fundamentalist religious organizations of being "soft on crime" or "soft on sex."

Here are some examples of the ways in which sex laws categorize and criminalize private, consensual adult sexual acts. The law usually differentiates between *fornication* (intercourse between unmarried heterosexual adults) and *adultery* (extramarital intercourse), with adultery being the more serious crime. In a few states, cohabitation is illegal. The statutes pertaining to fornication and adultery are infrequently enforced today; when they are, people on welfare or members of minority groups are usually involved. Not too long ago, in some states these laws applied only to biracial sexual couples, but such laws have now been abolished as discriminatory and unconstitutional. Punishment for those convicted of fornication or adultery is generally a fine, but in a handful of states, a jail sentence may be given, depending on the discretion of the judge.

An interesting subcategory of fornication is called *seduction*, which is legally defined as a situation in which a woman is enticed into sexual intercourse by a promise of marriage. Only the male can be prosecuted under this statute (who said the law was always fair?). Although laws against seduction are rarely enforced, they can carry prison sentences of five years or longer.

The laws pertaining to nonmarital heterosexual coitus have an interesting past. Adultery was originally frowned on because it violated the sanctity of the family and made it difficult to determine a child's paternity (complicating inheritance decisions). In addition, a married woman was considered the property of her husband; so if she engaged in extramarital sex, her husband's property rights were violated. A similar line of reasoning held in cases of seduction: the male who seduced a woman violated a verbal contract (his promise to marry her) just as surely as if he had pulled out of a business venture. The result of his broken promise was judged to be "damaged" or "used" property, which no longer had the value of the original, unsullied merchandise. His punishment? Take delivery of the "used" property (that is, marry her, thus fulfilling the original "contract") or go to prison.

In many states, most forms of noncoital sexual activity are considered illegal even if they are done in private by consenting adult partners. As astonishing as it may seem in an age when oral sex is statistically the norm rather than the exception, a pleasant interlude of cunnilingus or fellatio can, theoretically, lead to arrest and imprisonment. Anal sex is similarly banned. In general, the statutes refer to these acts as "crimes against nature," going back to the view that heterosexual intercourse (with its reproductive potential) is the only "natural," healthy, nonsinful way of having sexual relations.

While a few states permit these sexual practices in legally recognized marriages, many states do not. Needless to say, homosexual acts are banned by most states on the same gounds, and the likelihood of prosecution for gay men or women is substantially higher than it is for heterosexuals. Most arrests of homosexuals, however, are for public solicitation and not for specific sexual acts.

Although prostitution is legal in many countries across the world, where it is generally regulated by some form of government licensing and health checkups, it is illegal in all U.S. jurisdictions except for several counties in Nevada. Since prostitution often involves private sex acts between consenting aduts, we have chosen to discuss it in this context. This discussion does *not* apply to prostitutes who are minors, since they may not have the capacity for giving true consent to the behavior; nor does it apply to females who are forced into prostitution by one means or another.

The laws dealing with prostitution vary considerably from state to state. In some jurisdictions, the customer of a prostitute is technically not engaging in any illegal act, although he may be violating another statute by participating in a "crime against nature" or fornication. In some areas, however, patronizing a prostitute is a legal violation that can result in a fine. In states such as New York, where this type of law is in effect, it has been justified on the grounds that it is unfair to penalize the prostitute but not her customer.

In a number of cities, including Dallas and St. Louis, the police regularly use policewomen "decoys" on street corners where prostitution abounds. An unwary male approaches the decoy (usually by car), engages her in conversation, and — once he offers money in return for a sexual act — the decoy signals to her hidden police confederates, who move in to make the arrest.

The prostitute is liable to arrest, although it is primarily the street girl and the male hustler who are vulnerable. This is because many laws against prostitution are actually directed against loitering in a public place for purposes of prostitution or solicitation; the call girl or call boy doesn't usually risk violating these statutes.

In most cases of women arrested for prostitution, conviction leads to a minimal fine which is simply one of the expenses of doing business. Pimps often have a service contract with a lawyer who promptly arranges bail for any of their arrested "girls" so that their time off the streets is limited.

It is a tried but true political fact of life that in many cities, major campaigns to "wipe out prostitution" are primarily carried out to generate favorable publicity for the police and a mayor seeking reelection. These campaigns also result in a prominent surge in arrest statistics, making it look as though the war against serious crime is nearing a victorious conclusion. However, it is unlikely that prostitution — the "oldest profession" — will be wiped out soon. It seems equally unlikely that prostitution will be either decriminalized or legalized in the United States at any time in the near future since public opinion still runs strongly against this reform.

Sex between Adults and Minors

In sharp contrast to the variations in law regarding sex between consenting adults, sex between an adult and a child is strongly condemned in all jurisdictions in Europe and in the United States, Canada, and most other countries. Sexual activity between an adult and a child is now widely regarded as a form of child abuse. The specific crime may be rape, statutory rape, sexual assault, child molestation, impairing the morals of a minor, or incest, depending on the precise circumstances.

For the purpose of this discussion, we should point out that adults who have sexual contact with children or who exploit them in other sexual ways (by forcing them into prostitution or by using them in the production of pornographic materials) are generally dealt with harshly by our criminal justice system. The severity of sentencing depends, to some extent, on the age of the child and the nature of the sexual act. As a general rule of thumb, an act with an older teenage child is not judged as harshly unless force or injury was involved. The convicted sex offender is likely to receive a stiff prison term, particularly if it is a repeat conviction. In some jurisdictions, the offender may be given an indeterminate (open-ended) sentence. Depending on subsequent behavior and willingness to participate in treatment, the offender may be paroled in a shorter or longer time, or may be imprisoned indefinitely.

The Paraphilias

Many paraphilias involve sexual acts that violate the law in one form or another. For example, exhibitionism falls under laws against "indecent exposure" and, if genital exposure is accompanied by masturbation, it also violates statutes prohibiting sex acts done in public. Peeping involves an unwanted invasion of privacy. The cross-dressing of the transvestite is sometimes criminal under "public nuisance" laws (although females in male clothing are not arrested on these grounds). Sadistic acts may violate laws on assault; the fetishist may steal to enlarge his fetish collection; and bestiality is generally illegal as a "crime against nature."

People with a paraphilia are in an unusual state of limbo in a legal sense today. Often, they will be categorized as *sexual psychopaths,* a term that is not a psychiatric diagnosis but a legal label. This designation permits them to be given an indeterminate sentence and also suggests that they are a menace to society. While this may be true of some paraphiliacs, it is difficult to see how someone with zoophilia is menacing or what danger is posed by the ordinary transvestite. Similarly, the majority of voyeurs, fetishists, and exhibitionists are unlikely to endanger others. The problem is that there is no way to predict what subsequent behaviors will occur after treatment or after "successful" prison rehabilitation. With every case of a sexual psychopath who is arrested a second or third or fifteenth time, the public demands better protection, stricter law enforcement, and harsher sentencing. Until more reliable information is developed to understand the nature of these conditions, there is no effective answer to such concerns.

CHAPTER SIXTEEN

Coercive Sex: The Varieties of Sexual Assault

A FIFTEEN-YEAR-OLD female hitchhiker is kidnapped by a truck driver and raped repeatedly over an eighteen-hour period. She is stabbed eight times and left in a field for dead. She is a victim of sexual assault.

A seventy-five-year-old widow is returning to her apartment from the grocery store, struggling with a heavy bag of food. A teenage boy on her street offers to carry it for her. When she unlocks her apartment door, he pushes her inside, rips off her clothes, and rapes her. She is a victim of sexual assault.

A twenty-two-year-old secretary is repeatedly pinched and ogled at work. Invitations are made for sexual trysts at lunchtime, with a veiled threat that "if you don't do what I want, I'll find someone else who will." The secretary *is* finally fired by his boss, and files a lawsuit against her. He is a victim of sexual assault.

A twenty-six-year-old man is being held in the county jail on marihuana charges. Before his hearing, he is gang-raped by six other prisoners, who stuff a towel in his mouth and hold him down by force. The prison guards refuse to remove him from the cell. The next day, he is found hanging from the ceiling, an apparent suicide. He, too, is a victim of sexual assault.

These true cases provide a glimpse of the types of coercive sex in our society. Other aspects of sexual victimization include a child trapped in an incestuous situation, a teenager pushed into prostitution, the thousands of women who are sexually harassed and intimidated at work, and the wives who are beaten if they are not ready for sex when their husbands demand it. This chapter provides some historical, legal, social, and psychological perspectives about coercive sex.

RAPE

Rape is an emotion-laden subject surrounded by myths and misunderstandings. While it is defined as a sexual act, it is primarily an expression of violence, anger, or power. Its victims can be male or female, very young or very old, rich, poor, mentally retarded, disabled, or able-bodied. The victimizers — those who rape — are also a diverse group that defies neat classification or description.

Although there are many definitions of rape, it is legally defined in most jurisdictions as sexual assault with penile penetration of the vagina without mutual consent. Strictly speaking, penile penetration of the mouth or anus without mutual consent is not rape but falls under other basic laws against sexual assault.

Historical Perspectives

The word "rape" comes from a Latin term (*rapere*) which means to steal, seize, or carry away. In ancient times, rape was one way to procure a wife — a man simply overpowered a desirable woman and then brought her into his tribe. The man then had to protect his property and his honor by preventing others from seizing or raping his wife. This appears to have provided the origins for the first laws against rape in which rape was viewed as a crime against property or honor but not against women.

According to the Code of Hammurabi, a set of laws established in Babylonia about 4,000 years ago, a man who raped a betrothed virgin was to be put to death. If a man raped a married woman, however, both the rapist and his victim were regarded as guilty and were executed by drowning. A similar distinction with a slightly different twist was found in biblical injunctions about rape (Deuteronomy 22: 22–28): a married woman who was raped was seen as a willing accomplice, so she and her rapist were killed; a virgin was considered guilty only if she was raped in the city, since it was assumed that her screams would have led to her rescue. In contrast, a virgin who was raped in a field outside the city walls was spared, since no one could hear her screaming. If she was betrothed to someone, her rapist was stoned to death — if not, he had to marry her (whether or not she liked this arrangement didn't seem to matter).

Later laws against rape continued to specify varying circumstances by which rape was judged as more or less serious. Penalties were highest if the woman was a virgin or of high social class. Under William the Conqueror (1035–1087), a man who raped a virgin of high social standing was punished by castration and blinding. Guilt, however, was determined by trial by combat, so unless the victim had a companion willing to risk his life by fighting the accused rapist, she had no way of establishing her case.

By the twelfth century, jury trials replaced combat as a means of determining guilt or innocence. Yet all were not equal in the eyes of the law — a nobleman or knight could easily blame one of his men for a rape he committed and save his vision and chances for fatherhood. At the end of the thirteenth century, two additional changes appeared in English law concerning rape: the distinction between raping a virgin or a married woman was dropped, and the old custom of penitence through marriage was permanently banned. The essential elements of defining rape and its punishment had fallen into place. Seven centuries later, not too many changes have been made.

Despite the legal system, rape has not always been regarded as bad. In wartime, from thousands of years ago until today, victorious soldiers have raped enemy women. In literature, rape has sometimes been presented in heroic terms, as Ayn Rand did in *The Fountainhead*. In society, rape has often been practically defined in terms of the social positions of victim and victimizer: in the 1940s and 1950s, for example, a white male was rarely charged with raping a black woman in the South, but a black male charged with raping a white woman was dealt with swiftly and harshly. Even today, in most jurisdictions, a man is unlikely to be charged with raping a prostitute, and forced intercourse between husband and wife does not "count" as rape in thirty-eight of the fifty American states.

The last decade has seen a great increase in public awareness of rape. The women's movement played a major role in this process, raising issues and demanding improvement in services to rape victims, with the landmark book by Susan Brownmiller, *Against Our Will,* published in 1975, eloquently calling the public's attention to what had previously been a largely ignored problem. Today, almost all metropolitan police departments have specially trained teams (including policewomen) to work with rape victims; there are hundreds of rape hotlines and rape crisis centers for emergency and long-term assistance; many hospital emergency rooms have developed special procedures for treating rape victims; and trial procedures and laws about rape have been changed in important ways.

Myths about Rape

The most devastating myths about rape have cast women in the role of being responsible for the rapist's act. According to this view, women secretly "want" to be raped and really enjoy the experience. This nonsensical notion has led at least one rapist to give his name and phone number to his victim, so she could "get together" with him again. His conceit led to his immediate arrest. Lurking beneath the surface of this myth are some commonly held misconceptions: women find overpowering men irresistible; women's rape fantasies indicate a real-life sexual desire; and women dress and act provocatively to "turn on" men, who somehow are the hapless victims of their own reactions to this deliberate provocation.

Closely allied to this view of women as instigators is the idea that "she was asking for it, and she got what she deserved." Susan Brownmiller comments on this "explanation" of rape:

> The popularity of the belief that a woman seduces or "cockteases" a man into rape, or precipitates a rape by incautious behavior, is part of the smoke screen that men throw up to obscure their actions. The insecurity of women runs so deep that many, possibly most, rape victims agonize afterward in an effort to uncover what it was in their behavior, their manner, their dress that triggered this awful act against them. (Brownmiller, *Against Our Will,* 1975, pp. 312–313)

Most research shows that rapists look for targets they see as vulnerable (e.g., walking by themselves, appearing unfamiliar with where they are) rather than women who are dressed in a certain way or who have a certain manner of appearance. The "provocation" myth loses its believability when it is recognized that many rape victims are elderly women or young children. Furthermore, it is a little like believing people should dress in old, worn-out clothes in order to prevent being mugged, thus misplacing responsibility from the criminal to the victim.

Despite this, the woman is still frequently "blamed" for being raped. This view is based not only on her possible role as instigator but on the incorrect notion that a woman who resists *cannot* be raped. According to two old saws: "A girl can run faster with her skirts up than a man can with his pants down," and "You can't thread a moving needle." These bits of attempted "commonsense" humor completely overlook the terror of the rape victim, her fear of physical injury, mutilation, and death, and her shock and disbelief. Even when no weapon is in view, can a woman be certain that one is not hidden? In rape situations, many women hope that by seeming to cooperate with their assailant, they can avoid injury and get it over with more quickly. But, in an irony of our legal system, this intelligent way of coping is penalized — the case against her attacker depends in part on "proving" her physical resistance by cuts, bruises, and other signs that she put up a struggle. No such evidence is required to "prove" that a robbery occurred.

The last myth we will mention here (although there are many others) is the idea that women frequently make false accusations of rape. This view is so common that it has even found its way into contemporary literature: in John Updike's novel *Rabbit Redux* (1972) a woman observes, "You know what rape is? It's a woman who changed her mind afterward." While there have certainly been cases where a false cry of rape was made for some ulterior motive, as widely publicized in 1985 with the Dotson case, the belief that most women are capable of such an act is the ultimate view of women as emotional, vengeful "bitches." Yet the law in some states quietly upholds this view: unlike cases of assault or robbery, where the victim's word and evidence are enough to prove that a crime occurred, in rape cases another person's testimony or evidence — called "corroboration" — is required as proof.

Rape Patterns

Forcible rape is far and away the most common form of rape reported. Here, the act of penile penetration is achieved by force or the threat of force. Several subcategories of forcible rape can be distinguished, although most of these are not legally defined terms. The *solo rape* is carried out by one man, acting alone. The *pair rape* or *gang rape* — often a particularly terrifying form of rape — involves either two men or a group of men, sometimes with a female accomplice, who take turns raping the victim. An apparently rare form of gang rape involves several women raping a man. In another, more common version, a group of men rape another man rectally. This rape of men by men is infrequent among homosexuals and usually involves heterosexual men in prison.

Two other types of forcible rape should be mentioned: *date rape* and *mate rape*. In interviews with 300 women between the ages of eighteen and thirty, we found that about one woman in five had been forced into some form of unwanted sexual activity on a date or at a party. One woman in twenty-five had been raped in these circumstances, although very few of these cases were reported to the police. The following brief account from a twenty-one-year-old woman is typical of the explanations we have been given of these situations:

> I was out on my second date with Jerry and we'd been drinking and dancing and having fun. I agreed to go back to his apartment with two other couples. We drank some more, and the other people left, and we were messing around a little on his bed. When I said I had to get home, it was late, he got mad and pushed me down and raped me. I wasn't really hurt, but I was forced to do something I didn't want to. But I couldn't see that anything would be gained by reporting it.

Similar patterns of date rape were reported by Eugene Kanin (1969).

One of the primary problems of date rape is that many males believe the idea that women are so coy and illogical that "when they say no they mean maybe, and when they say maybe they mean yes." Thus, the male rejects his date's messages about sex partly because he wants to interpret them "his" way, rather than in their literal meaning. Furthermore,the male is often primed into forcing himself sexually on his date through the prior use of alcohol, which lowers his ordinary social inhibitions and may increase his sense of urgency to "score" and fulfill his "macho" self-image without recognizing what he is doing is actually a form of criminal behavior. In fact, in a recent study of 71 college males who had committed date rape, it was found that 76 percent attempted to seduce their partner by using alcohol and/or marihuana, whereas only 23 percent of a control group of age-matched college men who had never raped had done so.

Of particular interest in date rape research is a recent study that disproves the notion that males who coerce their dates into intercourse do so because they are otherwise unable to find willing sex partners. In 1985, Kanin found that men who admitted to having committed date rape actually had averaged almost twice as much heterosexual activity over the year preceding their rape as an age-matched group of controls. Kanin concluded that date rapes are largely a result of a process of sexual socialization in which some males develop an exaggerated sexual impulse and put a premium on attaining sexual "conquests." Finding themselves aroused by noncoital sexual involvement with a date, but frustrated in their ultimate goal of "scoring," they are driven by their high expectancy of success to disregard the female's limit-setting as not genuinely meant, so that they push on by any means available to attain their goal without regard for their partner's feelings or wishes. Thus, in a sense the date rapist behaves as he does to "prove" how good he is at getting what he wants.

In addition, some men think they've paid for sex by picking up the tab on a date — they see women who refuse their sexual advances as backing down from their side of the bargain. Thus, two of the best ways of avoiding potential date rape situations are for women to be prepared to pay their own way and not to drive anywhere with a man who has been drinking too much. Brownmiller notes that date rapes "look especially bad for the victim in court, if they ever get to court."

Mate rapes are probably even more common but can be charged in only 12 states at present because rape laws generally exempt a husband from raping his wife on the assumption that their marriage provides firm evidence of her consent to sexual relations. It is interesting to note that marriage does not exempt people from conviction for other acts of physical violence against their spouse.

Marital rape is estimated to be a far more common form of family violence than had previously been realized. In one study, rape by a husband reportedly occurred more than twice as often as rape by a stranger, with one out of eight married women saying that they had been victimized in this manner. Other researchers suspect that the real incidence of mate rape is much higher, noting that many women are either unwilling to report being forced to have sex by their husbands or don't think of this as a "real" form of rape. While some cases of forced sex in marriage might not qualify as rape according to a court of law, in other instances the victim (usually, but not always, the wife) is beaten and battered or otherwise abused. A recent study of marital rape found that there is little evidence that wives who have been raped by their husbands provoke these assaults by refusing reasonable sexual requests. Instead, it appeared that the husbands liked violent sex and used physical force to intimidate and control their wives.

One of the most troublesome aspects of mate rape, of course, is that the victim not only must live with the memory of her traumatic experience but also must live with her rapist, never being sure when she will be assaulted again. One thirty-three-year-old woman thus described it:

> If he's mad at me, he loves to drag me into the bedroom and force me to have sex. It's a punishment session, he says. I learned a long time ago that physically resisting just goaded him on and left me bruised and bloody. So I don't fight back now, but I don't see how anyone could call this violence a form of making love.

In addition to forcible rape, there is also a somewhat smaller category of *nonforcible rape.* Included here is *statutory rape,* defined as intercourse with a girl below the age of consent (even if she agreed to or initiated the sexual contact). While a woman could theoretically be charged with statutory rape, cases of intercourse between a woman and an underage boy are generally prosecuted under charges such as "contributing to the delinquency of a minor" or "carnal abuse." Also included in the category of nonforcible rape are rapes where the woman's capacity to consent is impaired because of mental illness or retardation, drugs or alcohol, or deceit. Other varieties of nonforcible rape involve some form of coercion: a blackmailer who extracts sexual payment, a professor who demands sex in return for a better grade, a sex therapist who "diagnoses" or "treats" his female patients by having sex with them, a prospective employer who makes it clear that a job offer depends on sexual submission.

Rape Victimology

Rape is an act of violence and humiliation in which the victim experiences overwhelming fear for her very existence as well as a profound sense of powerlessness and helplessness which few other events in one's life can parallel. (Hilberman, 1976, p. 437)

Rape is a crime against the person, not against the hymen. (Metzger, 1976, p. 406)

You have to stop being a victim of the rape. The person who raped you didn't do it for sexual pleasure; he did it to have power over you. And if you let him have power over you for the rest of your life, he's really won. A lot of women remain victims the rest of their lives. (Anonymous rape victim, quoted in the *St. Louis Globe-Democrat,* March 20, 1983)

According to FBI statistics, there were 82,000 forcible rapes reported in the United States in 1981. Most researchers and law enforcement authorities believe that reported rapes are only a small fraction of actual rapes, so it is possible that the number of annual rapes is over one-half million. A recent estimate suggests that one woman in six will be the victim of an attempted rape in her lifetime and one in twenty-four will be the victim of a completed rape. In light of these statistics, it is particularly important to understand the effects of rape on the victim.

The rape victim, whether female or male, young or old, emotionally composed or in a terrified state of shock and disbelief, needs careful medical attention. Physical injuries are common — and not always visible or obvious — and some are so serious that they present a life-threatening emergency.

In addition to the detection and treatment of physical injuries, the victim should be provided with information about testing for and possibly treating sexually transmitted diseases to which she or he may have been exposed. Female rape victims who could become pregnant should also undergo a pregnancy test and should be informed of the pregnancy prevention options available to them. These include the use of DES (diethylstilbestrol), insertion of an IUD (which prevents implantation), menstrual extraction, and abortion.

Finally, if the victim consents, the medical examination may be used to gather evidence for possible legal proceedings. For this reason, it is advisable that the woman not bathe or shower (or otherwise clean herself up) before being examined if she wants to report the rape to legal authorities.

For many rape victims, a major question is whether to report their rape to the police or not. While reporting a rape may seem logical, many women have hesitated or decided against it for any or all of the following reasons: (1) fear of retribution by the rapist, who may get out on bail; (2) an attitude of futility — "the police probably won't catch him, and even if they do, he'll probably get off"; (3) fear of publicity and embarrassment; (4) fear of being mistreated by the police or trial lawyers; (5) pressure from a family member against reporting; and (6) occasionally, unwillingness to ruin a friend or relative's life by sending him to prison. In addition, the victim of a date rape may be afraid of facing the adverse judgment of mutual friends if she reports the incident, and victims of marital rape may fear the social and economic consequences if their husband is convicted and sent to jail.

These concerns are, by and large, founded in fact. In the past, police often scoffed at a woman's story and asked humiliating questions like "Did you enjoy it?" or "Do you like sex a lot?" As Gager and Schurr (1976, p. 68) comment, "Such questions have little to do with finding the rapist and much more with human curiosity or satisfying the officers' vicarious sexual urges." Similarly, in many instances a report of rape never gets to trial, even if the rapist is identified — the district attorney can simply decide that the case is unfounded or unprovable.

Even when reporting a rape leads to identification and arrest of a suspect, the trial itself may be an anguishing ordeal for the victim. Typically, the woman is made to feel that *she* is on trial rather than the accused man. The defense lawyer may try to show that she consented to sexual activity; if she waited for more than a few hours to report the rape, her motivation and truthfulness may be questioned; if she showered or changed her clothes, there may be insufficient evidence; and in some instances, her past sexual behavior may be questioned on the presumption that a woman with many sexual partners is likely to have consented rather than have been raped.

Fortunately, there have been some major advances in police investigations of rape cases as well as in the legal process. In many states, for example, it is no longer permissible for a defense attorney to introduce the woman's past sexual behavior into the trial and women are not required to "prove" that they attempted to resist the rape by signs of physical injury. Police have also generally improved their sensitivity toward dealing with rape victims. In most American cities, police departments now have specially trained teams for dealing with victims of sexual assault.

Unfortunately, despite these advances, the outcome of rape trials is still a far too subjective matter. Recent research funded by the National Center for Prevention and Control of Rape, a division of the National Institute of Mental Health, shows that American juries are far more likely to believe that a female has been raped if she appears chaste and has a traditional life-style. In this study, which involved interviews with 360 jurors in sexual assault trials immediately after they were over, Barbara Reskin, a professor of sociology at the University of Michigan, and her colleagues found that if the jurors questioned the victim's "moral character" they were far more likely to vote against convicting the defendant. Rape victims were most likely to be believed by jurors if they were married and were assaulted in their own homes. Conversely, if the rape victim was portrayed as sexually active or morally "loose," having had an illegitimate child or using marihuana, for example, the jurors were apt to not take her testimony seriously. Furthermore, while jurors were relatively biased against defendants who seemed to be losers or who had a scruffy appearance, those who were good-looking and could show that they had sexual access to a woman were likely to get the jury to side with them more easily. Despite the fact that many states now prohibit defense attorneys from questioning rape victims about their sexual histories, women continue to be put through a major ordeal on the witness stand in such cases, with attorneys often asking questions that they know will be objected to simply to plant a seed of doubt in the minds of the jury. In one trial attended in Reskin's study, "A woman who was raped in her own home at 2 A.M. while she was sleeping was asked by the defense attorney if she had been wearing a bra."

One additional point should be noted about legal aspects of rape. Although a criminal case can be filed and prosecuted only by the city, county, or state, the rape victim herself may choose to file a civil action in which she sues her assailant for personal injury, pain and suffering, or punitive damages. In a civil suit, unlike criminal proceedings, the woman can hire her own attorney. In addition, since civil proceedings require that the suspect be proven guilty only by "a preponderance of evidence" rather than the more stringent standard of guilt used in criminal proceedings ("beyond a reasonable doubt"), there may be a greater likelihood of winning the case. However, a civil suit will not succeed in putting even a convicted rapist in prison: it will only obtain a monetary judgment.

The psychological impact of rape can be profound from the first moments of the attack and for years afterward. The rape victim reacts with a sense of isolation, helplessness, and total loss of self. How the victim handles the severe stress of this crisis usually falls into a recognizable pattern. The *acute reaction phase* usually lasts for a few days to a few weeks. The victim typically reacts with shock, fear, disbelief, and emotional turmoil. Guilt, shame, anger, and outrage are commonly seen in those women who are able to talk about their feelings. Other women, who adopt a more controlled style, have an apparent calmness that may indicate that they are forcing an attitude of control or are denying the reality or impact of the experience. This phase is usually followed by a *post-traumatic "recoil" phase,* which can last weeks or months. The victim undergoes a limited degree of coming to grips with herself and her situation. Superficially, she may seem to be over the experience. She tries to relate to her family and friends. She returns to her everyday activities and tries to be cheerful and relaxed. But deep down inside, she has not really grappled with her fears, her self-doubts, and her feelings about the experience.

The final phase, a *long-term regrowth and recovery process,* varies considerably depending on the victim's age, personality, available support systems, and how she is treated by others. Frightening flashbacks and nightmares are common. Fears about being alone, suspicious men, and sexual activity surface with distressing frequency. Proper counseling or psychotherapy may be needed to deal with these fears and the depression that often occurs.

One recent study by Nadelson and her co-workers of women interviewed an average of twenty-two months after being raped found that three-quarters of the women reported changes in their lives that they directly attributed to the rape experience. Almost half of the women reported some form of fear, anxiety, or symptoms of depression; many also had trouble sleeping, feelings of vulnerability, and fear of walking alone, even during the day. The most common symptom still present almost two years after the rape was a general suspiciousness of others.

After a rape, some women avoid any type of involvement with men in sexual or social situations. Other women take just the opposite approach, with mixed results.

> After I was raped, I had intercourse with my husband as a ritual gesture. (I had learned as a child to get back on my bike after falling, lest I never mount again.) Intercourse was easy. It didn't matter. I was an abandoned house. Vacated. Anyone or anything could enter. (Metzger, 1976, p. 406)

Women who have been raped may face a number of sexual problems as a consequence. Sexual aversion or vaginismus (involuntary spasms of the muscles around the outer vagina in response to attempts at intercourse) are the most dramatic responses to the trauma of rape, but other women have difficulty with decreased sexual desire, impaired vaginal lubrication, loss of genital sensations, pain during intercourse, and anorgasmia. One study indicates that even though rape victims may have the same *frequency* of sexual activity one year after their rape as nonraped women, their sexual satisfaction is significantly reduced. Another study found that more than half of a group of women who were victims of either rape or incest had post-assault sexual dysfunctions, with fear of sex, lowered sexual desire, and difficulty becoming sexually aroused accounting for the majority of problems. The husband or sexual partner of the rape victim may also encounter sexual difficulties. Erectile dysfunction is not unusual, and *his* sexual desire may be affected by anger or disgust.

The male partner of a rape victim often experiences a psychological crisis, too, in which shock, blame, and a sense of guilt emerge. This may cause him to become overprotective or to try to take charge of the legal and medical decisions that his partner faces. In some instances, the male may become preoccupied with thoughts of vengeance as a means of dealing with his own turmoil and discomfort. In other cases, the male may try to show that his love for his partner is still intact by pushing for sexual intimacy, not recognizing that the woman needs to be the one to decide — based on her own feelings and reactions, and not his — when to resume sexual activity.

What is most helpful is a willingness on the male's part to allow for open communication, giving his partner a chance to express her anger, anxieties, or other feelings in an atmosphere of acceptance and empathy. This openness should also include his being able to accept her silence, if that is what she requires. In addition, if the couple has children, they also will be likely to sense that something significant has happened. As Grossman and Sutherland (1982/83, p. 32) point out: "What children imagine is usually more frightening to them than knowing the facts. It is helpful if they are given the opportunity to deal with their feelings."

As the couple adjusts to the post-rape period, the male may benefit from counseling, too. In fact, the long-term process of recovery from rape often can be facilitated by counseling for the couple, since there is some evidence that in stable, committed relationships, the male is a prime source of support for his partner.

How a rape victim resolves all of these problems is not well understood at present. For many, the counseling experience provides a useful opportunity for working through feelings of anger, worthlessness, depression, or fear. Others seem to handle things most comfortably on their own; although until more information is available, it is not certain that their adjustment is as satisfactory.

The Rapist

Information about men who commit rape almost exclusively depends upon studies of convicted rapists. The information obtained from these studies cannot be applied to all rapists because the less intelligent, less affluent rapist is most likely to be arrested and found guilty. Many of these studies are also done many months or even years after the rape was committed, and the rapist may not be accurate in recalling the details of his act.

Summarizing findings from a number of sources, it is possible to outline some general facts about convicted rapists:

- Eighty-five percent have a prior criminal record.
- Eighty percent never completed high school.
- Seventy-five percent are under thirty years old.
- Seventy percent are unmarried.
- Seventy percent are strangers to their victims.
- Sixty percent are members of racial minorities.
- Fifty percent were drinking heavily or drunk when they committed the rape.
- Thirty-five percent have previously been convicted of rape.

However, convicted rapists are not all alike. Their motivations for raping vary considerably, and their methods of finding a victim, overpowering her, and sexually tormenting her are not the same.

In some cases, the rapist methodically commits a long string of carefully planned assaults. Such rapists are not always deranged outcasts from society: in 1983, for example, Dr. Edward F. Jackson, a respected practicing physician and hospital board member, was convicted of twenty-one rapes over a seven-year period and pleaded guilty to an additional fifteen rapes in a separate case. Jackson, who had kept a list in his car of the names of the women he had assaulted and the dates of his attacks, was sentenced to 191 to 665 years in prison. The much-publicized case of Ted Bundy, a one-time law student who was convicted of the rape and murder of a number of women, also shows how "respectable" some rapists may seem.

Other rapists act impulsively, with no apparent premeditation of their act. In Colton, California, a forty-year-old man who discovered four teenage boys raping a twelve-year-old girl in a shed in his backyard "apparently got in line and participated," according to police (*The New York Times*, March 23, 1983). And, in a case that received major national publicity in 1983, a twenty-one-year-old woman who stopped in a bar in New Bedford, Massachusetts, for some cigarettes and a drink was grabbed by a group of men and brutally raped for several hours while the bar's other patrons stood by watching, laughing, and cheering without attempting to call the police (*The New York Times*, March 17, 1983).

One of the key advances of the last decade in studying rapists has been the realization that rapists are not oversexed men and that rape is usually an expression of power or anger and not an act of sexual desire. Notably, most rapists do not lack available sexual partners. This is not to say that rape has no sexual meaning or motivation; in most rapes, however, the aggressive components are so predominant that the sexuality of the act becomes secondary.

Groth, Burgess, and Holmstrom — a psychologist, nurse, and sociologist — studied 133 rapists and 92 rape victims to better understand the dynamics of the rape situation. They found that forcible rape could be classified as either power rape or anger rape. None of their rape cases showed sex as the dominant motive. According to these researchers, *power rape* occurs when the rapist tries to intimidate his victim by using a weapon, physical force, and threats of bodily harm. The power rapist is usually awkward in interpersonal relationships and feels inadequate as a person. Rape becomes a way for him to reassure himself about his strength, identity, and sexual adequacy.

In *anger rape*, the rapist brutalizes his victim and expresses rage and hatred by physical assault and verbal abuse. The motive behind this type of rape is often revenge and punishment against women in general and not the victim specifically. The anger rapist usually gets little or no sexual satisfaction from the rape and may have difficulty getting an erection or being able to ejaculate with his victim.

Groth, Burgess, and Holmstrom later described a third pattern, *sadistic rape*, where sexuality and aggression are fused together and the suffering of the victim is the primary source of the rapist's satisfaction. The victim of a sadistic rape may be tortured or deliberately injured by cigarette burns, bites, or whipping. Sex murders, with grotesque mutilations of the victim's body, are extreme cases of sadistic rape. Groth estimates that about 5 percent of rapes are sadistic rapes, 40 percent are anger rapes, and 55 percent power rapes.

Two separate studies shed additional light on the psychological makeup of the rapist. In each study, a group of rapists was compared to a group of nonrapists in terms of erection measurements while listening to taped descriptions of rape and of mutually consenting sexual scenes. Both studies showed that rapists developed erections while listening to descriptions of rape, but nonrapists did not. In response to descriptions of mutually consenting intercourse, the erection responses of rapists and nonrapists were similar. Interestingly, the rapists did not show greater sexual arousal to forced or violent sex than to consenting sex. These findings suggest that nonrapists may have internal controls, such as fear or empathy for the victim, that inhibit their sexual arousal to descriptions of rape situations, while rapists either lack such internal controls or have learned to overcome them. It should be noted, however, that these studies involved only a small group of rapists, and it is uncertain if these findings apply more widely.

It is also not clear if the patterns described above apply to the motivations and dynamics of date rape. In this situation — which may actually be the most common form of rape — the sexual anticipation component may be a primary factor, although the power issue is also involved. Further research is required in this area.

Male Sexual Dysfunction during Rape

A fascinating study of 170 convicted rapists showed that sexual dysfunction during rape attempts is a frequent occurrence. After eliminating 69 cases where no data were available or where an evaluation of sexual function was inapplicable because of successful resistance by the victim, an interrupted assault, or no attempt made at penile penetration, 101 cases were left. Of these cases, erectile difficulties occurred in 27 men, premature ejaculation occurred in 5 men, and ejaculatory incompetence was seen in 26 cases.

These findings have two important implications. First, they confirm the view that rape is not primarily an act of sexual desire. Fifty-eight percent of the rapists in this study were sexually *dysfunctional,* indicating that either desire or arousal had gone awry. Ejaculatory incompetence is a particularly infrequent male dysfunction in the general population, and its high rate of occurrence in rapists may signify that their preoccupation with expressing power or anger inhibits their sexual responsiveness. Second, a very practical point arises from this study. In some rape cases, the woman's testimony has been discredited if no sperm were detected on her body. Now two new findings will prove useful in prosecuting rapists who in the past may have been acquitted — many men do not ejaculate during a rape and others may ejaculate prematurely before touching their victim.

Treatment of the Rapist

Since rape is a crime, rather than a medical diagnosis, most convicted rapists are sent to prison. Often, there is little attempt at rehabilitating the rapist; instead, the prison term is regarded purely and simply as punishment. Given this practice, it is not surprising that approximately three-quarters of convicted rapists become repeat offenders.

Various past attempts to provide psychological counseling to men convicted of rape have not improved the situation very much. One of the problems with these "psychotherapy only" programs has been that they have not succeeded, by and large, in quelling the inner compulsion to rape that some of the rapists claim they feel. Recently, this has been approached (in a limited number of cases) by the combined use of the drug medroxyprogesterone acetate (MPA; also known by its trade name, Depo-Provera) and psychotherapy. The use of MPA results in a sizable reduction in circulating testosterone, which in turn causes a marked drop in a man's sex drive. This enables psychotherapists to be more effective in helping a man to reorient his sexual and aggressive impulses, although the effect is a temporary one that lasts only for as long as the drug is taken.

Dr. Fred Berlin of Johns Hopkins University, who has pioneered in the use of MPA for rapists and other sex offenders, states that 17 out of 20 men treated with this drug were able to self-regulate their sexual behavior while receiving the medication. However, almost all of the men who stopped taking the drug subsequently relapsed, and it is not as yet clear how this experimental treatment could be monitored effectively even if it ultimately proves useful. Another, more compelling criticism is that MPA is hardly the answer to what is basically a crime of violence. Since there is no guarantee that MPA will prevent a man from functioning sexually, nor that it would reduce a rapist's hostility toward women, it is even possible that the use of MPA might make some rapists more hostile, and thus more likely to break the law, than they may have been before. In any event, it is clear that more information is required on the long-term effectiveness of the use of MPA in treating rapists before this method will be widely accepted.

INCEST

Incest (from the Latin word for "impure" or "soiled") refers to sexual activity between a person and a close relative, such as a parent, a brother or sister, a grandparent, or an uncle or aunt. Although brother-sister incest is probably most common, most cases of incest reported to the authorities involve an adult-child interaction. Therefore, we will discuss this form of incest in more detail. All states require the reporting of such cases of suspected incest under child abuse laws, with the underlying assumption that a child is unable to consent in a meaningful way to a sexual interaction with an adult.

The incidence of incest can only be guessed at since reported cases make up only a small fraction of the overall number. Estimates suggest that perhaps 50,000 children are abused sexually by their parents or guardians each year, with an even larger number victims of rape or molestation at the hands of other family members. In one study, in 32 percent of rapes involving children, the offender was a relative.

Research on incest has been plagued by the limitations of working with clinical samples (those seeking treatment) or samples drawn from prisons. One distortion that this has introduced is the idea that father-daughter incest is the most frequent pattern. In reality, father-daughter incest seems to be far less common than brother-sister incest, but sexual activity between siblings is almost never reported or brought into treatment. The Playboy Foundation survey done in 1975 found that about 4 percent of men and women had ever had sexual contact with a sibling, but only 0.5 percent of the women had sexual contact with their fathers and an even smaller percent of men described participation in parent-child sex. The Kinsey surveys also found that brother-sister incest was far more common than parent-child sexual relations. Recent data from a sex-therapy clinic confirm this finding.

David Finkelhor, who is associate director of the Family Violence Research Program at the University of New Hampshire, has recently developed a list of the factors that appear to be important predictors of childhood sexual abuse. The single most important predictor among these factors was having a stepfather, which more than doubled a girl's vulnerability to sexual victimization. Next in importance was having a mother who was punitive or highly negative about sexual matters (e.g., scolding or punishing a child for asking sexual questions or masturbating). The other vulnerability factors Finkelhor noted were having a mother who had not graduated from high school, not having a good relationship with the mother, having ever lived apart from the mother, not receiving physical affection from the father, having a family income under $10,000, and having two or fewer close friends in childhood. In a study of 796 college students, Finkelhor found that if none of these risk factors was present in a student's background there were virtually no reports of childhood sexual victimization. Strikingly, however, among those with five or more of these risk factors, two-thirds had been sexually abused.

Myths about Incest

Several myths about incest continue to be widely believed. The origins of these myths are not entirely clear but they continue to influence people's thinking about incest.

Myth: *Incest occurs primarily in poor, uneducated families.*
Fact: Incest is not bound by family wealth or education. While incest in middle-class or well-to-do families may be handled privately without being reported to the courts or social agencies, solid evidence shows that families in all walks of life can be affected.

Myth: *Incest is usually committed by a father who is a sexual degenerate.*
Fact: Most studies show that fathers who commit incest are neither "oversexed" nor fixated on children as sex objects.

Myth: *Claims of incest by a child are usually made up.*
Fact: This myth can be traced back to Freud, who suggested that reports of sexual activity with a parent were actually based on fantasies due to Oedipal desires. Unfortunately, most children's reports of incest — no matter how shocking and unbelievable they may seem — are likely to be true.

Patterns of Incest

Incest occurs in a wide variety of forms, and it would be foolish to regard them all as equivalent. Some cases of incest, for example, are one-time occurrences producing so much guilt or anxiety for either participant that they are never repeated. Other cases involve long-term interactions in which both parties seem to be interested (with no physical force used), one party is overtly coerced and terrorized, or multiple incest occurs, as when a father molests several daughters.

Other variables to consider when defining the incest situation include the child's age at the outset of the relationship, the openness or secrecy of the activities, the types of sexual activity involved, and the impact of the interaction on family dynamics. Often, the incest behavior begins as a kind of teasing, playful activity with prolonged kissing, wrestling, and surreptitious genital touching. Over time, these activities can develop into overt genital sexuality, without any physical force being used.

Commonly, the daughter is made to feel that the father and mother's happiness, their love for her, and the stability of the family rests on her willingness and her silence. Unlike many other forms of sexual abuse, incest often leads to very complex, ambiguous feelings in the daughter. It is not uncommon for the daughter to experience some sexual pleasure and a feeling of importance and power in her family. Often these feelings are intertwined with negative feelings such as sexual displeasure, pain, and guilt. (Gottlieb, 1980, pp. 122–123)

At other times, incest begins abruptly and forcefully. The father may be drunk or may have had a vicious argument with his wife, or decides to use sex to "punish" his daughter or to "teach her what she needs to know." The child is likely to fight back and may be physically injured.

Interestingly, most men who become involved in incest are shy, conventional, and devoted to their families. Many profess to be strongly religious, although privately they are apt to voice scorn or disregard for ordinary taboos. However, the dissimilarities among fathers who commit incest are far more noticeable than the similarities. They may be heavy drinkers or teetotalers, construction workers or professionals, highly educated or elementary school dropouts. Although many seem to be mild-mannered, this outward appearance can be quite deceptive. While some are confirmed sociopaths, with little or no regard for others and a glib, believable explanation for everything, and others are clearly psychotics who suffer from a vastly distorted sense of reality, many incest offenders do not have a diagnosable psychiatric disorder.

This variability in the profile of male incest-offenders is mirrored by the considerable variability in their incest acts. Most commonly, incest behavior begins when the child is eight to twelve years old, but we have seen a number of cases where incest was initiated while the child was still in diapers. In some incest families, the father selects only one child (usually the oldest daughter) as his victim, but it is common for several siblings to be victimized — sometimes sequentially over the years, but sometimes simultaneously. This variability is also evident in terms of the frequency of incestuous contacts (ranging from daily acts to those that occur only once or twice a year), the type of sex acts involved, and other characteristics such as whether there is an overtly sadistic element to the forced sexual encounters.

Father-son incest is considerably less common than father-daughter incest; consequently there is far less information available about this pattern. However, there are several pertinent observations that can be made about this virtually neglected form of incest.

1. In contrast to the situation with father-daughter incest, where the likelihood is that the daughter has been the sole child to be victimized, when a boy is the incest victim of his father the odds are much higher that there is at least one other sibling who also has been victimized.
2. Fathers who sexually abuse their sons are not necessarily, or even typically, homosexuals. In many cases, these fathers have never had an adult sexual experience with another man.
3. Boys who are sexually victimized by their fathers tend to be somewhat younger than daughters who are victimized.
4. A substantial number of fathers who sexually abuse their sons were themselves the victims of incest as children.

Wives of husbands who commit incest were often themselves the victims of sexual abuse as children and tend to be dependent, disenchanted women who withdraw from the family either through depression or outside diversions. The mother may actually force the daughter into assuming her role, relieved at having the daughter as a "buffer" between her and her husband and sometimes pleased to have to deal with her husband's sexual advances no longer. Even after incest is discovered by a mother, in more than two-thirds of cases she does not try to help or protect her child. Yet the mother is not the primary culprit in most cases of incest: she may be defenseless to stop her husband, unable to control her daughter, fearful of her husband's physical retaliation, and worried about having her family break up if her husband is put in jail.

Recent evidence has begun to show that incest is particularly common in reconstituted families, those in which remarriage occurs after divorce or the death of a spouse. Statistics gathered by researcher Diana Russell support this view: in interviews of 930 San Francisco women Russell found that only one out of forty was sexually abused by her biological father, but one out of six reared in reconstituted families had been sexually abused by a stepfather. The explanation for this difference may be that there is less of an incest taboo between non-blood relatives. In addition, in many reconstituted families the stepfather is thrust into close daily contact with an adolescent stepdaughter, a situation possibly contributing to sexual arousal. Lacking proper internal controls that ordinarily arise from the protectiveness of parenting a child from infancy on, the new father may find less to prevent such arousal from being translated into behavior. In fact, it appears that in at least some instances men have married divorced women primarily so they could gain sexual access to their children.

The incest victim, whether female or male, is subjected to a number of intense pressures that are apt to create considerable internal conflict. Almost invariably, the father makes threats to frighten the victim into maintaining secrecy. These threats run the gamut from promises of physical retaliation against the victim or other family members to predictions of breakup of the family if the incestuous activity is revealed. At the same time, many of the fathers who commit incest at least occasionally provide some form of positive reinforcement to the child they are victimizing, such as gifts, monetary rewards, or special privileges. This, together with the sense of guilt the child characteristically experiences over pleasurable sexual sensations during the incestuous acts, combines to give the child victim a sense of complicity in the situation that makes disclosure seem all the more difficult. In fact, many children who are incest victims acknowledge that the fear of being blamed for instigating the forbidden acts is one of the major factors in their continuing silence. Unfortunately, it is also common to find that when a child incest victim has finally gotten up enough courage to tell an adult, the adult's reaction is one of disbelief ("Oh, that can't *possibly* be true. You've got to stop letting your imagination run wild").

There is relatively little research information available on brother-sister incest. Reported cases usually involve an older brother (in his late teens or early twenties) and a considerably younger sister, but most cases probably involve siblings who are close in age. The brother is usually the dominant partner in sibling incest, but we have seen almost a dozen cases where the reverse relationship was true. Although most cases of brother-sister incest seem to involve mutual consent, in a few instances one sibling blackmails the other into providing sexual gratification, as shown by this description of her experiences given to us by a twenty-six-year-old woman:

> When I was fourteen, my older brother (who was sixteen) found out that I was doing drugs. Apparently he snuck around for awhile and got a whole set of "evidence" together, and then he confronted me with it one night when our folks were at a movie. He told me I had two choices: either give him a blow job, or he'd tell my parents what I was doing.

A recent survey of 796 college students found that of those who reported sibling incest experiences (15 percent of females and 10 percent of males), one-fourth of the experiences were categorized as exploitive.

Mother-son incest is rare. In one sample of 203 cases of incest in the nuclear family (that is, mother, father, and children — not other relatives), only two cases of mother-son sexual activity were described. According to Meiselman (1978, pp. 299–300), "In the great majority of reported cases in which the son initiates incest with his mother, the son is schizophrenic or severely disturbed in some other way prior to incest." In cases of mother-initiated incest, the mother is usually psychologically disturbed. Mother-son incest typically involves genital fondling without intercourse if the child is young, but with boys over age ten, coitus is the most typical activity.

Mother-daughter incest seems to be the rarest form of nuclear family sex. While father-son incest is encountered more frequently, it too is exceptionally rare and accounts for less than one percent of cases overall.

The Aftermath of Incest

Most researchers and clinicians agree that incest is an intensely damaging psychological experience. It can lead to drug abuse, prostitution, suicide attempts, and a host of other problems. One research team concisely summarizes this viewpoint:

There is a striking similarity in the reported reactions of incest participants: The children take over the responsibility and the blame from the initiating parent. The betrayal of parental responsibilities and the failure of responsible adults leads the child to feel he or she is fundamentally bad and unworthy of care or help. Sexuality, tainted with guilt and fear, becomes exaggerated as the only acknowledged aspect of attraction or power. (Summit and Kryso, 1978, pp. 248–249)

Perhaps the most striking, but not surprising, finding in incest victims is the long-term persistence of a variety of sexual problems. Sexual difficulties usually bring an adult woman into psychotherapy, where she finally is able to reveal a childhood incest situation ten to twenty years after the fact. In many cases, the woman has been unable to form close, intimate, trusting relationships with men because she expects betrayal, rejection, or punishment.

There are a few studies, however, that suggest that incest victims may not be harmed by their experience and can become healthy, well-adjusted adults. If such cases exist in any numbers, they would be the last to be identified or reported by traditional means. Nevertheless, it seems likely that an incestuous relationship between an adult and a child will create major conflicts for the child, even if these are eventually overcome. In addition, and perhaps of even more importance, is the fact that even in cases where incest has no demonstrably harmful effects on the child, it is still morally wrong because the child is not capable of truly giving a free and informed consent to such behavior. As Finkelhor points out, children are not knowledgeable enough about sexuality and its personal and social meanings to give a valid consent to sexual contact with an adult; furthermore, children don't have the freedom, either legally or psychologically, to give a meaningful consent to such behavior. While these remarks do not apply to cases of adult-adult incest, we believe that adults are morally and ethically bound to refuse to have sexual contacts with children and that failure to follow this ethical imperative should be regarded as an act of serious consequences.

If You Have Ever Been
Involved in Incest

Reading about incest can stir up old, hidden memories for anyone who has been involved in sexual contact with a relative. The
descriptions of incest presented here may or may not match
your own feelings and experiences; for example, you might be
surprised to learn that others were harmed by such involvement
if you never felt you were, or you might not have realized that
incest is illegal. Certainly, no two people cope with an incest
experience in exactly the same way. However, since current
studies show that many people who were involved in incest as
children never told anyone about their situation and have resid-
ual feelings of guilt, resentment, anger, or poor self-esteem that
seem to be linked to the incest experience, it may be helpful to
consider the following options.

Some people feel completely comfortable with a "forgive and
forget" attitude and have no desire to talk about this topic with
anyone. Others who feel they were in consensual, nonexploitive
incest relationships may not even see a need to forgive or forget,
since they may view the entire experience as a positive one.

On the other hand, people with a background of incest who
are having sexual problems or difficulty forming intimate relationships are likely to benefit from professional help in dealing
with these issues. Even though the incest may have occurred
decades ago, even if it was a single episode rather than a re-
curring pattern, consulting a psychologist, psychiatrist, social
worker, or sex therapist can help you determine whether counseling on this issue can help. In fact, in these circumstances many
people discover that the opportunity to bring the incest experience and their long-range reactions to it out in the open is a key
step in gaining more control over their lives.

If you fall in between these two groups — those who are most
comfortable leaving things as they are and those who seek
professional help — there are two other options you can also
consider:

1. Some incest victims have felt tremendous relief in confronting the person who initiated the incest behavior years later,

in adulthood, to explain how they felt and to obtain an acknowledgment or even an apology. This approach can help you feel more in control and less a victim, but it can backfire if you let your anger drown out all your other feelings, or if the person you confront denies your accusations or even asserts that *you* initiated the sexual contact.

2. Short of confrontation, which may not always be possible (for instance, if the other person has died or is mentally incapacitated), it can be helpful to confide in someone else — a spouse, sibling, parent, lover, best friend, or member of the clergy, for example — so that you're not forced to bottle up your feelings and carry around this burdensome "secret" for the rest of your life. In many instances, the very act of disclosure to someone you trust can be a tremendous relief. But only you can judge whether you are comfortable with this approach.

One other aspect of incest should be mentioned here. If your child ever tells you that he or she has been approached sexually by a family member, do *not* dismiss it as a "misunderstanding" or something that the child "imagined." Realize that younger children usually won't have a vocabulary to describe just what happened — so they may say something like, "Uncle Joe was trying to do funny things to me." All children should be taught that they have the right to say "No" to adults who ask them to do something they don't want to do; they should also know that any form of genital contact with adults is strictly off limits.

CHILD PORNOGRAPHY AND SEX RINGS

The most recent variety of sexual abuse of children to attract national attention has been the use of children in the production of pornographic photographs, movies, or videotapes. While the number of children involved in such activities is unknown, it is certain that thousands are exploited in this fashion every year. In some instances, very young children apparently have no idea they are posing for pornographic purposes — they may actually be posed with teddy bears or dolls. In other cases, what begins as nude modeling quickly escalates to posing in nude scenes staged to resemble sexual activity to finally enacting "live" sex so that movies can be shot or photographs can become more realistic. The allure of payment for such work is usually the principal motivating factor, but in some cases "cooperation" is obtained by threats, blackmail, or kidnapping. There are also cases where an adolescent seeks out a chance to "be in the movies," where not only the money but the thirst for adventure and "stardom" comes into play.

Many of the "stars" and "starlets" for such productions come from the ranks of the estimated 700,000 to one million children who run away from home each year. Usually, the runaway child has no realistic plan of survival and only limited financial resources; lacking friends, family, lodging, and the ordinary restraints of everyday routine, such children are perfect victims for the porn recruiters and pimps who prowl bus depots and hamburger stands looking for their victims. In addition to money, drugs are often provided as enticement or pay. One of the unpleasant ironies of this situation is that many runaways flee their homes because of sexual abuse, only to find themselves further entwined in coercive sex once they're on their own.

Recently, additional information has come to light suggesting that sex rings in which adults exploit children — often, but not only, for purposes of producing pornography — are far more common than previously imagined. In an important study published in the *American Journal of Psychiatry,* Ann Burgess and her colleagues described 11 such sex rings that involved 66 children aged six to sixteen. The fourteen adults who led these sordid activities included a scout leader, a school bus driver, a respected coach, a teacher, and a grandfather. Both "solo" sex rings, in which one adult was sexually involved with small groups of children, and "syndicated" rings, in which several adults formed an organization for recruiting children, producing pornography, and providing direct sexual services to a network of adult customers, were noted. Children were sometimes recruited for these rings by their siblings or friends and were often given drugs by the adult ring leaders to "reward" them for their participation. The adult leaders simultaneously frightened children into not revealing their involvement by threats; they maintained the children's participation in part by convincing them that such activities were normal. In addition, the children themselves typically exerted pressure on their fellow group members to maintain secrecy.

Burgess and her co-workers noted that "the sexual abuse of the children by the adult is compounded by the adult's supporting the children's exploitation of each other." Often, the older, stronger children abused the younger, smaller ones on a regular basis, mimicking the sadistic, humiliating practices of the adult ring-leaders. In all of the sex rings studied, adult pornographic books and magazines were shown to the children to "instruct" and "educate" them. Not surprisingly, three-quarters of the children had identifiable patterns of problems with psychological and social adjustment after their involvement was discovered. While no systematic studies have yet been done on the long-term effects of the exploitation of children by such sex rings, clinical observation suggests that the impact may be profound.

The effects of participation in child pornography are serious and lasting. A psychoanalyst warns, "Children who pose . . . begin to see themselves as objects to be sold. They cut off their feelings of affection, finally responding like objects rather than people" (*Time*, April 4, 1977, p. 56). As with other victims of sexual coercion, they are prone to sexual problems as adults. Worst of all is that "sexually exploited children tend to become sexual exploiters of children themselves as adults" (Baker, 1980, p. 304).

Recently, a government report recommended the following legislative steps to combat the problem of child pornography:

1. Require film processors and laboratories that receive what appears to be child pornography to turn the material over to local law enforcement bodies or to the state's attorney.
2. Require photographers wishing to film a child nude or seminude to receive and maintain possession of a signed release from the child's parent or guardian authorizing such photography.
3. Amend the civil code to provide for licensing of all children used in commercial modeling or performing, with carefully worded prescriptions and substantial sanctions against the use of such children in sexually explicit activities. (General Accounting Office, 1982)

Unfortunately, these steps will not do much to deter the activities of the sex rings described above. These are more likely to be combated effectively by educating children and parents about sexual exploitation by adults rather than pretending "it can't ever happen to *my* child."

SEXUAL HARASSMENT AT WORK

Many women who work outside the home have been victimized by another sort of sexual coercion. Although sexual harassment at work is less shocking to most people than rape or incest, it is a social problem of considerable size. Traditionally joked about or viewed as trivial, it has now become an important issue of sex discrimination in both a legal and practical sense. Although there has been relatively little research on the subject, cases of male sexual harassment at work have also come into view.

Sexual harassment at work can appear in a number of different forms. One version is in the attempt to seek employment. Here, the prospective employer makes it clear that hiring the applicant depends on her sexual availability — and a "sample" is requested as a sign of "good faith." Jokes about the "Hollywood casting couch" as a means for aspiring starlets to gain entry to the world of entertainment are based on fact. Even with far less glamorous jobs, the person doing the hiring has economic power to assist him in making such a request. To a woman who has been unable to find a job and needs to support her family, economic realities may make giving in the simplest and most logical thing to do.

A more common situation for sexual harassment occurs when a boss or supervisor makes sexual compliance a condition for keeping a job, getting a promotion, or obtaining other work-related benefits. Again, the person who does the hiring and firing has economic power. The coercion is even stronger than the pre-employment situation because if the woman is fired for her unwillingness to cooperate sexually, her boss will have to make up a reason "for the record," and this may hurt her chances for future employment. In some situations, the request for sexual interaction is direct, blatant, and threatening. In other cases, while the sexual invitation may be direct, the threat is unspoken — the woman is left to decide what will happen to her if she refuses to play along.

Sexual harassment can occur in virtually any setting. A recent survey of nurses found that more than 60 percent had experienced sexual harassment in the preceding year, and surveys of female law students, medical students, and military personnel have also reported alarmingly high rates of sexual harassment. A survey of twenty thousand federal employees found that 42 percent of the women and 15 percent of the men reported having been sexually harassed at work in the preceding 24 months. In many places little significance is attached to this issue and victims have no formal path for voicing grievances or having a full investigation conducted, although some employers have issued directives to combat this problem. The U.S. Navy, for example, has ordered commanders to deal with sexual harassment with "swift and appropriate" discipline. Because there are few, if any, effective safeguards against reprisal, however, the victim of sexual harassment may have too much to lose to report the problem.

According to Catherine MacKinnon, a lawyer who has written the most detailed book on the subject, when a woman refuses the sexual advances of her boss, retaliation may come in any number of ways. She may be demoted or have her salary cut, unfavorable reports may be put in her personal file, she may be denied requests for vacations, she may be passed over for promotion, she may be given tedious or unpleasant assignments, or her work conditions may be made undesirable. In the last category,

> She may be constantly felt or pinched, visually undressed and stared at, surreptitiously kissed, commented upon, manipulated into being found alone, and generally taken advantage of at work — but never promised or denied anything explicitly connected with her job. . . . Never knowing if it will even stop or if escalation is imminent, a woman can put up with it or leave. . . . Most women are coerced into tolerance. (MacKinnon, 1979, p. 40)

Although a survey by *Redbook* in 1976 answered by more than 9,000 working women revealed that almost nine out of ten experienced some form of sexual harassment at work, the impact of this type of sexual coercion has been studied only to a limited degree. The experience seems to be a degrading, humiliating one in which the woman usually feels a sense of helplessness similar to the feeling reported by rape victims. As one woman explained:

> I'd heard before about women being told to "put out or get out." But when my boss, who was my father's age, tried to get me to sleep with him, I thought at first it was a nice compliment. Nothing too shocking, and something I assumed would pass. I wasn't prepared for his change in behavior. Suddenly, nothing I did was right. I was constantly pressured to change my mind. He began to grab at me, and one day he pulled me down on his lap. I felt cheap, confused and used — although I hadn't done anything wrong. Later, I felt anger, and I went to see a lawyer.

A 1981 survey of almost 2,000 business executives done by *Redbook* and the *Harvard Business Review* found that men and women in top management positions disagree considerably on the extent of the problem of sexual harassment at work, with two-thirds of the men believing the scope of the problem "is greatly exaggerated" while only one-third of the women agreed with this statement. Male and female executives, however, were nearly unanimous in feeling that "unwanted sexual approaches distract people from the job at hand" and nearly three-quarters were in favor of management issuing a statement to all employees disapproving of sexual harassment.

The legal definition of sexual harassment is now quite specific:

> Unwelcome sexual advances, requests for sexual favors, and other verbal or physical conduct of a sexual nature constitute unlawful sexual harassment when (a) submission to such conduct is made either explicitly or implicitly a term or condition of an individual's employment, (b) submission to or rejection of such conduct by an individual is used as the basis for employment decisions affecting such individual, or (c) such conduct has the purpose or effect of unreasonably interfering with an individual's work performance or creating an intimidating, hostile, or offensive working environment. An employer is responsible for sexual harassment by its agents or supervisory employees regardless of whether the employer knew or should have known of their occurrence. An employer is responsible for acts of sexual harassment in the workplace committed by "non-supervisory

employees" where the employer knows or should have known of the conduct and no immediate and appropriate corrective action was taken. (*EEOC Rules and Regulations*, 1980)

As more and more women have become aware of their legal rights in situations of sexual harassment, a large number of lawsuits have been filed with claims based primarily on sex discrimination under Title VII of the Civil Rights Act of 1964. Since few states have laws banning (or defining) sexual harassment at work, many of these cases have not been decided favorably for the woman even when it was clear that sexual harassment did occur. Here, the situations were viewed as "personal" instances of harassment rather than sex discrimination. Some landmark cases in the last few years, however, have now been decided in the woman's favor, and it may be anticipated that such judgments will become more realistically decided in the future. Recently, a male employee of the Wisconsin Department of Health and Social Services won a major sexual harassment suit against his female superior; a federal jury of five women and one man awarded him $196,500. New regulations published by the U.S. EEOC in 1980 state that employers now have an "affirmative duty" to prevent and eliminate sexual harassment at work.

SEXUAL HARASSMENT AT SCHOOL

Sexual harassment is not confined to the workplace, of course. Examples of sexual harassment are common in many situations in which there is a hierarchy of power. Even the halls of academe are not immune from this form of abusive behavior, as these examples show:

> *A twenty-two-year-old woman:* My economics professor kept asking me to schedule meetings with him to review the work I was doing on my senior thesis. Whenever he could, he would drape his arm on my shoulder while we were talking and he would remind me that I needed his support to get an honors grade and to get into grad school. One day, he tried to rub my breasts — and when I pulled away from him, he just came out and said, "Look, either you put out for me or I won't put out for you." When I went to

the dean to complain, I was told that without proof of my accu-
sation, nothing could be done.

A twenty-seven-year-old woman: One of my professors seemed
to take a strong interest in my work and asked if I would help
him grade some exam papers from his intro course. While I was
working in his office, he calmly peeled off his clothes and said,
"Let's take a break." I made a fast retreat, believe me, but I was
shaken up by the experience. Then he called me at home to
say that I shouldn't have been upset, he was only trying to be
friendly.

The scope of sexual harassment on college campuses is not
known for certain, but Bernice Sandler of the Association of
American Colleges estimates that one out of five college coeds is
subjected to sexual harassment. A 1983 survey done at Harvard
University found that one-third of female undergraduates and
41 percent of female graduate students had encountered some
form of sexual harassment, although outright assaults accounted
for only a small fraction of these incidents. Whatever the precise
numbers may be, it is clear that sexual harassment of students is
a problem of considerable magnitude that has not attracted publicity (or action) to date.

DEALING WITH
SEXUAL HARASSMENT

Deciding how to handle sexual harassment in its various forms
is not an easy task. As the preceding discussion shows, victims of
sexual harassment are typically in a precarious position because
they have less power than the person who is harassing them.
Because of this, they may worry about whether their accusations
will be believed and whether the harasser will be able to get back
at them in some manner. Thus, as with other forms of sexual
coercion, it appears that the overwhelming majority of cases of
sexual harassment are never reported, leaving the harasser free
to victimize others repeatedly. Being informed about options for
handling this problem can help turn things around. Here, then,
are a number of practical pointers:

1. If you have been the victim of actual or attempted sexual assault or rape by a boss, supervisor, teacher, or co-worker, you can file either civil or criminal charges against the offender.

2. In cases of sexual harassment that have not included an actual assault, you can confront the person who is harassing you in a number of different ways.

   a. Consider writing a letter to your harasser telling him (or her): (1) what the facts are as you see them (e.g., "On March 14, 1983, when we met in your office to go over my draft of the Smith contract, you put your arm around me and tried to kiss me, and then asked me to come to your apartment so we could 'work more intimately together' "), (2) how you feel about what happened (e.g., "Now I am upset when I see you and worried that you won't evaluate my work objectively"), and (3) what you want to happen next (e.g., "I am willing to forget what happened if our relationship is a purely professional one from this point on").

   b. An alternate approach would be to confront the person who is harassing you, either in person or by telephone, although this is more likely than a letter to produce an emotional response.

   c. Another possibility is to have an attorney write to your harasser for you, telling him to immediately stop such behavior or to run the risk of a subsequent lawsuit. (An attorney's letter may be more effective than your own in letting the harasser know you mean business.)

3. Keep careful documentation of each incident of harassment, including memos that include the dates, times, and specific details of offensive actions. Note the names of any witnesses, since they may be of considerable help in substantiating your case.

4. File a grievance with the appropriate person (for example, someone at the dean's office at school, or the personnel office at work, or with a union representative, if you're a union member).

5. If confronting the offender and filing a grievance don't work, seek help from sympathetic co-workers. You may discover others who have been victims of sexual harass-

ment at the hands of the same person. Consider forming a group to discuss and deal with issues of sexual harassment.

6. If you aren't able to remedy the situation by the above steps, or if you have been unjustly fired or discriminated against in any other way, you can file a complaint with the Equal Employment Opportunity Commission (EEOC). You may be entitled to unemployment compensation and back pay, in addition to damages.

7. Don't let yourself feel guilty. As Backhouse and Cohen (1981, p. 165) put it: "The most important consideration in dealing with sexual harassment is to protect yourself, to refuse to feel guilty or in any way responsible for your problems. . . . You are a victim and are in no way to blame for being the target of this heinous behavior."

THE CULTURAL UNDERPINNINGS OF SEX VICTIMOLOGY

We live in a society that trains and encourages females to be victims of sexual coercion and males to victimize females. This statement is harsh but true, and it has important implications for what must be done to prevent sex victimization in its many forms.

Without repeating the detailed discussion of gender roles presented in chapter 8, we can summarize the situation by saying that females are generally socialized for passivity and dependence, while males are programmed for independence and aggressiveness. This fundamental difference lies at the heart of sexual victimization, which is primarily an act of power and control.

> The fact is that families generally are given the job of socializing children to fill prescribed roles and thus supply the needs of a power society. . . . Ingrained in our present family system is the nucleus of male power and domination, and no matter how often we witness the devastatingly harmful effects of this arrangement on women and children, the victims are asked to uphold the family and submit to abuse. (Rush, 1974, p. 72)

The teenage boy is quick to learn that he is expected to be the sexual aggressor. For him, it is acceptable — even "manly" — to use persuasion or trickery to seduce his prey. He is also taught (by our society, if not in his home) that females do not really know what they want, that when they say "no" they mean "maybe," and when they say "maybe" they mean "yes." He may also have heard that bit of male folk-wisdom that says — in reference to some "uppity" or unhappy female — "What she needs is a good lay." Given this background, it's not surprising that what men see as being an "active, aggressive (and desirable) lover" may quickly be transformed into sexual assault in its various forms.

Most women have been taught as children not only to be passive ("nice," "polite," "ladylike"), but also to be seductive and coy. They are usually not trained to deal with physical aggression (unlike boys, whose play activities develop this capacity), but *are* trained to deal demurely with sexual situations. Thus, the female in a situation of sexual coercion is ill-prepared to act against it. Faced with a physical threat, she often becomes psychologically paralyzed. Faced with unwanted sexual demands, she is likely to question what it was about her manner, dress, or behavior that produced the attention — she blames herself and feels guilt instead of taking more positive action. This hesitancy is frequently misread by the male who sees it as a sign of weakness and a chance that she will give in. His past experience may prove him right — how many women "give in" in various undesired sexual situations just is not known.

There are no perfect solutions that can wipe out sexual coercion, but a significant part of the problem can be addressed in two fundamental ways. First and foremost, as this discussion implies, is to change traditional gender role socialization that puts females in the position of being vulnerable to sexual abuse. Second, in-depth attention is required to identify the conditions that push men into the "victimizer" role. Only when a clear understanding of the causes and motivations underlying coercive sex is at hand will it be possible to develop effective strategies for dealing with this problem on a large-scale basis.

CHAPTER SEVENTEEN

Increasing Sexual Satisfaction

THERE IS little question that personal dissatisfaction with
sex is commonplace in our society today. Half of all American marriages are troubled by some form of sexual distress
ranging from disinterest and boredom to outright sexual
dysfunction. A high proportion of unmarried adults voice similar complaints about sex: they can't find partners who are "right"
for them, they feel that sex is pressured rather than relaxed and
pleasing, their own erotic responsivity is inhibited or frustrated
in one fashion or another. Even adolescents — supposedly at the
peak of their sexual appetites and unencumbered by puritanical
attitudes about sex — frequently complain that sexual matters
trouble them. While it is true that some of these people need sex
therapy to overcome the difficulties they are facing, we feel confident in saying that many can learn to increase their personal
sexual satisfaction on their own if they'd only go about it in a
sensible way. The trouble is that this rarely happens, since most
people respond to the sexual problems in their lives by means
that only serve to entrench them more deeply in a downward
spiral of sexual uncertainties and anxieties. This chapter presents a variety of specific suggestions and practical pointers for
anyone who is interested in greater sexual pleasure.

COMMON PROBLEMS
LIMITING SEXUAL SATISFACTION

Our work as sex researchers and therapists has led us to believe that the four types of sexual problems people encounter most frequently are inhibitions and guilt, performance anxiety, erotic boredom, and blind acceptance of sexual misinformation or myths. In fact, these four problems collectively account for more than 80 percent of the sexual dissatisfactions in modern American society. For this reason, it seems logical to examine the dimensions of these specific areas in some detail before proceeding to look at ways in which they can be overcome.

One might think, at first blush, that our contemporary world — with its sometimes blatant openness about sex — is hardly a place where sexual inhibitions would linger in the minds of many people. But inhibitions and guilt about sex continue to be ubiquitous even as their causes and directions may have shifted somewhat from earlier times. Today, for example, not only do we still have the "sex is dirty" legacy of some religions to contend with and the innate sense of secrecy about sex that is linked with its customarily private, hidden practice in our culture, we also have a whole set of sexual inhibitions that stem from our anxieties about our personal attractiveness. While concerns about personal attractiveness have been with us for a long time, their intensity has been heightened by our everyday exposure to visual media — television, most notably — and our current cultural obsession with physical fitness (or at least the *appearance* of physical fitness). Women who don't happen to have the physical attributes of a Jane Fonda, a Tina Turner, a fashion model, or a *Playboy* centerfold are apt to judge themselves harshly as far as their erotic allure goes, often translating this personal sense of body-image deficiency into tangible sexual behavior patterns. In recent years, many males have been smitten with a similar malady — unless they literally can measure up to the physical appearance of Sylvester Stallone, Bubba Smith, or Burt Reynolds, they also feel unattractive and thus somehow deficient as lovers. While there were undoubtedly some young men in the 1940s who desperately wanted a physique like Johnny Weissmuller's or Buster Crabbe's, the matter just didn't seem as inexorably linked to sex appeal. Males are affected in another, related fashion even more powerfully: the need to "measure up" in terms of penis size is one of the most common anxieties sex therapists hear about from males.

Just as the man who feels inadequate because he thinks his penis is too small is apt to measure his self-worth in diminished terms, which often lead to a lack of sexual self-confidence and behavioral hesitancy, so too the woman who feels sexually unattractive because she considers herself to be overweight or too flat-chested is apt to hold back in sexual encounters because she feels she isn't really a sexual person. While there are, of course, many who do not fit this mold — including some who attempt to cope with such anxieties by overcompensating in their sexual behavior, as though they can prove their self-worth by the frequency or variety of their sexual encounters — not fitting current cultural models of attractiveness continues to be a common source of sexual inhibitions.

> My wife insisted that we make love in the dark for the first six years we were married. I thought it was her notion of propriety, so I reluctantly went along with it. It was only recently that I found out that the real reason was that she was embarrassed by how she thought her body looked. When we finally got this out in the open and I was able to convince her that I found her body great — a terrific visual turn-on — our sex life improved about 300 percent.

Another frequent type of sexual inhibition that continues to confound sex therapists is that associated with a sense of shame or repulsion about oral sex. Rather than being able to relax and enjoy this form of intimate caress, many people find that oral sex is highly problematic, as these comments indicate.

> *A twenty-one-year-old single woman:* My boyfriend wants to go down on me just about every time we make love and it really makes me uncomfortable. I don't mind giving him a blow job — in fact, it's a turn-on for me — but whenever his lips or tongue get near my genitals I freeze up.

> *A thirty-eight-year-old married man:* My wife and I have a pretty good sex life overall, I guess you'd say — we have sex four or five times a week, and she always has an orgasm. But I rarely can get

up the courage to give her oral sex, even though I know she really loves it. I guess that deep down inside, no matter how much I *think* I'm liberated about sex, I still think that eating her is somehow dirty.

A forty-two-year-old divorced woman: I was never able to endure oral sex when I was married, either as the giver or receiver, and I haven't been able to change now, after my divorce. It's disturbing to me because it certainly seems like every man I go with now expects it and looks at me like I'm crazy when I can't do it. But I just find it to be a dreadful act, no matter what all the sex manuals and research surveys show.

What's going on here? we might ask. Are these inhibitions based mainly on aesthetic considerations (not liking the taste or odor of oral sex, for instance)? Are they more fundamentally a reflection of the persisting view that they are a perversion, either banned by religious edict or by the idea that the only "natural" sex acts are those with reproductive potential? Are they mainly a statement of propriety, maintaining a sense of personal decorum and control? Do they tell us something about status and power — for instance, is the twenty-one-year-old woman quoted above, who finds performing fellatio a turn-on but freezes up with cunnilingus, really telling us something about her perception of appropriate gender roles, with her view of the world one in which the female must service her partner sexually but renounce any sexual pleasure herself as unfeminine or tawdry? In our view, it is too simplisic to assume that there is a single explanation that would apply to every person with inhibitions about oral sex. The key fact to keep in mind, though, is that whatever their source (which may, as a practical matter, be undiscoverable), there are usually specific techniques that can be used to overcome such inhibitions, first reducing their intensity, then extinguishing them, and ultimately replacing them with a new set of attitudes that can not only lead to toleration of the previously anxiety-provoking act but also to enjoyment and gratification from it.

Sexual inhibitions may be much more global in scope than those mentioned so far even when they don't reach the magnitude of an outright phobia. These global inhibitions don't always preclude sexual participation either, although they typically reduce the potential for erotic satisfaction quite drastically. Here are three brief examples of the variety of inhibitions that may take on global dimensions.

Case 1: Mr. H. was a 48-year-old man who had been married for 22 years when he found himself suddenly involved in a short-lived affair with one of his neighbors. Guilt-ridden by the experience, which he ended voluntarily, he found that subsequent sexual relations with his wife became emotionally draining and a purely mechanical exercise. While he was able to function physically during sex, he felt that it was a relatively detached, impersonal experience, and complained that his foray into infidelity had shattered the intimacy and pleasure of his marriage.

Case 2: Ms. W. was a 23-year-old married woman whose sexual interactions with her husband were consistently troubled by fantasy scenes in which she was making passionate love to a young priest who had been in her parish when she was a teenager. She had used this fantasy as a frequent accompaniment to masturbation when she was younger, when it brought her great pleasure, but now she felt unable to keep this imagery from intruding into her sexual arousal. As a result, she became generally unresponsive to her husband's lovemaking, although she went through the motions — "doing everything I can to hurry him up," as she put it — so as to preserve their marriage.

Case 3: Mrs. B. was a 31-year-old woman who had been divorced and remarried. At age 10 she had been sexually abused by an uncle, who had sworn her to secrecy under the threat that he knew of a crime her father had committed and would put him in jail if she told anyone. As a teenager, she had avoided sexual activity completely, and when she married for the first time, at age 24, she was still a virgin. Her sex life was something that she barely tolerated, saying, "I never feel anything." Only after she was able to tell a psychiatrist about being abused sexually as a child and was placed in a therapy group with women who had had similar experiences was she able to overcome her sense of guilt and remorse about sex.

As these examples suggest, sexual inhibitions are often reflections of guilt in one form or another. The guilt may either stem from long-past events, as in Case 3, or it may be a result of recent circumstances. Thus, it is not unusual to find sexual inhibitions occurring among infertile couples who have never previously had such difficulties, as they seek to "explain" their infertility as a form of retribution for a past misdeed. Likewise, some people who develop a sexually transmitted infection find that their reaction to it includes a passing phase of temporarily reduced sexual ardor and responsivity.

Performance anxieties constitute the next category of common sexual problems that we will consider. In the next chapter, we will examine the fear of sexual performance (or, more accurately, the fear of *not* performing sexually) as a direct cause of sexual dysfunction, but it is important to recognize that performance pressures can affect sexuality in a variety of other ways. Central to all such effects is the way in which anxiety about performing properly becomes an overriding block to the free, spontaneous flow of sexual feelings and thoughts. This is because awareness of the performance anxiety typically produces so much preoccupation with the anxiety itself that the person fighting to overcome the anxiety becomes less fully involved in the sexual interplay in what is ultimately a paradoxical cognitive juggling act. As the anxious person worries about how to be sexually spontaneous and responsive (how to be a "good lover"), he or she focuses on each component part of the lovemaking — where a hand is placed, how rapidly the partner is breathing, whether a minute shift in position is required, how much erection is present — dissecting the action so deliberately that enjoyment is virtually impossible. An analogous behavior with a different genesis but a similar result is found among persons with anorexia nervosa, the self-starvation disease. These individuals are so anxious and preoccupied about food and eating that they frequently engage in such "dissecting" at mealtime — either literally, as in the ritual of cutting a small piece of food, such as a lima bean, into a dozen or more pieces before they can eat it, or obsessively, by actions such as chewing each morsel of food at least 200 times before swallowing it. They too manage to remove the enjoyment from their participation in the activity that provokes their anxiety, and by working at the task of eating manage to also reduce their appetites.

The type of performance anxiety alluded to above is categorically different from another version of performance pressure that acts to heighten sexual passion. Pressures that stem from

attempting sex in forbidden places or with forbidden, illicit partners are more apt to add to excitement than to dampen arousal.
Thus, it isn't simply that anxiety is an automatic sexual turnoff.
But when anxiety leads to the dissecting phenomenon — like the
proverbial centipede who suddenly becomes unable to walk
when asked how he manages to coordinate the motion of all
his legs at the same time — it commonly results in sexual non-
response.

The fear of sexual performance is not limited to anxiety about
physical responsivity as measured by the swiftness or durability
of erection or vaginal lubrication or the intensity or frequency
of orgasms. It also reflects anxiety about the totality of one's
sexual response — how much passion, tenderness, intimacy, and
sensitivity a person shares with his or her partner. For this reason, a person who is having no apparent problems in the physical side of sexual response may nevertheless be deeply distressed
by an internal perception of inadequate or inappropriate sexual
performance. In some instances, of course, this may be an accurate reading of a rather soulless, mechanical coupling. But in
other cases, judging that sex should always be a supreme act of
blissful closeness, a sort of communion with one's partner, the
person who has felt surging, passionate sexual impulses may feel
anxious about having been too selfish and not concerned
enough with his or her partner's feelings and response. In one
of the intriguingly paradoxical twists of sexual thinking, this
viewpoint manages neatly to devalue sexual sponaneity in comparison to a more civilized, synchronized style of lovemaking.

Fears of sexual performance often become most problematic
by leading to avoidance of sexual interaction or at least minimizing the amount of sexual interaction that occurs. This can result
in one member of a couple mistakenly interpreting the situation
as a form of rejection by his or her partner. This is most unfortunate because the underlying motive has nothing to do with
rejection; instead, the avoidance is an ego-saving defense that
generally helps a person feel more in control and less guilty
about being inadequate.

A forty-three-year-old married man: Whenever my wife tried to get
romantic, my first reaction was, "God, she probably won't have
an orgasm with me — she'll have to use that damn vibrator

again." It was like a slap in the face; I wasn't man enough to satisfy her. So gradually I found myself making excuses to avoid having sex. You know, like "I've got to do the checks now," or "There's a ballgame on the tube I want to watch." The result was that she thought I didn't love her or find her attractive anymore, but what she didn't realize, and what I never told her, was that sex was making me feel so inferior that it wasn't any fun.

Avoidance of sex also occurs because of plain, old-fashioned boredom with the erotic side of the relationship. Despite the fact that it seems as if it should be obvious when boredom is the principal cause of sexual problems, and despite the fact that coming up with ways of counteracting this sort of boredom should also be simple and straightforward, people are remarkably predictable in avoiding the recognition or solution of this type of problem in their lives. Perhaps for this reason, although no one has yet succeeded in studying erotic boredom as a cause of sexual dissatisfaction, most sex therapists see it as a commonplace cause of a wide variety of sexual problems.

If partners are genuinely gratified with their sexual interaction, boredom sets in only if they become satiated with too much of a good thing or affected by too much sameness in their sexual routines. On the other hand, partners whose sexual interaction is not particularly satisfying are prone to become bored much more easily. Whether this is because a lack of consistently reinforcing gratification makes them somewhat detached and disengaged from their sexual encounters, or whether boredom results from too much of an effort expended in working at having good sex (with a poor physical/emotional return on the energy put forth), or whether boredom is stricly an attitudinal problem has yet to be decided. Indeed, it may be that erotic boredom reflects a different level of reality: namely, that many couples who spend time together gradually come to dispense with so many of the romantic, pleasuring, attentive little gestures and behavioral nuances that marked the earlier stages of their relationship (both sexually and nonsexually) that the quality as well as quantity of their sex lives dissolves before their eyes.

Sexual boredom is also apt to reflect a fundamental gap that most people have about sex — the gap between our sexual expectations and our sexual realities. In our expectations, just as in our fantasies, sex is not only immensely passionate, explosively gratifying, and intensely consuming, it is also free of such problems as fatigue, runny noses, bad breath, flaccid sex organs, apathetic (or nonexistent) orgasms, and bad feelings between partners. In effect, our expectations cannot possibly be matched by our real-life sexual experiences in sustained relationships because these expectations are usually fictionalized, idealized expressions of wishful thinking. It's not surprising that this is so, since we are inundated with fictionalized depictions of sex in movies and books that somehow lead us to believe that everyone else is having supercolossal sexual ecstasy as a regular diet, while we are consigned to what is, by comparison, mundane and unremarkable.

The aspect of sexual boredom that has been discussed more than any other is the question of whether long-term familiarity with one's partner is more or less bound to reduce eroticism. Sociobiologists argue, for example, that the male is inherently promiscuous, going from partner to partner to ensure the most widespread distribution possible of his gene pool. The conclusion, rather than the explanation, of the sociobiological argument is supported indirectly by many sex researchers, who note numerous surveys showing that males are more likely than females to have extramarital affairs. We suspect that these data may be misleading since they do not take into account the substantial number of women in their samples who are full-time housewives, since it now seems that women who work outside the home are far more apt to engage in extramarital sex than those who do not. It is unlikely that this is a biologically mediated phenomenon. Instead, we suggest that it shows how much past social constraints have caused women to submerge their sexual feelings. (Admittedly, though, the complexity of this issue is considerable, since there are indications that women enter affairs seeking emotional closeness more than sexual gratification, whereas men do just the reverse. *Caveat emptor.*)

Is having sex with the same partner over a long time a preordained ticket to erotic boredom? Here are comments from three married men that may be of interest.

> *A thirty-seven-year-old man:* I've been married 16 years now and been completely faithful to my wife. I'd have to say, in all candor, that our sexual relationship now is much better than it was when

we were first married. To me, the trick has been to be able to be creative about our sex lives — not with whips and chains, or gymnastic positions — but with keeping away from routines; and staying attuned to one another's needs. For instance, we've learned to spend a lot of time giving each other sensual massages as part of our foreplay. It's an almost indescribable, tantalizing delight . . . and something we never did back when we were kids.

A forty-four-year-old man: Let's face it, my wife isn't as attractive as she once was, she's let her body go to pot, and her idea of good sex is spreading her legs for me twice a week. It's like she's doing me a favor. Is it any wonder I say our sex life is boring? In fact, I have more of a thrill when I masturbate. So I go to a classy hooker once a month as my way of getting even. Didn't someone once say, "Living well is the best revenge"?

A forty-nine-year-old man: Although I've been happily married for 27 years, I've certainly noticed a definite deterioration in our sex life. We get along really well, we do more together now than we ever did, but sex is the one area in our lives where we seem stuck in a rut. I'm not sure who's to blame for it, since I think we're both involved, but I guess we've both given up too easily, accepting a lousy sex life as a given. I'm not proud to admit it, but the way I've coped with this for years is by having occasional affairs. At least that way I feel like I can still turn *someone* on.

Women are less likely to complain about sexual boredom — perhaps their expectations in this respect are more realistic than men's — but they are much more vociferous about the lack of intimacy and communication in their lives. On the other hand, a significant number of women voice feelings such as the following:

A twenty-eight-year-old woman: Tom and I have only been married for five years, but I've already found a discouraging pattern emerging. Tom's so wrapped up in his work that we have very little time together, and when he's not working he wants time to relax. Sex seems like something we only get to on weekends, and even then it's not very special. It's like he can't be bothered to make the effort, and whenever I try to talk to him about it, he brushes me off by making it into a joke. "You know I'm the world's best lover," he'll say. Well, the world's best lover better wake up soon or his wife will go elsewhere for some fun.

A thirty-eight-year-old woman: My husband complains about our sex life all the time. We don't have sex often enough, we don't try new things, we don't have any passion in our lives. Well, he's right. But I don't think I'm the problem — he better take a look in the mirror one morning and think how much fun it is to get passionate with his forty-pound beer belly, his cigar breath, and his body odor. Worst of all, he seems to think that just because HE's ready to screw, I should be instantly flat on my back with a smile on my face. It's like he doesn't think about my sexual feelings at all, or what I might like to feel turned on. Sure our sex life is boring, but it's not so hard to figure out why.

Beyond boredom, performance anxieties, and inhibitions as common contributants to sexual dissatisfaction is the broad range of myths and misinformation people cling to in their attempts at being sexually knowledgeable and sophisticated. We have already alluded to a number of these myths in previous chapters: for instance, the notion that males reach the peak of their sexual prowess at age eighteen and forever after are doomed to an insidious decline of their sexual interests and capabilities; the belief that when a female lubricates vaginally it is a sure sign that she's ready to have intercourse; and the idea that women are inherently less sexual than men are all examples of the genre. The danger of such myths, of course, is wrapped up in how willing people are to accept such "common knowledge" as fact, using it as an excuse to explain their behavior or feelings without regard for what the correct facts may be. Intriguingly, it's common for someone who has latched onto such an erroneous explanation of the way things are sexually to struggle against being informed of the real facts by a therapist. It often seems as though the person prefers the comfortable explanation provided by the myth and is reluctant to try a different viewpoint even if it means enduring sexual frustrations.

One of the curious elements of sexual myths is that they are so easily believed despite the lack of evidence to substantiate their existence. Perhaps this is because most of these myths sound eminently sensible and scientific — almost as though it's the way the world *should* be. A case in point of relatively recent vintage is the idea that feminist women are creating a new wave of impotence in their partners, as solemnly reported by a psychiatrist (to much media attention) a few years back. The implicit assumption supporting this notion is the idea that feminists are more masculine and aggressive than "traditional" women and thus are more likely to frighten their partners, who are unused to such aggressiveness from their bedmates, into poor sexual performance because they have had their own masculine prerogatives undermined. While this nonsense may have been of some political-philosophical solace to the likes of Phyllis Schlafly or members of the Moral Majority, who would prefer to see feminism discredited on any grounds possible, the facts are that (1) there has been no surge of impotence corresponding to the spread of feminism; (2) male partners of women with full-time careers outside the home actually have *lower* rates of impotence than the male partners of traditional housewives; and (3) more "masculine" women, as rated by a battery of sex-role tests, are as sexually responsive and satisfied as more "feminine" women. Despite these facts, though, the myth of the "new" impotence has gained some credence and, like any myth, has undoubtedly become a sort of self-fulfilling prophecy for some men.

Why are people so susceptible to sexual myths and misinformation? In part, it is undoubtedly because these poorly conceived or sloppily researched issues receive a great deal of play in the media. Having listened to a self-proclaimed "expert" expound on the nature of sexual behavior for three minutes on a news program or a TV talk show, or having read a newspaper or magazine article espousing a particular theory about sexuality as gospel, the average consumer of such information has little reason to doubt its veracity and is quite likely to accept it uncritically as true. Unlike the situation with media economic pundits, whose predictions about the stock market or interest rates can be matched to what actually happens over time, the media expert on sex is not easily judged by the public except on the basis of very subjective factors such as appearance, title, and so on. Since many of the experts who make media appearances do so in conjunction with promoting a book they have written, and since the fact that Dr. So-and-So has written a book seems to confer an extra aura of authenticity to his or her pronouncements, this is a particular sore spot in the field.

To be sure, many myths about sex are essentially harmless

curiosities (e.g., the notion that bald men make poor lovers). In other instances, though, the seemingly innocuous myth is grasped so fervently or applied so inappropriately in a person's life that it leads to anguish or turmoil. Perhaps worst of all, people who latch onto myths about sex often use them as excuses for giving up on changing their sex lives for the better.

ENHANCING YOUR
SEXUAL SATISFACTION

Recognizing problems is one thing, but doing something about them is something else entirely. What follows, then, is a list of specific suggestions for people who feel that their sex lives are less gratifying or fulfilling than they might be. While neither totally exhaustive nor a substitute for sex therapy when problems are too large or too complicated to be handled alone, these suggestions have been developed on the basis of our experience advising thousands of couples about their sexual satisfaction. These suggestions have been numerically listed as a matter of convenience only; they are not rank-ordered in terms of their importance or likelihood of being helpful.

1. *Always remember that good sex begins while your clothes are still on.* This doesn't mean that sensuous stripteases are necessary for good sex. Instead, it's a reminder that the buildup of sexual desire and the atmosphere that will be carried into the bedroom is largely dependent on what happens outside the bedroom not just in the moments before deciding whether to make love but over the hours or days preceding a sexual interlude. In this sense, "getting in the mood" is not a deliberate act of playing romantic music or dining by candlelight (although there's nothing wrong with such activities) but is part and parcel of an ongoing nonsexual closeness between partners both psychologically and physically. Being able to express tenderness and affection, both verbally and nonverbally, without its being a direct invitation to sex is a particularly important ingredient here.

2. *Take time to think about yourself as a sexual being.* One of the problems people often have with sex is that they compartmentalize it as something that happens to them in isolation from the rest of their lives. If you can spend time thinking about your sexuality in its many different dimensions, you can better appreciate that as long as sex is only a discrete activity, like a set of tennis, it will be a somewhat fragmented, disjointed experience rather than part of the fabric of your life and your relationship. In addition, if you think of yourself as sexual you are, in effect, learning to accept your sexuality as a fact about yourself instead of simply seeing it as an impulse or urge that arises from time to time.

3. *Take responsibility for your own sensual and sexual pleasure.* As much as you might like to think that someone else will turn you on and give you joyful paroxysms of sexual pleasure, in actuality we are each responsible for our own eroticism. While males seem to have an easier time accepting this aspect of sex, females often slip into a culturally dictated role of passivity in which they expect that their partner will sweep them off their feet and be the "expert" when it comes to sex, relieving them of any responsibility except to respond on command. This notion of sex as something one person does *to* the other — or the somewhat kinder notion that sex is something one person does *for* the other — both can lead to problems. By taking responsibility for your own sensual and sexual needs, you actually are paying your partner a terrific compliment: in effect, you're saying to them, "I care enough about you to want to keep you from having to guess at what I want, what I like, and what can make me happy."

4. *Talk about sex with your partner.* One of the most amazing things to us about sexual behavior is how reticent most people are when it comes to talking about sex with their lovers. It's as though talking about sex would somehow spoil it or take away its spontaneity. Instead, what we see over and over again are couples where one person doesn't have the foggiest notion of what the other wants or likes sexually, as well as couples whose well-intended caresses

fall short of the mark because they're too much, too soon, too little, too light-handed, too far off the mark — all matters that could be easily corrected by a few words, discreetly murmured, at the right time.

A cautionary note is in order here. When we suggest that talking about sex with your partner is apt to be beneficial, we don't mean that this should be done immediately after you've had sex together, in a sort of postmortem or critique. In fact, it's best to avoid sounding critical entirely, since that almost invariably gives your partner the message "Look what you're doing wrong," which is most likely to set them on the defensive. Instead, we're suggesting that talking about sex can help establish a number of useful understandings ranging from how you can handle intense sexual feelings when your partner isn't much interested to developing a repertoire of code words or signals that convey sexual messages if this seems useful.

5. *Make regular time for togetherness with your partner.* Many couples find their sexual desires on the decline because they forget to create unencumbered pockets of private time together. It's all too easy to relegate our primary relationships to a back seat as we look after the kids, try to get ahead at work, spend time socializing with friends, and pursue our hobbies — but the cost of this sort of benign neglect is apt to be high. If sex is always last on your agenda, it's safe to say that you'll be tired out and frazzled by the time you think about it, if you even think about it at all. This certainly doesn't set the stage for a satisfying, relaxed lovemaking session. Likewise, if you don't spend much quality time with your partner outside the bedroom, it's hard to create a sense of closeness and affection the instant you're ready for sex. This doesn't mean that every time you're together with a bit of privacy you have to seize the opportunity for sexual action — in fact, it's much more sensible to let your feelings and needs dictate just what you do with this time together — but it doesn't hurt to have the option available more often than once or twice a week.

6. *Don't let sex become routine.* For too many couples, sexual dissatisfaction is a direct reflection of boredom resulting from an absence of variation or creativity in their sexual

interaction. To avoid too much "sameness" in sex, don't always try to get amorous at the same time of day, change the scenery on occasion (for instance, try sex someplace other than your bedroom whether it means checking into a hotel for a night or sending the kids to sleep at friends' homes so you can make love in front of the living room fireplace), and vary the action. This can be done in many different ways, including:

- Don't always have one partner be the aggressor and the other be passive; try switching these roles at times, or even try having *both* partners be aggressors.
- Initiate sexual opportunities with unexpected kinds of invitations. This might be a phone call home from the office saying "I'm feeling kind of romantic, so tonight before dinner I'd really like to make love," a love-note pinned to a pillow, or a verbal invitation at an unusual time.
- Introduce new options into your sexual interactions in ways that are comfortable to you and your partner. This might mean experimenting with the art of sensual massage, watching erotic videotapes together, trying new sexual "toys" such as vibrators, or creating mutual sex fantasies which you proceed to perform. It can also include types of sexual activity you've rarely or never tried before, if they are appealing to you both.
- Try different positions. This strategy may provide an additional source of turn-on not just during intercourse but also during other kinds of sexual play.
- Don't be trapped into thinking that sex has always got to include intercourse to be meaningful or gratifying. By occasionally omitting intercourse from a lovemaking session, you may even discover other pleasures that are equally arousing.
- Use different tempos at different times. There's no rule that says that good sex must always be a long, leisurely, teasing process. At times, a throw-off-your-clothes, rush-into-bed quickie can be the height of passionate abandon, while at other times a slower-

paced session suits your mood better. Just remember, too much sameness can reduce pleasure in sex as in other matters — dining on *pâté de foie gras* and butter-dipped lobster every day would soon become incredibly boring for most of us.

One of the greatest threats to long-term sexual satisfaction is complacency — taking it for granted. Use your best creative thinking to avoid letting sex become just a dull, routine affair.

7. *Fantasy is one of the best aphrodisiacs you can find.* We've already devoted an entire chapter to the topic of sex fantasy, so we're simply using this occasion to remind readers that fantasy can be used in a wide variety of ways to enhance excitement and to bring variety to a person's sex life. Whether fantasy is used as a prelude to a sexual encounter — as a means of whetting your sexual appetite — or whether it is incorporated into your lovemaking sessions directly, your imagination can help you transform ordinary sex into something far more stimulating. In some instances, fantasy can be quite useful in getting beyond minor sexual inhibitions, too.

8. *Working at sex doesn't work.* Herein lies the tale of many a person's sexual woes. The natural tendency that most of us have is to try to deal with sexual anxieties or frustrations by working harder to overcome them. This approach usually backfires — although it may be eminently successful in dealing with other problems in our lives — because working at sex usually makes it so goal-oriented that it loses its fun and its spontaneity.

This doesn't mean that you should ignore any sexual problems you're having, hoping they will magically disappear. It may take some effort on your part (and some understanding from your partner) to make the sorts of changes in your sex lives that we're discussing here, changes that are apt to produce beneficial results. The distinction between making the effort to do something different and *working* at sex may seem slim, but there is more than just a semantic difference. Working at sex almost invariably means carrying a mental checklist into every sexual encounter and then watching yourself to see if you're performing properly. In effect, it involves pushing yourself into the spectator role we've previously mentioned. Implementing changes in your sex life, in contrast, may involve making an effort to plan different types of opportunities, but leaving the performance checklist out of the process. If there's no goal that *must* be attained right now, there's far less pressure to perform and far less of a need to be a spectator to check up on how you're doing.

9. *Don't carry anger into your bedroom.* This straightforward admonition is ignored by more couples than we like to think about, with the typical result being that sex gets turned into a power struggle or a contest. This is not to say that sex can't be used to end an argument; sometimes, as we all know, the sweetest, most delirious erotic moments come as we rush into our partners' arms in an act of making up after a fight. But those who try to settle a score from an unresolved issue that has provoked considerable anger by withholding sex or demanding sex are asking for trouble in the long run, no matter how satisfying it may be to gain a temporary sense of revenge. The best way to deal with this problem, of course, is to try to minimize the amount and duration of anger that you and your partner feel by recognizing its source and trying to handle it while it is still at the level of hurt or frustration, as we discussed in the chapter on "Intimacy and Communication Skills." If your anger is not so easily prevented or dissipated, however, it is usually best to acknowledge it and temporarily avoid sex unless both partners are willing to engage in a relatively impersonal version. There's nothing wrong with impersonal sex once in a while, but if it becomes the rule rather than the exception in a relationship, chances are that it will lead to less sexual satisfaction all around.

10. *Realize that good sex isn't just a matter of pushing the right buttons.* It's tempting to think that there's some magical formula for producing foolproof sexual responsivity in one's partner, but this just isn't true. In fact, we tend to overemphasize the mechanical aspects of sex and underplay

the emotional components that are probably far more important in terms of how we actually experience sex. People who get preoccupied with matters of sexual technique frequently find that sexual satisfaction becomes more and more elusive to them, which is not surprising because sex for them becomes more and more depersonalized. While careful attention to mechanical detail may help straighten your golf swing or improve your bowling scores, it's often more of a hindrance than a help when it comes to sex.

11. *Keep some romance in your life.* While working at sex usually produces dismal results, as we've already said, working at being romantic is another matter entirely. Creative ingenuity and plain old-fashioned thoughtfulness can go a long way in this direction. Preserving romance is an ongoing process that involves everything from those little nonsexual exchanges of affection that are so easy to forget about to the out-of-the-ordinary special gestures like a love poem written for an anniversary, flowers sent for no reason at all, or a surprise gift. These gestures are not just ceremonial; they make the recipient feel special and loved. In fact, the lavish attention we bestow on our loved one during courtship has a great deal to do with the passionate feelings that are present at that time; thus, it's no surprise that as we fall into nonromantic complacency in our relationships, our sexual interests and passions are apt to dwindle, too. (This observation applies to nonmarried couples in long-term relationships as much as it does to married couples. Marriage isn't the only condition in which romance disappears from a relationship.)

12. *Don't make sex too serious.* In the Polynesian islands, natives would be horrified at how effectively our society has managed to squeeze much of the fun out of sex by making it into serious business. A byproduct of our goal orientation and performance anxieties, our work ethic, and the guilt we seem to attach automatically to having a good time, the seriousness with which we regard sex is a definite drag on our sexual satisfaction. If people could only take themselves a bit less seriously in the bedroom and allow them-

selves to experience sex as a pleasurable pastime or adventure rather than a task with predetermined objectives (e.g., excite your partner, reach orgasm), they might be surprised to find that sex was more enjoyable. And they also might be able to accept their sexual shortcomings and foibles without a fear that the world was coming to an end. Consider, for a moment, the difference between going on a leisurely, relaxed walk in the woods with no goals or expectations versus having the same experience after having first decided that you want to spot a yellow-bellied warbler. In the latter instance, if you never see the elusive warbler you're apt to be disappointed and rate your experience a failure, whereas in the unstructured experience you could simply notice the scenery and enjoy what you were doing.

13. *Don't always wait to be "in the mood" before agreeing to have sex.* As revolutionary as this may sound, it's a little silly to think that you always must feel sexy before getting into a sexual situation. For one thing, it's perfectly reasonable to accommodate your partner's needs when you're not feeling particularly in the mood — after all, this isn't very different from what you might do if your partner was hungry and asked if you'd make them a sandwich. You wouldn't agree to make the sandwich only if you were hungry too, would you? Second, while you may not feel very sexy at the moment your partner has invited you to fool around, if you give yourself a chance to get into the situation you may be surprised to find that your feelings can change quite rapidly. But if you keep the opportunity from ever developing except when you feel in the mood, you prevent yourself from being able to enjoy this facet of your sexual appetite — that is, having your appetite arise when you least expect it to.

14. *Realize that if you and your partner don't see eye-to-eye on what you enjoy sexually, it isn't tantamount to destroying your relationship.* For a variety of reasons — aesthetic and otherwise — not everyone shares the same sexual tastes and preferences. If your partner won't participate in some form of sex that you find quite appealing, the important thing to

remember is that he or she is rejecting the activity, *not* rejecting you. Try to talk out the problem without using an accusatory tone, since it may be possible to find some compromise solution. For example, a woman may be willing to try fellatio if she has her partner's assurance that he won't ejaculate in her mouth or if there's prior agreement to the fact that she'll do it only when she wants to, not when she's asked. If talking about the problem doesn't result in any progress, a visit to a qualified sex counselor or therapist may be helpful.

15. *If sexual problems occur in your life, realize that you're like most other people. But if sexual problems persist over time, don't just wait for them to go away — get help while treatment is still apt to be easy.* There's little question of the fact that the vast majority of people have times when they're temporarily out of commission sexually. This may be for physical reasons — such as fatigue, infection, acute illness — or for reasons related to factors such as high levels of stress, grief, depression, preoccupation with school or job, family problems, and so forth. Most of the time, these prove to be relatively transient episodes of sexual distress that disappear within a matter of weeks. There's no reason to get yourself all worked up about such an occurrence if it happens to you or your partner, but if difficulties persist for longer than a few months without showing definite signs of improvement, it's time to seek out professional advice. The longer such a problem lingers the less likely it is to disappear by itself; furthermore, a professional will usually be able to make a rapid diagnosis of what's going on. In fact, sometimes a sexual problem can be the first sign of a medical condition that requires treatment. In any event, the sooner the cause of the problem is identified and appropriate treatment is started, the easier it generally is to cure it.

16. *Keep your sexual expectations realistic.* If you're looking for every sexual encounter to be sheer ecstasy for you and your partner, you're setting yourself up for failure. If, instead, you can accept the fact that sex isn't always the great passionate joining of souls that Hollywood would have us believe it to be — that it's sometimes rather feeble, awkward, and even unsatisfying — then you won't be a

prisoner of unrealistic standards. In real life every orgasm isn't a stupendous, earth-shattering event — some are more like small, ordinary twitches. Keeping in mind that we are human beings, rather than machines, may not only help us accept the ups and downs of our sexuality but may also help us demystify our sexual experiences.

CHAPTER EIGHTEEN

Sexual Dysfunctions and Sex Therapy

LIKE MANY other body processes, when sexual function goes along smoothly, it is usually taken for granted and given little thought. But if sexual function is a problem in one way or another, it can be a source of anxiety, anguish, and frustration that often leads to general unhappiness and distress in personal relationships.

This chapter begins with a description of sexual dysfunctions — conditions in which the ordinary physical responses of sexual function are impaired. Our attention then turns to the causes of these dysfunctions. Next, we consider another category of sexual problems that are not dysfunctions in the ordinary sense of that term, since they are disorders of sexual desire. The concluding portion of the chapter discusses the methods and effectiveness of sex therapy in dealing with these problems.

MALE SEXUAL DYSFUNCTION

For most men in most societies, sexual adequacy is considered a yardstick for measuring personal adequacy. The man who does not "measure up" sexually is often embarrassed, confused, or depressed over his plight, which he regards as reflecting poorly on his manhood. The sexually dysfunctional male may change his behavior to avoid sexual situations (fearing in advance that he will fail); he may cope with his dilemma by inventing excuses (blaming the dysfunction on his partner, for example); or he may try to overcome his problem by diligently "working" at sex, which usually makes the situation worse instead of better.

Erectile dysfunction, or *impotence,* is the inability to have or maintain an erection that is firm enough for coitus. Erectile dysfunction is classified as either primary or secondary: the male with *primary erectile dysfunction* has never been able to have intercourse, whereas the male with *secondary erectile dysfunction* has succeeded in having intercourse once, twice, or a thousand times before his dysfunction began. Secondary erectile dysfunction is about ten times more common than primary erectile dysfunction.

Erectile dysfunction can occur at any age and can assume many different forms. Total absence of erection is infrequent except in certain medical conditions. More typically, the male with erectile dysfunction has partial erections that are too weak for vaginal insertion (or anal intercourse). Sometimes, there are firm erections that quickly disappear if intercourse is attempted. In other instances, a man with erectile dysfunction may be able to have normal erections under some circumstances but not others. For example, some men with erectile dysfunction have no problem during masturbation but cannot get erections during sexual activity with a partner. Other men have solid erections during extramarital sex but only feeble erections with their spouses. The reverse of this pattern is also common: some men who have no sexual difficulty with their wives are unable to function during attempts at extramarital sex.

Isolated episodes of not having erections (or of losing an erection at an inopportune time) are so common that they are nearly a universal occurrence among men. (For this reason, Masters and Johnson [1970] classified a man as secondarily impotent only if his erection problems occurred in at least 25 percent of his sexual encounters.) Such isolated episodes do *not* mean that a man has a sexual dysfunction; they may reflect a temporary form of physical stress (having the flu, being tired, having over-indulged in food or drink), or may relate to other problems like tension, lack of privacy, or adjusting to a new sexual partner. If the man does not take such incidents in stride and becomes deeply upset by his "failure" to respond the "right" way physically, he may set the stage for difficulties in later sexual situations because he is worried about his ability (or inability) to perform.

Fears of sexual performance — "Will I lose my erection?" "Will I satisfy my partner?" — are likely to dampen sexual arousal and cause loss of erection. The stronger and more insistent such fears become, the greater is the likelihood that they will become self-fulfilling prophecies, and the man will experience an actual inability to get and keep an erection. On a long-term basis, performance fears may lead to lowered interest in sex (avoidance), loss of self-esteem, and attempts to control the anxiety by working hard to overcome it (which usually reduces sexual spontaneity and causes sex to be even more of a "performance" instead of just being *fun*.) In addition, fears of performance often cause one or both partners to become spectators during their sexual interaction, observing and evaluating their own or their partner's sexual response. By becoming a spectator, a person usually becomes less involved in the sexual activity because of the distraction of watching and evaluating what is going on.

The spectator role, which can affect men and women, is not only found in cases of erectile dysfunction. When a person slips into the spectator role because of performance fears, the reduced intimacy and spontaneity of the situation combined with pre-existing fears usually stifle the capacity for physical response. This cycle tends to feed on itself: erectile failure leads to performance fears which lead to the spectator role, which results in distraction and loss of erection, which heightens the fears of performance. Unless this cycle is broken, there is a strong possibility that sexual dysfunction will be firmly established.

Men react to erectile dysfunction in various ways, ranging from great dismay (probably the most typical response) to studied nonchalance (the least typical). While there are some men and women who see sex as more than a throbbing, erect penis and do not judge the satisfaction of a sexual encounter on the basis of having intercourse alone, for most people the practical limitations of erectile dysfunction are bothersome. One thirty-four-year-old man relates his personal feelings in dealing with it:

> After a while, the problem becomes so predictable that you start to make excuses in advance. It's as though you lose any chance of having sexual pleasure because you become preoccupied with the

notion of failure. And that failure hits you right in the gut — you don't feel like much of a man.

The partner of a man with erectile dysfunction may blame herself for not being attractive enough to turn him on or not being skilled enough to arouse his passion, or she may fear that she is pressuring him and causing his difficulties. On the other hand, the partner may blame the man in various ways for his sexual problems. We have encountered women who accused their husbands of extramarital sex, being homosexual, or not being in love with them as explanations of erectile dysfunction. Sometimes the impact of erectile dysfunction can alter the fabric of a close relationship by introducing strain, doubt, irritability, and frustration, all of which have effects outside the bedroom.

Premature ejaculation, or rapid ejaculation, is a common sexual dysfunction but difficult to define precisely. Older definitions that used a specific duration of intercourse as the dividing line ("less than two minutes," for example) or that specified a minimum number of penile thrusts before ejaculation have now been discarded. This is fortunate because some men actually tried to time themselves with a stopwatch to determine if they were normal, and others tried to hurry their thrusting ("just four more thrusts, dear") although this usually speeds up ejaculation instead of delaying it.

In *Human Sexual Inadequacy*, an attempt was made to define premature ejaculation in terms of the interaction between sexual partners, not just the male alone. (Prior to 1970 premature ejaculation had frequently been classified as a form of impotence, a belief that reflected a poor understanding of the underlying physiology.) A man was considered to ejaculate prematurely if his partner wasn't orgasmic in at least 50 percent of their coital episodes, but it was acknowledged that this definition was still lacking. Specifically, it couldn't be applied to situations in which a woman was infrequently orgasmic or never had orgasms during intercourse, and it was an arbitrary way of estimating normality at best. Later, in *The New Sex Therapy*, Helen Kaplan suggested that premature ejaculation occurred if the male didn't have voluntary control over when he ejaculated — although most sex therapists agree that total voluntary control over the timing of ejaculation is the exception rather than the rule.

The American Psychiatric Association has recently side-stepped this issue by defining premature ejaculation in terms of "reasonable voluntary control." "The judgment of 'reasonable control' is made by . . . taking into account factors that affect duration of the excitement phase, such as age, novelty of the sexual partner, and the frequency and duration of coitus." Another view, offered by the psychologist Joseph LoPiccolo, suggests that premature ejaculation does *not* exist if both partners "agree that the quality of their sexual encounters is not influenced by efforts to delay ejaculation."

Despite the shortcomings of these definitions (or the definitions of this shortcoming), it is usually not too difficult to decide when rapid ejaculation is problematic in a sexual relationship. Although Kinsey and his co-workers suggested that rapid ejaculation was a sign of biological competence, noting, "It would be difficult to find another situation in which an individual who was quick and intense in his responses was labeled anything but superior . . . however inconvenient and unfortunate . . . from the standpoint of [the] wife," today most sexologists disagree with this idea. Kinsey's belief may have influenced his finding that 75 percent of men ejaculated within two minutes of vaginal entry, but it now seems unlikely that this figure is accurate. While it is certainly true that some people see sex as primarily for the male's pleasure — and some females may actually be grateful to "get it over with" quickly — these ideas, which were once widespread, seem to have been replaced today by a more egalitarian view of sexual interaction except among the least educated and lowest socioeconomic levels.

Clearly, the male who persistently ejaculates unintentionally during noncoital sexual play or while trying to enter his partner has a problem. While this extreme situation is found in fewer than 10 percent of cases, it is likely to be particularly distressing. More typically, the premature ejaculator is able to participate in a variety of sexual activities and only loses his ejaculatory control soon after intercourse begins. Premature ejaculation may occur in some situations and not in others. For example, a man may have this problem only during extramarital sex.

Some men are not bothered at all by ejaculating rapidly. Many others question their masculinity and have low self-esteem. Fears of performance often seem to heighten the lack of ejaculatory control and can occasionally lead to erectile dysfunction by the "fears-spectator-failure-greater fears" cycle described earlier. Erectile difficulties can also occur if a premature ejaculator struggles to control his sexual arousal by using distraction (thinking about the office or counting backwards from 1,000): if he succeeds too well in distracting himself from involvement in the sexual interaction, he may lose his erection as well as the urgency to ejaculate.

While many of the female partners of men with premature ejaculation are understanding and accepting of the involuntary nature of the problem, others "feel angry and 'used,' leading them to seek professional guidance, to seek another lover, or to avoid sex," according to the psychologist Michael Perelman. Because most males have a tendency to ejaculate more quickly if it has been a long time between sexual opportunities, avoidance is likely to worsen the problem and may worsen the relationship too. Similarly, if the man tries to reduce his arousal by shortening the time of noncoital play, his tactic is not only ineffective but also may backfire by further convincing the woman of her partner's selfishness.

Although premature ejaculation is less frequent than erectile dysfunction among male patients at the Masters & Johnson Institute, we believe it is probably the most common sexual dysfunction in the general population. We estimate that 15 to 20 percent of American men have at least a moderate degree of difficulty controlling rapid ejaculation, but less than one-tenth of this group consider it to be enough of a problem to seek help. Some men find that they can overcome premature ejaculation on their own by using a condom to cut down on genital sensations; others discover that a glass or two of an alcoholic beverage may reduce their ejaculatory quickness; and others find that controlling ejaculation is no problem "the second time around" — that is, once they have already had one orgasm and attempt intercourse within the next two or three hours. Over-the-counter creams and ointments that "desensitize" the penis deaden sensations. If they help control rapid ejaculation at all, they do so at the cost of not feeling very much or by the power of suggestion.

At the opposite end of the spectrum, *ejaculatory incompetence* is the inability to ejaculate within the vagina despite a firm erection and relatively high levels of sexual arousal. It must be distinguished from *retrograde ejaculation,* which is a condition where the bladder neck does not close off properly during orgasm so that the semen spurts backwards into the bladder, where it is mixed with urine. *Retarded ejaculation* can be thought of as the opposite of premature ejaculation; here, although intravaginal ejaculation eventually occurs, it requires a long time and strenuous efforts at coital stimulation, and sexual arousal may be sluggish.

Ejaculatory incompetence is an infrequent disorder mainly seen in men under age thirty-five. The most common pattern (about two-thirds of patients) is *primary ejaculatory incompetence* or never having been able to ejaculate in the vagina. *Secondary ejaculatory incompetence* refers to men who have lost the ability to ejaculate intravaginally or who do so infrequently after a prior history of normal coital ejaculation. In either the primary or secondary versions of this dysfunction, ejaculation is usually possible by masturbation (about 85 percent of patients in our series) or by noncoital partner stimulation (about 50 percent of patients in our series). In about 15 percent of our cases, men with ejaculatory incompetence had never experienced ejaculation except through nocturnal emissions. Rarely, ejaculatory incompetence can be situational, occuring with one partner but not another.

Ejaculatory incompetence can be a source of sexual pleasure because it permits prolonged periods of coitus. A few of our patients have told us that they were regularly able to sustain an erection for one or two hours of intercourse — much to the delight of their partners, many of whom marveled at this staying power. However, once the woman discovers that her partner is unable to ejaculate intravaginally, a new reaction most likely sets in. She may assume that the man does not find her attractive, is not enjoying the experience, or is "withholding" orgasm as a sign of selfishness. If reproduction is a goal of the sexual partners, ejaculatory incompetence can be even more frustrating and may lead to accusations and arguments that can threaten even the best relationships, as this example from our files shows:

> *A twenty-seven-year-old married man:* I'm sick and tired of being psychoanalyzed by my wife because of this problem we're having. I want a baby just as much as she does, but my penis doesn't seem

to understand. But that's no reason to accuse me of being homosexual.

Retarded ejaculation is seen in all age groups from adolescence on and is probably two or three times more common than ejaculatory incompetence. Although it may also be a source of sexual enjoyment, sometimes the prolonged periods of coital thrusting required to bring about ejaculation are uncomfortable both physically and psychologically for the female, whose own sexual needs may have been amply met in a briefer time frame. The woman may become resentful of the sexual demands placed on her by her partner. Her feeling corresponds to the male whose female partner needs to be stimulated coitally for a long time to reach orgasm.

Here again, it is important to distinguish between the fairly consistent pattern of sexual dysfunction and the occasional episodes when a man cannot ejaculate intravaginally or requires a long period of vaginal containment and thrusting to ejaculate. Occasional difficulty with ejaculation is *not* a sign of sexual disturbance and is often related to fatigue, tension, illness, too much sex in too short a time, or the effects of alcohol or other drugs. In addition, a male may be unable to ejaculate with a partner he's not very emotionally involved with (e.g., when he's just having sex because he feels it's expected of him).

Finally, painful intercourse, or *dyspareunia,* is generally thought of as a female dysfunction but it also affects males. Most typically, the pain is felt in the penis but it can also be felt in the testes or internally, where it is often associated with a problem of the prostate or seminal vesicles. Causes of painful intercourse in males are discussed later in this chapter.

FEMALE SEXUAL DYSFUNCTION

Until fairly recently, it was presumed that women were less sexual than men. Females with a sexual dysfunction were therefore not seen as incomplete or "unable to measure up," as men with sexual problems were. In the past two decades, traditional views of female sexuality were all but demolished, and women's sexual needs became accepted as legitimate in their own right. But as part of this process, female sexual responsivity became something of an expected accomplishment, with women — helped along by numerous magazine articles, how-to books, and TV talks shows — suddenly put on the spot with performance pressures of their own. As a result, females began to develop more awareness of the existence of sexual dysfunctions. The woman who sees herself as "unresponsive" in one way or another often becomes embarrassed, confused, or depressed just as men do. She too may try to cope by avoidance, inventing excuses, or studiously "working" at sex to find the "right" technique to unlock her sexual potential.

Vaginismus is a condition in which the muscles around the outer third of the vagina have involuntary spasms in response to attempts at vaginal penetration. Females of any age can be affected, and the severity of the reflex is highly variable. At one extreme, vaginismus can be so dramatic that the vaginal opening is tightly clamped shut, preventing not only intercourse but even insertion of a finger. Less severe, but still considerably distressing, is when any attempts at coitus — no matter how gentle, relaxed, and loving — result in pelvic pain. In its milder versions, vaginismus may allow a woman to have intercourse but only with some discomfort. The frequency of vaginismus in the general population is unknown, but judging from our patients, it accounts for less than 10 percent of cases of female sexual dysfunction. We estimate that 2 to 3 percent of all postadolescent women have vaginismus.

Although some women with vaginismus are very fearful of sexual activity, which may impair their sexual responsiveness, most women with this dysfunction have little or no difficulty with sexual arousal. Vaginal lubrication occurs normally, noncoital sexual play may be pleasurable and satisfying, and orgasm is often unaffected. Women with vaginismus usually have normal sexual desire and are upset by their inability to enjoy intercourse. Vaginismus can be particularly troubling to a couple who wants to have children, and it is often this consideration that pushes the couple to seek help.

The male partner of the woman with vaginismus may be completely baffled about why sexual difficulties arise. Often, he has no specific knowledge of the involuntary muscle spasms involved and either thinks that he is doing something that hurts his partner or sees her as deliberately avoiding intercourse by "tensing up." If he thinks he is hurting her, he may become more and more passive in sexual situations. Erectile dysfunction may develop, especially if he assumes the blame for the situation. If, instead, he blames the woman, he may lose patience after awhile and become resentful or openly hostile or may simply seek other sexual partners.

Vaginismus may be suspected from a woman's history (e.g., the woman may have had difficulty using a tampon or a diaphragm), but identification of this dysfunction can only be made with certainty by a careful pelvic examination. Unfortunately, not all physicians are well versed in detecting sexual problems, and women are sometimes mistakenly told "everything's normal" when vaginismus is unquestionably present.

Before the publication of *Human Sexual Inadequacy* in 1970, the term *frigidity* was generally used to describe a number of female sexual difficulties ranging from not having orgasms to not being interested in sex to not becoming sexually aroused. As this term lacked diagnostic precision and was increasingly used in a negative, disparaging way, portraying women as "cold" or "rejecting," many sexologists abandoned its use. Masters and Johnson (1970) and Kaplan (1974) substituted the term *orgasmic dysfunction* to describe women who have difficulty reaching orgasm; *anorgasmia* is currently used as a synonym.

As with many sexual dysfunctions, there are several categories of anorgasmia. *Primary anorgasmia* refers to women who have never had an orgasm. *Secondary anorgasmia* refers to women who were regularly orgasmic at one time but no longer are. *Situational anorgasmia* refers to women who have had orgasms on one or more occasions but only under certain circumstances — for example, women who are orgasmic when masturbating but not when being stimulated by their partner. Women who are orgasmic by a variety of means but do not have orgasms during intercourse are described in a subcategory of situational anorgasmia called *coital anorgasmia*. Finally, *random anorgasmia* refers to women who have experienced orgasm in different types of sexual activity but only on an infrequent basis.

As indicated by these definitions, there are many forms of anorgasmia. Within these classifications, the diversity is even greater. Some anorgasmic women get little pleasure out of sex and see it as an obligation of marriage or a means of maintaining a relationship. Other anorgasmic women find that sex is stimulating and satisfying. One woman told us, "Since I've never had an orgasm, I don't really know what I'm missing, but I certainly know when I'm having fun." Many women with orgasmic difficulties voice opinions somewhere between these extremes, as shown by these comments from our files:

> *A twenty-two-year-old single woman:* I enjoy sex, but I keep pushing to reach for an orgasm, and the worry that it won't be there keeps gnawing away inside me. I'd feel a lot better if I knew I could come every time.

> *A thirty-one-year-old married woman:* I've always been able to have orgasms when I masturbate, but it's never happened with my husband. After eight years of marriage this has really become a strain on our relationship — for him, because he feels he's inadequate; for me, because I'm missing a special kind of sharing.

> *A nineteen-year-old college student:* There's so much talk about orgasms that I've been wondering what's wrong with me, that I don't have them. I used to enjoy sex a lot, but lately it's a bad scene because I just get reminded of my own problems.

Not having orgasms can create fears of performance that propel a woman into the spectator role, dampening her overall sexual responsiveness, just as is true for a man. Anorgasmia can also lead to less self-esteem, depression, and a sense of futility.

It must be stressed that the stimulation of intercourse *alone* is not always sufficient for female orgasm to occur. Many women usually require additional stimulation (e.g., stroking the clitoris) during intercourse to have coital orgasms. Unfortunately, some people believe that "genuine" coital orgasms are only the result of penis-vagina contact, which is simply not the case.

There is some controversy today about the number of women who are anorgasmic during intercourse. A number of studies, viewed as a group, suggest that about 10 percent of women have never experienced coital orgasm. Perhaps another 10 percent have coital orgasms on an infrequent basis. Kaplan suggests that these women should not necessarily be thought of as having a problem, since she believes that not having orgasms during intercourse is within the normal range of female sexual response.

Hite voices a similar opinion, pointing out that many of the women she surveyed preferred noncoital orgasms to those occurring during intercourse. To us, their approaches are flawed for several reasons: (1) if the same line of reasoning was applied to males, premature ejaculation would be viewed as "within the normal range," since it appears to affect a similar number of men; (2) many coitally anorgasmic women are easily able to begin having orgasms during intercourse with the aid of short-term sex therapy; and (3) there is no reliable research evidence that a sizable fraction of women are incapable of coital orgasm. Furthermore, since many women who have coital anorgasmia are distressed by their situation, what is gained by telling them that everything is normal and it's no cause for concern?

We would like to make it clear that we believe that people should seek sex therapy with attainable goals and a genuine sense of need. A woman who is content not to have orgasms or who is unconcerned about how she reaches her orgasms should not be pushed into sex therapy, and we would be the first to tell her this directly. We also believe that it is usually a mistake to undertake sex therapy for female anorgasmia when the primary motivation is to satisfy the male partner.

Anorgasmia in its different forms is far and away the largest category of female sexual dysfuntion, accounting for about 90 percent of cases in most large studies. Many women, however, are not orgasmic during every sexual encounter yet do *not* have a sexual dysfunction. Lack of orgasm must be viewed in terms of the individual's desires, the skill and sensitivity of her partner (as well as his attractiveness, cooperation, etc.), the circumstances of sexual activity (privacy, timing, comfort, and so on), and other factors too numerous to mention here. A woman who sometimes has orgasms should be classified as having orgasmic dysfunction only if her orgasmic frequency is so low that it is a source of distress or dissatisfaction.

The male partner of an anorgasmic woman may feel sympathetic while also feeling threatened, since many men assume it's their responsibility to make their partner orgasmic. If he sees his role as "tutor" or "coach," he may become impatient or angry if his partner doesn't reach orgasm in response to his attentions. If he concentrates on romance and carefully orchestrated sexual technique, he may become resentful if the female does not achieve orgasm. Some men give up trying and become resigned to the situation, while others are convinced that their partners are deliberately withholding orgasm. If a man discovers that his partner has been faking orgasm, he is particularly likely to be upset or angered.

Although premature ejaculation in men has been widely discussed, its female counterpart — rapid orgasm — has been almost completely ignored by sexologists. This is probably because this condition is relatively rare. In more than two decades of our research, we have encountered only a handful of women who complained of reaching orgasm too quickly. The primary problem for these women is that once orgasm occurs they have little interest in further sexual activity and often find that it is physically uncomfortable. In contrast, most women who have rapid orgasms remain sexually interested and aroused (often going on to additional orgasms) and thus consider it an asset, not a liability. The male partner of a woman who reaches orgasm rapidly is also likely to view it in a positive light, either seeing it as a sign of a very responsive partner or giving himself a pat on the back for his lovemaking skills.

Dyspareunia, or painful intercourse, in women can present a major stumbling block to sexual satisfaction. In this condition, which can occur at any age, pain can appear at the start of intercourse, midway through coital activities, at the time of orgasm, or after intercourse is completed. The pain can be felt as burning, sharp, searing, or cramping; it can be external, within the vagina, or deep in the pelvic region or abdomen.

The incidence of dyspareunia is unknown. We have found that about 15 percent of adult women experience coital discomfort on a few occasions per year. We estimate that 1 to 2 percent of adult women have painful intercourse on more than an occasional basis.

Dyspareunia detracts from a person's sexual enjoyment and can interfere with sexual arousal and orgasm. The fear of pain may make the woman tense and decrease her sexual pleasure; in many cases, the woman may avoid coital activity or abstain from all forms of sexual contact. The partner of a woman with dyspareunia may be either very understanding and sensitive to her feelings, or resentful and demanding in spite of her discomfort.

CAUSES OF SEXUAL DYSFUNCTION

It is customary to classify the causes of sexual dysfunction as either *organic* (physical or medical factors such as illness, injury, or drug effects) or *psychosocial* (including psychological, interpersonal, environmental, and cultural factors). The precise cause of a specific dysfunction in a given person cannot always be identified, and in some instances, it may be a combination of several different factors.

It is generally estimated that 10 to 20 percent of sexual dysfunction cases are caused primarily by organic factors. In another 15 percent of cases, organic factors may contribute to the sexual difficulty, although they may not be the direct or sole cause of the disorder. Given these facts, it is important for a person seeking treatment for a sexual dysfunction to have a thorough physical examination as well as appropriate laboratory testing of blood and urine samples to identify or rule out organic conditions that might be affecting sexual function. We will now review each sexual dysfunction to examine organic factors that may be important. Many of these disorders are discussed more fully in the next chapter.

Erectile dysfunction can result from many medical conditions. *Diabetes* (a condition in which the body improperly handles blood sugar regulation) and *alcoholism* are the two most prominent organic causes of erectile dysfunction. Together, they probably account for several million cases in the United States alone. Other organic causes of erectile dysfunction include spinal cord injury, multiple sclerosis, or other neurological disorders; infections or injuries of the penis, testes, urethra, or prostate gland; hormone deficiencies; and circulatory problems. Both prescription medications (such as drugs for high blood pressure) and street drugs like uppers (amphetamines), downers (barbiturates), and narcotics sometimes cause difficulty with erection. Overall, as many as one-half of all cases of erectile dysfunction may be of organic origin.

Premature ejaculation rarely results from organic causes. In over 500 cases of premature ejaculation seen at the Masters &

Johnson Institute, we have found only one instance where an organic condition proved to be of importance.

With *ejaculatory incompetence,* organic causes can be eliminated as a possiblility if ejaculation occurs in noncoital situations. In cases of complete inability to ejaculate, drug use and neurological disorders are sometimes found, accounting for about one out of twenty cases. Drug use and alcoholism account for about 10 percent of cases of retarded ejaculation.

Painful intercourse in males can be due to several different organic problems, although psychosocial factors appear to cause at least half of such cases. Inflammation or infection of the penis, the foreskin, the testes, the urethra, or the prostate are the most likely organic causes of male dyspareunia. A few men experience pain if the tip of the penis is scratched or irritated by the tail of an IUD (the stringlike portion that protrudes through the cervix into the vagina). Other men develop painful penile irritation when exposed to a vaginal contraceptive foam or cream.

Vaginismus is most frequently a psychosocial problem rather than an organic one. However, any of the organic causes of female dyspareunia can condition a woman into vaginismus as a natural protective reflex. Even when the underlying organic problem is detected and successfully treated, the vaginismus may remain, particularly if it has been present for a long period of time.

Anorgasmia is linked to organic causes in less than 5 percent of cases. Severe chronic illness of almost any variety can impair female orgasmic response. Specific disorders that sometimes block orgasm include diabetes, alcoholism, neurological disturbances, hormone deficiencies, and pelvic disorders such as infections, trauma, or scarring from surgery. Drugs such as narcotics, tranquilizers, and blood pressure medications can also impair female orgasm. *Rapid orgasm* in women has no known physical causes.

Female dyspareunia can be caused by dozens of physical conditions, although psychosocial factors may be as frequent as organic ones. Any condition that results in poor vaginal lubrication can produce discomfort during intercourse. The chief culprits here seem to be drugs that have a drying effect (e.g., antihistamines — used to treat allergies, colds, or sinus conditions; certain tranquilizers; and marihuana) and disorders such as diabetes, vaginal infections, and estrogen deficiencies. Other causes of female dyspareunia include:

1. *Skin problems* (blisters, rashes, inflammation) around the vaginal opening or affecting the vulva;
2. *Irritation or infection of the clitoris;*
3. *Disorders of the vaginal opening,* such as scarring from an episiotomy, intact hymen or remnants of the hymen that are stretched during intercourse, or infection of the Bartholin glands;
4. *Disorders of the urethra or anus;*
5. *Disorders of the vagina,* such as infections, surgical scarring, thinning of the walls of the vagina (whether due to aging or estrogen deficiency), and irritation due to chemicals that are found in contraceptive materials or douches;
6. *Pelvic disorders* such as infection, tumors, abnormalities of the cervix or uterus, and torn ligaments around the uterus.

It has been much more difficult to develop a clear understanding of how psychosocial factors "cause" sexual dysfunction. Much of the research to date has found *associations* between factors such as developmental traumas, psychological traits, behavior patterns, and relationship difficulties and the existence of a sexual dysfunction, but research of this sort cannot prove what *causes* sexual dysfunctions. Furthermore, many people whose histories are loaded with potentially devastating psychosexual events have completely normal sexual function, while others who have unremarkable histories turn up with sexual dysfunctions.

Despite these problems, we can identify some psychosocial factors that are currently thought to contribute to the origins of sexual dysfunction. Since many of these are nonspecific — that is, they may lead to a number of dysfunctions in either men or women — we will consider them in terms of several broad categories.

Many authorities have suggested that developmental factors such as troubled parent-child relationships, negative family attitudes toward sex, traumatic childhood or adolescent sexual experiences, and gender identity conflicts may all predispose one toward developing later sexual dysfunctions, either singly or in combination. For example, a child who is brought up believing that sex is sinful and shameful may be handicapped in later sexual enjoyment. Children who have been repeatedly and severely punished for touching their genitals or for innocent sex play with other boys or girls are also liable to become fearful about sex in any form and may have difficulty developing a positive view of sex as an intimate, pleasurable, desirable activity.

A traumatic first coital experience — either physically or psychologically painful — is another common problem found in the backgrounds of many people with sexual dysfunctions. Such an experience can raise fear about sexual encounters, lead to avoidance, or cause considerable guilt. Another variation is shown in this comment from one of our patients:

> *A forty-eight-year-old coitally anorgasmic woman:* When I was twenty-one and still a virgin I had been looking forward to my wedding night in a romantic, idealized way. But the wedding day was exhausting, my husband and I both had too much to drink, and when we tried to make love for the first time, instead of being blissful and tender, it was hurried and disastrous. It seems as though we were never able to catch the spark of loving sex after that — it's always been disappointing and unpleasant for me.

One other developmental factor will be mentioned briefly. In *Human Sexual Inadequacy,* it was noted that a rigid religious background during childhood seemed to be associated with many sexual dysfunctions. What was striking about these cases was not the specific set of religious teachings (since these did not always condemn sexuality) but that sex was strongly regarded as evil and dirty in these rigidly religious families. Since 1970, when these findings were published, we have gathered more information in this area. We can now say that rigid religious upbringing seems to be a common factor only in certain dysfunctions: vaginismus and primary anorgasmia in women, and ejaculatory incompetence and primary impotence in men. Furthermore, by interviewing many individuals from similar backgrounds who did *not* have sexual dysfunctions, we have become even more confident in stressing that it is generally *not* the religious beliefs that are troublesome, but the severely antisexual attitudes that are forced upon the child.

Personal factors comprise another substantial category of psychosocial causes of sexual dysfunction. We have already noted that fears of performance often suppress sexual function. Other types of anxiety, including fears of pregnancy, venereal disease, rejection, losing control, pain, intimacy, and even success can also block the pathways of sexual response.

Other feelings can also affect sexual responsiveness. Guilt, depression, and poor self-esteem are encountered frequently in association with sexual dysfunctions. Sometimes, though, it is difficult to know which came first, the feeling or the dysfunction. It is natural for people who have sexual problems to become depressed about them or to feel less good about themselves. Thus, identifying such a feeling does not always mean it caused the dysfunction.

Other personal factors that can play a part in sexual dysfunction are lack of sexual information and blind acceptance of cultural myths. Uncertainty about the location of the clitoris or lack of awareness of its importance in female sexual response are prime examples of lack of sexual information. Believing that the capacity for sexual function disappears with aging or that the pace of sexual activity must be set by the male are examples of how cultural myths can translate into personal attitudes and behavior.

Although a number of different studies have attempted to correlate sexual dysfunctions with particular types of personalities, no solid evidence exists documenting such a relationship.

Interpersonal factors also are of tremendous importance in most sexual dysfunctions. The most common problem is poor communication, both in sexual and nonsexual areas of the relationship. Communications problems either can lead directly to a sexual dysfunction (through misunderstandings or defensiveness) or can play a key role in perpetuating a dysfunction. As we have stressed throughout this book, sex is a form of communication, and effective communication is extremely important in sexual relationships. Other interpersonal factors frequently involved in sexual dysfunctions include power struggles within a relationship, hostility toward a partner or spouse, preference for another partner, distrust or deceit, lack of physical attraction to a partner, and gender role conflicts (which often become power struggles). Conflicts in the sex value systems of partners or

"What do you mean, no?"

widely different sexual preferences in terms of timing, frequency, or type of sexual activity may also contribute to dysfunctions.

It must be recognized, however, that such problems do not *always* lead to sexual difficulties. Some couples find that sex is most enjoyable when they are angry at each other. Other couples with terrible communications have fantastic sexual relationships. As always, we must be careful not to oversimplify.

Before concluding our discussion on the causes of sexual dysfunction, a few additional words are in order. Until the detailed studies of Masters and Johnson were released, it was generally thought that sexual dysfunctions were invariably due to deep-seated personality problems that originated in childhood. Today, most sexologists recognize that many people with sexual dysfunctions have completely normal personalities, no signs of emotional illness, and simple, straightforward explanations for their problem.

Yet there are still some major differences between the ways in which psychoanalysts and behaviorists explain sexual dysfunction. The traditional psychoanalytic viewpoint has been that the dysfunction is not the primary problem but is instead a symptom of a deeper psychological disturbance. Analysts have suggested that ejaculatory incompetence, erectile dysfunction, and premature ejaculation result from castration anxiety and unresolved Oedipal wishes that are usually present at the unconscious level. Similarly, the analytic viewpoint generally suggests that vaginismus and anorgasmia also reflect unresolved Electra conflicts, as well as unconscious hostility toward men because of penis envy. In essence, old fears of being punished for sexual play that were supposedly learned in early childhood are reawakened (in the unconscious) by adult sexual encounters and cause psychological conflict, anxiety, and dysfunction.

In sharp distinction, the learning theory viewpoint sees sexual dysfunction as a conditioned, or learned, response. A man may develop erectile dysfunction if his partner constantly criticizes his performance or if he feels guilty after every sexual encounter. A woman may be anorgasmic because she was conditioned to believe that sex was "wrong" or shameful, or because she was taught that "nice" girls don't enjoy sex. Premature ejaculation may originate from early sexual experiences in which ejaculating quickly was desirable. Such experiences might include situations with a risk of being discovered by someone else, as in sex in a parked car; sex with a prostitute who typically encourages speed so she can see more customers; or group masturbation, where ejaculatory speed was seen as a sign of virility. If the conditioning is powerful enough, it cannot be easily "unlearned" even when circumstances change.

The learning theory model also points out that some dysfunctions are maintained by positive reinforcement. That is, the dysfunction may result in increased tenderness or attention from a partner, or it may give a person the upper hand in a power struggle. Furthermore, behaviorists generally believe that a precise understanding of causation is less important in treating a sexual dysfunction than recognizing the conditions that *maintain* the difficulty, since these are the ones that need to be changed.

Perhaps as sex research becomes more sophisticated it will be possible to approach the question of causation with more certainty. At present, the understanding of this area remains limited.

DISORDERS OF SEXUAL DESIRE

Since the mid-1970s, sex therapists have become increasingly aware of a new category of sexual problems that are not, strictly speaking, sexual dysfunctions. In these conditions, which are collectively referred to as disorders of sexual desire, the capacity for physical sexual response is usually preserved, and the problem is one of lack of willingness to participate in sexual relations due to either fear or lack of interest. If lack of interest in sex is the predominant problem, it is classified as *inhibited sexual desire* (ISD). If lack of participation in sex is mainly due to overwhelming fear it is classified as *sexual aversion.*

In deciding when a disorder of sexual desire is present, it is necessary to remember that while some people seem to be interested in sex at almost any time, others have low or seemingly nonexistent levels of sexual interest. Only when lack of sexual interest is a source of personal or relationship distress, instead of voluntary choice, is it classified as ISD. Schover et al. (1982) suggest that ISD is present if there is both a low rate of sexual activity and "a subjective lack of desire for sexual activity; desire here includes sexual dreams and fantasies, attention to erotic materials, awareness of wishes for sexual activity, noticing attractive potential partners, and feelings of frustration if deprived of sex."

People with ISD characteristically have low interest in initiating sexual behavior and are generally unreceptive to a partner's sexual advances, although they may reluctantly "give in" to their partner's wishes from time to time in order to preserve peace in the relationship. Generally, people with ISD are sexually functional (in physiological terms) but this disorder can also coexist with one or more sexual dysfunctions. ISD can be primary (lifelong) or secondary, and it can be either generalized (occuring all of the time) or situational. Although the exact incidence of this problem is unknown, it has been appearing frequently in sex therapy clinics around the country in the recent past, accounting for as many as three out of ten cases in some clinics.

The causes of inhibited sexual desire include both organic and psychosocial conditions. Hormone deficiencies, alcoholism, kidney failure, drug abuse, and severe chronic illness may each play a role. Ten to 20 percent of men with this disorder have pituitary tumors that produce excessive amounts of prolactin; the prolactin suppresses testosterone production and sometimes causes impotence as well as ISD. A majority of cases of ISD appear to be psychosocial in origin, reflecting problems such as depression, prior sexual trauma, poor body image or self-esteem, interpersonal hostility, and relationship power struggles. In some cases, ISD seems to develop as a means of coping with a pre-existent sexual dysfunction. For instance, a man with erectile dysfunction who develops a low interest in sex finds that this allows him to avoid the unpleasant consequences of sexual failure such as embarrassment, anxiety, loss of self-esteem, and frustration.

Women or men with ISD may be sexually functional or may have difficulties with sexual arousal and orgasm. In many cases, they seem to not be able to recognize early signs of sexual arousal in themselves and use a limited set of cues to define a situation as sexual. For example, such persons may ignore warmth and tenderness as possible signs of sexual feelings while waiting to be swept off their feet by a tidal wave of sexual passion. In addition, many people with ISD believe their initial desire is a good predictor of their ultimate response to a sexual situation, so if they are not feeling "turned on" at the first touch or kiss, they give up all hope of enjoying themselves.

ISD is not a source of difficulty in all marriages or relationships where it occurs. Sometimes a couple reaches an acceptable accommodation to the situation: for example, a person with ISD may agree to participate in sex when his or her partner requests it, regardless of personal interest. Alternatively, some couples reach a workable situation by allowing — or even encouraging — the partner with an intact sex drive to seek sexual activity outside the relationship. Most frequently, however, when only one person in the relationship has low sexual desire, it poses a major strain.

Sexual aversion is a severe phobia (irrational fear) about sexual activity or the thought of sexual activity, which leads to avoidance of sexual situations. It, too, affects both males and females. The intense fear or dread found in sexual aversion is sometimes expressed in physiological symptoms such as profound sweating, nausea, diarrhea, or a racing, pounding heartbeat. But in many cases, the phobia expresses itself in purely psychological terms: simply put, the person is terrified of sexual contact.

Perhaps surprisingly, people with sexual aversion are likely to be able to respond fairly naturally to sexual encounters — if they can get past the initial dread. Some patients with this disorder have told us they have more difficulty with undressing and touching in a sexual context than they do with participation in intercourse.

Between 1972 and 1985, we have seen 164 cases of sexual aversion at the Masters & Johnson Institute. The primary causes seem to be: (1) severely negative parental sex attitudes; (2) a history of sexual trauma (e.g., rape, incest); (3) a pattern of constant sexual pressuring by a partner in a long-term relationship; and (4) gender identity confusion in men. In the typical case of sexual aversion, the frequency of sexual activity drops to only once or twice a year — if that often. This can obviously become a major source of relationship stress, and the partner of the person with sexual aversion often becomes extremely angry and considers leaving the relationship. Fortunately, the success rate for treating sexual aversion is over 80 percent even in cases of long duration. Schover and LoPiccolo have also reported that cases of ISD and sexual aversion usually have successful treatment outcomes.

SEX THERAPY

Prior to 1970 the treatment of sexual dysfunctions and problems was generally the province of psychiatry. Treatment usually took a long time and successful reversal of the sexual distress was very uncertain. The traditional psychiatric model of individual treatment (one therapist working with one patient) was almost always used. Today, sex therapy is a field that includes practitioners of many different backgrounds—psychology, medicine (both psychiatry and other specialties), social work, nursing, counseling, and theology, just to name a few. There are also many approaches to sex therapy, some of which are described below.

The Masters and Johnson Model

Beginning in 1959, Masters and Johnson began their revolutionary program for treating sexual dysfunctions. It is considerably different from prior approaches in a number of ways. For example, they characteristically work with couples (instead of individuals) because they feel that there is no such thing as an uninvolved partner in a committed relationship in which there is any form of sexual distress. This does not mean that the partner always *causes* the problem, but points out that he or she is *affected* by it, as the relationship is affected. This strategy shifts the therapeutic focus to the relationship instead of the individual. Furthermore, it provides a more effective means of identifying the full dimensions of a problem. Masters and Johnson found that the input of both partners usually proved more useful than the one-sided view that an individual provided. Finally, this strategy provides an opportunity to gain the cooperation and understanding of both partners in overcoming the distress.

A logical extension of this approach is the use of *two* therapists — a man and woman working together as a *co-therapy team*. This team increases therapeutic objectivity and balance by adequately representing male and female viewpoints and gives each partner a same-sex therapist to whom he or she can (theoretically) relate more easily. The co-therapy team also provides a model for the patient couple in important ways: for example, they can easily demonstrate effective communication skills.

Another important element of the Masters and Johnson approach is the integration of physiologic and psychosocial data in assessment and treatment. In the past, many psychiatrists never examined their patients because they feared that this might trigger unwanted sexual feelings that could complicate the treatment relationship. Masters and Johnson recognized that it was important to identify organic conditions that might require medical or surgical treatment instead of sex therapy. They also found that explaining the anatomy and physiology of sexual response to patients often had important therapeutic benefits of its own.

Finally, the Masters and Johnson model involves a rapid, intensive approach to treating sexual problems. Couples are seen on a daily basis for a two-week period (the average duration of therapy is actually just under twelve days). This format permits a day-to-day continuity that appears to be beneficial in certain aspects of sex therapy, such as reducing anxiety or helping patients overcome mistakes. Couples are also urged to free themselves from their ordinary work, family, and social activities during the two weeks of therapy to be able to devote their undivided attention to their own relationship without outside distractions.

Against this general background, some additional treatment concepts of the Masters and Johnson approach are important:

1. *Therapy is individualized to meet the specific needs of each couple.* The couple's values and objectives are the primary determinants of exactly what is done. Therapists must avoid imposing their own values on their patients.

2. *Sex is assumed to be a natural function, controlled largely by reflex responses of the body.* Although many different factors can interfere with sexual function by disrupting these natural reflexes, sex therapy does not generally involve "teaching" the desired sexual response. Masters and Johnson focus instead on identifying obstacles that block effective sexual function and on helping people remove or overcome these obstacles. When this happens, natural function usually takes over promptly. (Sometimes removing the obstacles to natural function is not enough, particularly for people who have a lifelong pattern of sexual dysfunction. They may also need specific therapeutic attention to facilitate arousal or to improve sexual techniques. In this situation, there is some "teaching" going on.)

3. *Because fears of performance and "spectatoring" are often central to cases of sexual dysfunction, therapy must be approached at several levels.* Pressures to perform are removed initially by banning direct sexual contact. Couples are then helped to rediscover the sensual pleasures of touching and being touched without the goal of a particular sexual response (the "sensate focus" exercises). The therapists also help couples relabel their expectations so they do not judge everything they do as "success" or "failure." They also give people "permission" to be anxious, which helps them to talk about

their anxiety more openly. This open communication often reduces the severity of the anxiety.

4. *Determining who's "to blame" for a sexual problem is discouraged as a counterproductive exercise.* Instead, couples are assisted in finding out what makes them feel comfortable and relaxed as opposed to tense or nervous. In this approach, each person is urged to take responsibility for himself or herself rather than waiting for his or her partner to provide the "right" mood, the "right" touch, or the "right" style of making love.

5. *Helping couples see that sex is just one component of their relationship is stressed.* Often, when a sexual problem occurs in people's lives, it causes them to worry about sex so much that they devote a disproportionate amount of time to thinking and talking about this topic. One typical objective of therapy is to help couples achieve a balanced perspective toward sex in which it is neither the totality of their relationship nor the most neglected part of their relationship. In fact, a general truism of sex therapy is that when a couple's relationship improves outside the bedroom, it is apt to have positive results inside the bedroom, too.

In the Masters and Johnson approach, on the first day of therapy the couple begins by briefly meeting with their co-therapists, who introduce themselves and explain the events of the next few days. Following this overview, the patients are separated and a detailed history is taken from each partner by the same-sex therapist. After a break for lunch, during which the co-therapists get together to discuss what they have learned, a second history is taken but this time on a cross-sex basis — that is, the male therapist interviews the female patient and the female therapist interviews the male patient. Rounding out a busy day, each patient has a complete physical examination. Blood samples are obtained the following morning for laboratory evaluation of general health.

The second day of therapy is the time of the "roundtable session" in which the couple and both co-therapists meet together. Here, the therapists present their assessment of the sexual *and* nonsexual problems that the couple is facing and give an honest opinion of the chances of successful treatment. The patients are encouraged to comment on the therapists' impressions and to correct any factual errors that have been made. The therapists try to explain the most plausible causes of the sexual dysfunction(s) or problem(s) and begin to outline the type of approach for treatment. Generally, there is some discussion of sex as a natural function, how fears of performance originate, the effect of spectatoring, and the importance of communication skills. The roundtable usually concludes with suggestions to try the "sensate focus" exercises (described soon) in the privacy of their own home or hotel room.

Each initial history-taking session usually lasts one and one-half to two hours, and the cross-sex histories average about forty-five minutes apiece. The roundtable generally lasts about ninety minutes, although any of these times are quite variable, depending in part on how talkative the couple is. Daily sessions after the roundtable average about an hour in length.

From the third day, the patient couple and the co-therapists continue to meet in a four-way interaction, although occasionally the therapists may see each patient separately to discover if there are any individual concerns that one partner is hesitant to discuss in the presence of the other. Each partner is asked to describe the events of the previous twenty-four hours, with particular attention to communication patterns and interaction during the sensate focus assignments.

Interestingly, a majority of time in the therapy sessions is usually spent on nonsexual issues (such as dealing with anger, self-esteem, or power struggles), although there is a direct attempt to provide information about sexual anatomy and physiology while attending to the couple's other needs. Couples who have negative sexual attitudes are encouraged to adopt new viewpoints.

At the beginning of therapy, each couple is asked to refrain from direct sexual interaction involving genital contact. This approach helps to remove performance pressures and provides a framework for breaking the fear-spectatoring-failure-fear cycle that is often deeply ingrained. To learn more effective ways of sexual interaction, the idea of sensate focus is introduced.

In the first stage of sensate focus exercises, the couple is told to have two sessions in which they will each have a turn touching their partner's body — with the breasts and genitals "off limits."

The purpose of the touching is *not* to be sexual but to establish an awareness of touch sensations by noticing textures, contours, temperatures, and contrasts (while doing the touching) or to simply be aware of the sensations of being touched by their partner. The person doing the touching is told to do so on the basis of what interests them, *not* on any guesses about what their partner likes or doesn't like. It is emphasized that this touching should not be a massage or an attempt to arouse their partner sexually.

The initial sensate focus periods should be as silent as possible, since words can detract from the awareness of physical sensations. However, the person being touched must let his or her partner know — either nonverbally (by body language) or in words — if any touch is uncomfortable.

Although many people say, "Oh, we've touched lots of times before — can't we just skip this and go on to a more advanced level?", this first stage of sensate focus is critical in several ways. For example, it allows the therapists to find out additional information about how a couple interacts that supplements their histories in important ways. This stage also has a specific treatment value of its own, as shown by the fact that many times, men who have not had erections for years in attempts at sex suddenly discover a king-size erection, probably because the performance demand was removed. After all, they were told that sexual arousal was not expected but even if it occurred, it was not to be put into action. Finally, it provides an excellent means for reducing anxiety and teaching nonverbal communication skills.

In the next stage of sensate focus, which is generally introduced over the next few days, touching is expanded to include the breasts and genitals. The positions shown in the drawings are recommended but not required. The person doing the touching is instructed to begin with general body touching and to not "dive" for the genitals. Again, the emphasis is on awareness of physical sensations and not on the expectation of a particular sexual response.

At this stage, the couple is usually asked to try a "hand-riding" technique as a more direct means of nonverbal communication. The couple takes turns with this exercise. By placing one hand on top of her partner's hand while he touches her, the woman can indicate if she would like more pressure, a lighter touch, a

The "hand-riding" technique for conveniently giving nonverbal messages.

faster or slower type of stroking, or a change to a different spot. The male indicates his preferences when the situation is reversed. The trick is to integrate these nonverbal messages in such a way that the person being touched doesn't become a "traffic cop" but simply adds some additional input to the touching, which is still primarily done based on what interests the "toucher."

In the next phase of sensate focus, instead of taking turns touching each other, the couple is asked to try some mutual touching. The purpose of this is twofold: first, it provides a more natural form of physical interaction (in "real life" situations, people don't usually take turns touching and being touched); and second, it doubles the potential sources of sensual input. This is a very important step in overcoming spectatoring, since one thing the spectator can try is to shift attention to a portion of his or her partner's body (getting "lost" in the touch) and away from watching his or her own response. Couples are reminded that no matter how sexually aroused they may feel, intercourse is still "off limits."

These positions are suggested to couples for the sensate focus exercises at the stage that includes genital exploration. When the man is touching his partner, the position in the top drawing is used; when the woman is touching her partner, the position in the bottom drawing is used. Couples are urged to modify these positions to whatever they find comfortable and to experiment with other positions, too.

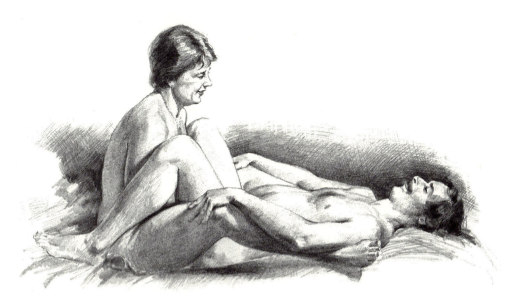

The next stages of sensate focus are to continue the same activities but at some point to shift into the female-on-top position without attempting insertion. In this position, the woman can play with the penis, rubbing it against her clitoris, vulva, and vaginal opening regardless of whether there is an erection or not. If there is an erection, and she feels like it, she can simply slip the tip of the penis a bit inside the vagina, all the while focusing on the physical sensations and stopping the action or moving back to simple nongenital touching or cuddling if she or her partner becomes goal-oriented or anxious. When comfort is developed at this level, full intercourse can usually occur without difficulty.

As simple as these techniques may sound, it is important to realize that they are used as part of a detailed program of psychotherapy, not just as a grab bag of "gimmicks." The remarkable thing is how dramatically and effectively these techniques work, even in cases where severe sexual dysfunctions have been present for ten, twenty, or thirty years.

The general features of the Masters and Johnson treatment model just described are supplemented by some additional methods used in the treatment of various dysfunctions.

In *erectile dysfunction,* it is important to help the man understand that he cannot "will" an erection to occur on demand any more than he can "will" his blood pressure to drop or his heart rate to increase. He *can* set the stage for his own natural reflexes to take over by not trying to have erections and by moving out of his performance fears. Not surprisingly, the man with erectile dysfunction often finds himself having firm erections during the early stages of sensate focus experience. While this can be reassuring to him, it is also important for the man (and his partner) to realize that losing an erection is *not* a sign of failure; it simply shows that erections come and go naturally. For this reason, the woman may be instructed to stop stroking or fondling the penis when an erection occurs, so the man has an opportunity to see that it will return with further touching. A related problem is that many men with erectile dysfunction try to rush their sexual performance once they get an erection, out of fear that they will promptly lose it. The "rushing" adds one more performance pressure, and the usual result is a rapid loss of erection.

When intercourse is attempted (only after the man has gained considerable confidence in his erectile capacity *and* has been able to reduce his spectatoring behavior), the woman is advised to insert the penis. This reduces pressures on the man to decide when it is time to insert and removes the potential distraction of his fumbling to "find" the vagina.

In treating *premature ejaculation,* the couple approach is particularly important since the condition may actually be more distressing to the woman than the man. In addition to discussing the physiology of ejaculation, the therapists introduce a specific method called the "squeeze technique" that helps recondition the ejaculatory reflex. When genital touching is begun, the woman uses the "squeeze" periodically. As shown in Figure 18, the woman puts her thumb on the frenulum of the penis and places her first and second fingers just above and below the coronal ridge on the opposite side of the penis. A firm, grasping pressure is applied for about four seconds and then abruptly released. The pressure is always applied front to back, never from side to side. It is important that the woman use the pads of her fingers and thumb and avoids pinching the penis or scratching it with her fingernails. For unknown reasons, the squeeze technique reduces the urgency to ejaculate (it also may cause a temporary, partial loss of erection). It should not be used, however, at the moment of ejaculatory inevitability—instead, it must begin at the early stages of genital play and continue periodically, every few minutes. The "squeeze" can be used whether the penis is erect or flaccid, but the firmness of the pressure should be proportionate to the degree of erection.

When the couple begins having intercourse, the woman is asked to use the squeeze three to six times before attempting insertion. Once the penis is fully inside her, she should hold still for a fifteen- to thirty-second period, with neither partner thrusting, and then move off the penis, apply the squeeze again, and reinsert. This time a slow thrusting pattern can begin. Once the man improves his ejaculatory control both partners are taught the "basilar squeeze," another version of the squeeze technique (Figure 19), so that intercourse need not be interrupted by repeated dismounting to apply a squeeze.

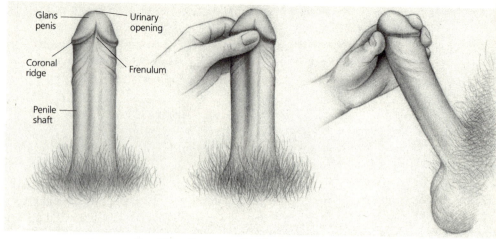

FIGURE 18.   The Squeeze Technique Used in Treating
Premature Ejaculation

Ejaculatory incompetence is treated by in-depth attention to the underlying psychological components combined with sensate focus experiences that seek to lead the man through a sequence of: (1) ejaculating by masturbation while alone, (2) ejaculating by masturbation in the presence of his partner, (3) ejaculating by manual stimulation received from his partner, and (4) having the partner stimulate the penis vigorously to the point of ejaculatory inevitability and then quickly inserting it in the vagina. In most cases, when the man has ejaculated in the vagina once or twice, the fears or inhibitions about this act disappear completely. In some cases where the treatment sequence has not worked, it may be helpful to have the man ejaculate (via manual stimulation) externally onto the woman's genitals. After he becomes used to seeing his semen in genital contact with his partner, intravaginal ejaculation may occur more easily.

Vaginismus is treated by explaining the nature of the involuntary reflex spasm to the couple and demonstrating the reflex in a carefully conducted pelvic examination with the male partner present and the woman urged to watch by use of a mirror. After this is done, the physician teaches the woman some techniques for relaxing the muscles around the vagina. The most effective method seems to be having her first deliberately tighten these muscles and then simply let go. It is much more difficult to relax

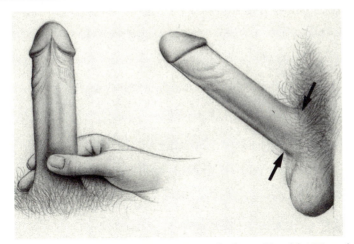

FIGURE 19.   The Basilar Squeeze Technique Used in Treating
             Premature Ejaculation

*Unlike the squeeze at the coronal ridge, the basilar squeeze can be applied
by either the woman or the man. Firm pressure is applied for about four
seconds and then released; the pressure should always be from front to back
(as shown by the arrows), never from side to side.*

on command. Next, the woman is given a set of various sized
plastic dilators. The smallest of these, slightly thinner than a
finger, is gently inserted in the vagina by the physician — often
to the amazement of the woman, who may have never been able
to insert *anything* in her vagina. She is then shown how to insert
the dilator herself, using plenty of sterile lubricating jelly, and is
asked to practice this at home several times a day, keeping the
dilators in place for ten to fifteen minutes at a time. Most women
with vaginismus find that within five or six days they are able to
use the largest dilator in the set which is similar in thickness to
an erect penis. If relationship issues have been dealt with adequately (often a key part of therapy), the transition to successful
intercourse is usually easy at this point. At this time, it is particularly important for the woman to insert the penis so she feels
in control.

The strategies used in treating *anorgasmia* depend greatly on
the nature of the dysfunction. Different approaches are needed
for a woman who has never had an orgasm as compared to a
woman who is easily orgasmic with masturbation, manual stim-

ulation, or oral-genital sex. Similarly, depending on the cause(s) of the anorgasmia, treatment strategies will vary widely from case to case. For example, a woman with a poor body image may be helped to find various ways of regarding her body more positively. A woman who is distracted from high levels of arousal by disturbing fantasies might be taught thought-blocking techniques, while another woman who cannot get beyond plateau may be encouraged to experiment with fantasies as an aid to boosting her into orgasm.

Common techniques used in treating anorgasmia include: (1) encouraging a woman to explore her own body, especially focusing on genital exploration and stimulation in a relaxed, undemanding fashion; (2) dealing with performance anxieties and spectatoring, with particular attention to reducing performance pressures from her partner; (3) fostering sexual communications so that the woman is able to let her partner know what type of touch or stimulation she prefers at a given moment; and (4) reducing inhibitions that limit the woman's capacity for arousal or that block orgasm. A woman helped by this last technique is often given "permission" to have sexual feelings and learns to overcome fears that orgasm involves losing consciousness or losing control of her bladder. In most cases, these strategies make it fairly simple for the woman to reach orgasm via masturbation or partner stimulation. To make the transition to have orgasms during intercourse, a "bridging" technique is used in which either partner stimulates the clitoris manually during coitus with active thrusting.

These methods have now been used for more than twenty years in the treatment of sexual dysfunctions, with approximately four out of five patients successfully treated. Between 1959 and 1973, patients were followed for five years after therapy to evaluate the permanence of treatment results. More recently, a two-year follow-up period has been used. About one couple in twenty returns to the Masters & Johnson Institute for additional therapy sometime after their two-week course of treatment.

Other Approaches to Sex Therapy

Important contributions to sex therapy have been made by a number of other workers using treatment methods that differ from Masters and Johnson's model. Most therapists, for example, see patients on a once-a-week basis rather than daily. Many therapists believe that a single therapist can work as effectively as a co-therapy team, which reduces the expense of sex therapy. Others have experimented successfully with group therapy formats or individual therapy. Hypnosis has also been reported to be useful in treating sexual difficulties. Here, we will consider some of the most common approaches used by other therapists.

Psychiatrist Helen Kaplan has written extensively on the treatment of sexual dysfunctions, integrating many of the methods of Masters and Johnson with principles of psychoanalytic therapy. In her view, human sexual response is best seen as *triphasic*, or consisting of three separate but interlocking phases: desire, arousal, and orgasm. She believes that desire phase disorders are most difficult to treat, because they tend to be associated with deep-seated psychological difficulties. She also states that "the standard sex therapy methods seem to be effective primarily for those sexual problems which have their roots in mild and easily diminished anxieties and conflicts." To deal with more complex cases, she uses a lengthened form of sex therapy which seeks a deeper level of insight and addresses unconscious conflicts. One of her underlying theories is that a sexual disorder usually results from multiple levels of causes — some more immediate and accessible, some more remote and hidden.

The details of Kaplan's treatment methods differ considerably in some ways from the Masters and Johnson methods. In the treatment of premature ejaculation, for example, she advocates use of the "stop-start" technique instead of the squeeze. In the "stop-start" method, introduced by James Semans in 1956, the female partner stimulates the penis manually until the man feels that he is rapidly approaching ejaculation, at which time she stops all stimulation until the sense of ejaculatory urgency disappears. Stimulation then begins again, and the stop-start cycle is repeated several times before the man is allowed to ejaculate.

Quite different from Kaplan's approach are the methods of

behavioral therapy as applied to sex therapy. Modern behavioral methods for treating sexual difficulties are generally traced back to Joseph Wolpe (1958) and enjoy widespread acceptance today. While many methods used by Masters and Johnson are very similar to behavioral techniques, there are some theoretical and practical differences in approach.

The operant approach to behavioral therapy is to carefully analyze the problematic behavior (e.g., the sexual dysfunction) and to use positive and negative reinforcers best suited to the individual case. Gradual exposure to imaginary scenes of sexual activity may be used before exposure to real-life sexual situations in sensate focus exercises. These techniques are both forms of *desensitization,* in which the controlled exposure to limited amounts of anxiety helps to eliminate the anxiety. *Relaxation training* (learning specific breathing and muscle exercises to reduce tension) and *assertiveness training* (learning how to say what you feel and what you need) are other methods used to reduce anxiety. *Directed masturbation,* which is used in the treatment of anorgasmia, is a nine-step program for helping a woman learn to masturbate to orgasm and then include her partner in her orgasmic response, first in manual stimulation and then during intercourse.

An interesting and logical way of approaching sexual problems has been suggested by a behavioral psychologist, Jack Annon. He uses a four-level model, represented by the acronym PLISSIT, to go from the simplest to more advanced levels of treatment. The four levels are: P = permission, LI = limited information, SS = specific suggestions, and IT = intensive therapy. This model makes use of the fact that in simple cases of sexual difficulties, reassurance and education are sometimes enough to solve the problem. Specific suggestions might include instruction in the squeeze technique or in sensate focus, without dealing with major psychosocial issues. Intensive therapy goes beyond these steps to deal with relationship conflicts, psychological problems, and other complex issues that may be present.

There are now so many different approaches to sex therapy that it is practically impossible to describe them all. Only brief mention of a few other notable methods can be made here due to space limitations. Lonnie Barbach has pioneered in the use of women's groups for treating anorgasmia. These groups have the particular advantages of being useful for women without partners and being relatively inexpensive and quite successful. The psychologist Albert Ellis has developed Rational-Emotive Therapy (R.E.T.) as a useful approach to sexual and nonsexual problems. R.E.T. helps people overcome irrational beliefs and unrealistic expectations that feed into their sexual dysfunction. Blending behavioral methods with unique strategies for dealing with emotional discomfort (such as shame-attacking exercises, risk-taking exercises), R.E.T. principles can be applied in individual, couple, or group therapy.

Some sex therapists follow the controversial practice of "body work therapy" in which the therapist has sexual activity with the patient. Many ethical objections have been raised to this practice, which we feel has such highly exploitive potential that it has no viable role in modern treatment methods.

A different approach, still controversial but more widely accepted, is the use of a "surrogate partner" in sex therapy. The surrogate is usually a trained woman who is paid to participate in therapy and sexual activity with a single male to improve his chances of success; there are also a few male surrogates who work with female partners. The surrogate does not function as a co-therapist but usually makes periodic reports to the therapists on the patient's progress and follows their instructions closely. Although some people have criticized this practice as being only a thinly veiled form of prostitution, others see it as an important means of helping people who are unable to find a partner who will accompany them in sex therapy. Nevertheless, given the current state of uncertainty regarding the possible transmission of the AIDS virus in such a situation, it seems advisable to us to avoid using surrogates in sex therapy at present.

THE EFFECTIVENESS OF
SEX THERAPY

As there are many different models of sex therapy and the specialty is still fairly young, evaluating the effectiveness of these approaches is difficult. Most studies to date have not had control groups, and many reports are based on a small number of cases.

Other problems are (1) lack of uniform definitions of the dys-functions, (2) differences in selection of patients, (3) differences in defining success or failure, and (4) lack of adequate follow-up (checking the results periodically after therapy is over).

The psychologists Bernie Zilbergeld and Michael Evans have criticized Masters and Johnson's data on the effectiveness of sex therapy on a variety of methodological grounds. Specifically, they have questioned the criteria Masters and Johnson used to rate treatment outcomes, suggesting that the results obtained may have reflected undue leniency in deciding how to classify cases. In addition, they suggested that Masters and Johnson's results may have been artificially inflated by selecting only the best candidates for treatment (and rejecting fairly large numbers of cases that appeared "difficult"). In actuality, however, fewer than one out of fifty couples who apply to the Masters & Johnson Institute are turned down for therapy, and the criteria used to evaluate the outcome of sex therapy at the Masters & Johnson Institute are relatively stringent. These criticisms also seem to lose sight of the fact that several other sex therapy programs are reporting rates of success that are similar to those of the Masters and Johnson program.

Some critics of sex therapy have suggested that it is a dehu-manizing, mechanistic process. The psychiatrist Natalie Shai-ness, for example, argues that sex therapy "debases" sex by detaching sex acts from feelings and emotions. The psychiatrist Thomas Szasz sees sex therapists as "determined to conceal moral values and social policies as medical diagnoses and treat-ment."

As most sex therapists agree, success or failure is a subjective commodity. Some sex therapy patients who are classified as "fail-ures" may see their treatment as very beneficial. Others, who attained enough change in their sexual function to be called "successes" by their therapists, may continue to feel unhappy or anxious. Sometimes two partners disagree about whether ther-apy was helpful. In all these situations, it is impossible to say that one viewpoint is right or wrong.

Although sex therapy offers no instant, magical cures, a vari-ety of studies show that it can be of significant help to many people. Furthermore, the gains made in sex therapy tend to be long-lasting rather than short-lived. Thus, while sex therapy is no cure-all and some people admittedly require a different type of professional help, sex therapy has managed to improve the lives of thousands of people with sexual distress.

CHOOSING A SEX THERAPIST

Unfortunately, sex therapy is an unregulated profession today. People can call themselves sex therapists even if their "training" consists only of watching a movie or reading a book. Judging from letters received at the Masters & Johnson Institute and from situations we have encountered with our patients, there are many sex therapists who are quacks, and there are hundreds of well-meaning persons who try to do sex therapy but simply are not qualified. The hapless victim of these self-proclaimed "therapists" may be bilked out of thousands of dollars. Worse yet, emotional problems may be created by improper therapy, sexual problems may worsen, and the lack of success may discourage a person from seeking further care.

To minimize the risk of falling prey to unqualified sex therapists, the following guidelines may be of assistance.

1. Turn first to sex therapy centers that are affiliated with universities, medical schools, or hospitals. Alternately, your local medical society, psychological association, or your family physician may be able to provide a list of qualified therapists. Two professional organizations, the Society for Sex Therapy and Research (New York, NY) and the American Association of Sex Educators, Counselors, and Therapists (Washington, DC) publish national directories. AASECT has a sex therapist certification program; SSTAR has tougher membership requirements which it believes are equivalent to certification.

2. Ask about the education and training of the sex therapist. Unless he or she is willing to discuss this with you, do not continue any further. In addition to having a genuine graduate degree from a recognized university, be sure to verify that the therapist has received postgraduate training in sex therapy that included personal supervision. Attendance at a weekend seminar isn't the same as in-depth training.

3. Avoid therapists who make unrealistic promises or guarantees of cure and therapists who tell you that part of your "treatment" includes having sexual relations with them.

4. Be sure the therapist is willing to discuss treatment costs, schedules, and plans in an open, straightforward fashion. (Many sex therapy clinics have a sliding-scale fee arrangement where charges are adjusted to the client's ability to pay.)

5. If you feel uncomfortable with the therapist even after you've begun, consider the possibility of switching to another therapist. While a certain amount of nervousness and hesitancy on your part is perfectly normal, it is often best to trust your instincts in this situation.

CHAPTER NINETEEN

Sexual Disorders and Sexual Health

S EXUAL HEALTH is tightly interwoven with total health: both depend on freedom from physical and emotional limitations. The preceding chapter emphasized emotional causes of sexual distress and ways of dealing with these problems. In this chapter, we will consider the physical conditions that affect sexuality. Our discussion will cover four major topics: sex and disability, the impact of illness on sexuality, drugs and sex, and infections involving the sex organs. Learning about these problems is important because it increases our understanding of the physiology of sex and helps us better grasp the emotional dimension of such situations.

SEX AND DISABILITY

A number of common myths and stereotypes about the sexuality of disabled people have been identified by the Sex & Disability Project of George Washington University. These myths, which "can drastically and unnecessarily curtail the sexual expression of disabled people," are as follows:

1. Disabled people are asexual.
2. Disabled people are dependent and childlike, so they need to be protected.
3. Disability breeds disability.
4. Disabled people should stay with and marry their own kind.

5. Parents of handicapped children don't want sex education for their children.
6. Sexual intercourse culminating in orgasm is essential for sexual satisfaction.
7. If a disabled person has a sexual problem, it's amost always the result of the disability.
8. If a non-disabled person has a sexual relationship with a disabled individual, it's because he or she can't attract anyone else.

These myths collectively show how uncomfortable our society is with the idea of disabled persons as sexual beings. Somehow, many people seem to think, the disabled should worry about more important things in their lives and not concern themselves with an area that is obviously intended for healthy, "normal" people to enjoy. Fortunately, many of those who are disabled have refused to be intimidated by this line of reasoning and have joined forces with a number of workers in the health care professions as activists in seeking more attention to the sexual needs and feelings of people who have one form of disability or another. In fact, this movement has led to the development of a sexual bill of rights for the disabled that covers the following items:

1. The right to sexual expression.
2. The right to privacy.
3. The right to be informed.
4. The right to have access to needed services such as contraceptive counseling, medical care, genetic counseling, and sex counseling.
5. The right to choose one's marital status.
6. The right to have or not have children.
7. The right to make decisions which affect one's life.
8. The right to develop to one's fullest potential.

Although we are not yet in an era when the disabled are totally free to exercise these rights, we have made considerable progress in this direction in the last decade. We hope that as attitudes toward the disabled continue to change and as our society accepts sexuality as a positive, enhancing aspect of all of our lives, we may finally see the disappearance of myths and stereotypes and a more tolerant, open, and informed attitude taking their place.

Spinal Cord Injuries

A dramatic example of the impact of a disability on sexuality can be seen in people who have had injuries of the spinal cord. These injuries occur in a variety of circumstances, such as auto or motorcycle accidents, stab or bullet wounds, crushing industrial accidents, diving accidents, and serious falls. The injury usually results in *paraplegia* (paralysis of the legs) or *quadriplegia* (paralysis of all four limbs) and loss of all sensations in the body below the level of the injury. Normal bowel and bladder control is usually lost, and persons with spinal cord injuries are apt to have a significant loss of sexual function as well.

Most spinal-cord-injured (SCI) men lose the ability to have normal erections in response to psychological arousal, although they may be able to have brief reflex erections (which they cannot feel) in response to mechanical stimulation such as pinching or rubbing close to the genitals. Most (but not all) SCI males also become infertile and lose the ability to ejaculate. Those who are still able to ejaculate generally do so with no pelvic sensations of orgasm, and in many of the cases retrograde ejaculation (a backflow of semen into the bladder) occurs. On the other hand, intercourse may be possible for about four out of five men with incomplete spinal cord injuries. Even when erections do not occur, a technique called "stuffing" can be used. This involves tucking the soft or semi-firm penis into the vagina, with the woman thrusting with her hips, taking care not to move so the penis falls out.

Women with spinal cord injuries generally have normal fertility and can have children. While they usually retain their interest in sex, they typically lose their genital sensations and orgasmic responsiveness, and their vaginal lubrication is greatly reduced. Some SCI women — as well as SCI men — report experiencing "phantom" (non-genital) orgasms which include both psychological feelings of orgasm (e.g., intense pleasure) and physical sensations in some unaffected areas of the body that resemble prior patterns of orgasmic response.

Many people cannot understand why a spinal-cord-injured

person who has no genital sensations would want to have intercourse. While the motivations of all cord-injured people are not
the same, the following reasons are commonly given by people
in this situation:

1. Having intercourse is a special act of sharing and intimacy
   quite apart from genital feelings.
2. Having intercourse can be intensely *psychologically* arousing.
3. Being able to engage in intercourse can provide a boost to a
   person's self-concept.

In addition, many cord-injured males mention that being able to
have intercourse makes them feel more like a man, while some
cord-injured females note that having intercourse enhances
their sense of femininity. There are also some cord-injured persons who have intercourse primarily to please their partner (just
as some people *without* cord injuries do, too).

While a number of people have pointed out that the SCI
woman is less handicapped sexually than the SCI man because
she is still able to have intercourse, this is an oversimplification
of a complex issue and perpetuates a view of females as sexually
passive. Many cord-injured women are distressed by their sexual
limitations and require counseling to help them reaffirm their
sexuality. Unfortunately, sexuality counseling is not included in
many rehabilitation programs for SCI women, forcing many to
seek out such services on their own or to do without their potential benefits.

The sexual abilities of the spinal-cord-injured and their ways
of coping with their situation vary greatly. While some adopt a
completely defeatist attitude and avoid all sexual opportunities,
others have a continuing interest in sex which they express
through active involvement with a partner. What must be
stressed is that there are many methods of sexual interaction
besides intercourse. Oral-genital sex, kissing, use of a vibrator
(which a quadriplegic can sometimes apply by holding the base
with his or her teeth), massage, and cuddling are only a few of
the available options for having intimacy and sexual pleasure. In
addition, many SCI people find that unaffected regions of their
body become extraordinarily sensitive and erotic, so that stimulation above the region of sensory loss can produce arousal and
sometimes orgasm.

In addition, it is now possible for the SCI man with permanently impaired erections to have a device implanted surgically that will permit him to have erections and to participate in intercourse. Although these devices can provide firm enough erections for intercourse to take place, they cannot restore sensation to the penis or normal ejaculation if these have been lost due to organic causes.

There are two basic types of penile implants. The simplest is a pair of fixed, semirigid rods which are placed inside the shaft of the penis. These rods produce a permanent state of semierection, so they may be embarrassing to some men. More complicated and expensive, but more realistic in appearance and function, is an inflatable device that also requires surgical placement. Two tapered inflatable cylinders are inserted into the penis and are connected by a tubing system to a fluid storage reservoir implanted in the lower abdomen. When a simple pump and valve in the scrotum is pinched, fluid moves into the cylinders, causing a natural-looking erection. Releasing the valve moves the fluid out of the cylinders back to the reservoir (returning the penis to the flaccid state).

Either device may seem somewhat "artificial" to the man or his partner. For many men facing a "hopeless" situation, however, penile implants can restore a sense of manhood and improve self-esteem considerably.

It is important to remember that erections are not all there is to sex. Some thought-provoking comments about sexuality and the physically impaired provide an interesting counterpoint to the seemingly "mechanical" approach of the penile implants:

- A stiff penis does not make a solid relationship nor does a wet vagina.
- Absence of sensation does not mean absence of feelings.
- Inability to move does not mean inability to please.
- Inability to perform does not mean inability to enjoy.
- Loss of genitals does not mean loss of sexuality (Anderson and Cole, 1975).

Blindness and Deafness

If a person is blind or deaf from birth (or from a very young age), the sexual learning that occurs during childhood and adolescence can be seriously hindered. The person who has been blind since birth, for instance, is deprived of the ability to see the various shapes of human bodies both clothed and undressed that most of us silently notice and learn from. As a result, the person with lifelong blindness may be quite uncertain about the anatomic relation of one part of the body to another. (One fifteen-year-old boy was quite surprised to learn finally that the female breasts weren't located just above the waist, where he imagined them to be.) The person who has been deaf since birth often has difficulty understanding abstract concepts like maleness, parenting, and intimacy. Since very few parents of deaf children can communicate with them effectively with sign language, any semblance of sex education in the home is difficult if not impossible.

Neither blindness nor deafness produces any physical limitations on the body's sexual responsivity in and of itself. However, the sexual ignorance of many deaf or blind persons, coupled with their all-too-common lack of social skills, may create a predisposition toward sexual and relationship problems. In addition, the distorted body-image and poor self-esteem that many deaf or blind persons experience can also contribute to later sexual difficulties. Since blind persons do not have the visual cues surrounding sexual behavior that most of us depend on, and deaf people are apt to have significant problems in communicating with others, it's not surprising that such difficulties might occur.

Fortunately, sex education curricula have been devised for the deaf and blind. Most of these programs emphasize both social and biological aspects of sexuality and make good use of innovative methods of teaching, as well. For instance, blind students may be given sex-related articles like condoms, tampons, birth control pill packages, and vibrators to handle, so they develop familiarity with these products by touch. In fact, in Europe the touch method is sometimes used for anatomy lessons with live nude models. Many deaf students are now being taught by sex educators trained in sign language, and special visual materials have been prepared to make sex education easier despite the existence of hearing impairment.

Mental Retardation

Until very recently, people who are mentally retarded have been viewed as either asexual (in a state of child-like innocence) or as totally impulse-ridden and unable to control their primitive sexual urges. These myths, along with the general public's distaste for seeing the retarded as persons in their own right, have combined to make the topic of sexuality and the retarded a threatening one that arouses indignation and sometimes anger. One reflection of this attitude that has recently been changed was the involuntary sterilization that was often performed on young retarded adults or even adolescents to ensure that they couldn't reproduce (even though there is no evidence that most forms of retardation are genetically transmitted).

It is important to recognize that not all mentally retarded people are alike in their learning capacity, emotional stability, social skills, or capacity for independent living. The mildly retarded, for example, are educable and often lead productive lives, holding jobs, marrying, and raising families. Their lives are often indistinguishable from the lives of "normal" people, and they are usually capable of learning responsible sexual behavior without difficulty. The moderately retarded do not fit into our society so easily and are more apt to behave in a child-like manner both sexually and in other areas of living. They may alarm people by masturbating or disrobing in public or acting in a fashion that appears to be reckless or aggressive when showing affection for others, and they often require a protected or restricted environment in order to live their lives as comfortably and safely as possible. Yet they, too, are capable of learning much about sexuality and reproduction; for example, they can be taught how babies are conceived, what menstruation is (and how to deal with it), how to channel their sexual impulses into appropriate (i.e., non-public) masturbation. They also can learn how to avoid taking advantage of others and how to avoid sexual abuse. The severely or profoundly retarded, who must generally be institutionalized for their own safety, present a different range of problems since they are much more difficult to teach and are often capable of only the most rudimentary forms of communication.

The primary way of helping retarded people learn to handle their sexual feelings and express their sexual desires is through appropriate sex education. Kempton notes: "The retarded need sex education more than anyone else because they cannot readily learn about sexuality from friends, books, or from observing the behavior of others, and they have more than usual difficulty with sorting out reality from unreality" (1978, p. 138). In addition, cause-and-effect relationships are not always clear to the retarded; for instance, they may not realize that having intercourse can cause pregnancy. While some parents of retarded children have worried that sex education simply puts ideas into their children's heads, the available evidence suggests that withholding sex information does not deter sexual behavior and providing sex education does not lead to irresponsible sexual acts — in fact, just the opposite is likely to happen.

A number of specific issues regarding sex and the mentally retarded are too complex to discuss meaningfully here. These issues include:

1. Providing privacy for the institutionalized retarded.
2. Providing appropriate contraceptive protection for the retarded. (In general, the IUD seems to be the best option for retarded women who are sexually active. Birth control pills are appropriate for use by the mildly retarded, but condoms, diaphragms, and foams are usually not practical since they are too easily forgotten and too difficult for many to use.)
3. Providing opportunities for long-term relationships and marriage, if desired.

SEX AND ILLNESS

A wide variety of medical illnesses cause sexual problems. Until fairly recently, these situations were generally overlooked by health care professionals because they were often: (1) uncomfortable discussing sexuality; (2) uninformed about sexual aspects of illness; and (3) untrained in providing useful ways of dealing with these problems. Fortunately, this situation is rapidly changing today because fundamental research in this area is increasing and most health care schools include specific education about sexuality in their curricula.

Illnesses affect sexuality in direct and indirect ways. Physically, an illness can disrupt the normal reflexes of sexual response, as seen in persons with multiple sclerosis. Sometimes the treatment of an illness creates a sexual problem: medications that cause sexual side effects or surgery that results in sexual impairment are common examples. In other instances, sexual problems arise from the general effects of illness such as weakness, fatigue, or pain. There is an emotional side too: people with an illness may assume that it is "wrong" for them to have sexual feelings or may believe erroneously that their illness will prevent them from enjoying sex or from functioning sexually.

Now we will describe the sexual problems encountered in some specific types of medical disorders.

Neurologic Conditions

Sexual behavior and sexual function are both controlled in important ways by the nervous system. The brain itself is the ultimate synthesizer of sensory input (touch, sight, sound, taste, smell) and transforms electric impulses sent to it by nerve fibers into perceptions of pleasure, pain, and human emotion. Similarly, impulses sent from the brain to other organs, including the genitals, translate sexual desire into sexual response. For these reasons, it is easy to see why many diseases or injuries of the nervous system can lead to sexual difficulties.

Multiple sclerosis (M.S.) is a disease that typically affects young adults and involves patchy damage to the protective covering of nerves throughout the body. Loss of erections and ejaculation eventually affects more than half of men with M.S., while difficulty reaching orgasm (or complete anorgasmia) affects a similar fraction of women. Early in the disease there may be a partial numbness in the genitals, and sometimes any type of touch to the affected body region produces irritating, unpleasant sensations. For this reason, painful intercourse is another common problem. The sexual effects of M.S. tend to vary over time, so a person may have difficulties for several weeks or months and then go through a period of relatively normal sexual function.

People with *brain tumors* are somewhat less likely to encounter sexual difficulties. Although pituitary or hypothalamic tumors commonly cause erectile dysfunction and decreased sexual desire, tumors in other regions of the brain do not usually lead to sexual problems. *Epilepsy* is another neurologic disorder with little physical effect on sexual function. Some people with epilepsy, however, feel inferior because of their condition and may avoid sexual opportunities in the mistaken fear that arousal will cause a seizure. Although *polio* can cause spinal cord damage, it affects muscular function, not sensations. For this reason, most people with paralysis from polio are not impaired in their sexual function, although their inability to move may create problems in coital thrusting.

Endocrine Conditions

The most common type of endocrine disorder is *diabetes,* which affects approximately 4 percent of the U.S. population. About 50 percent of diabetic men have erectile dysfunction, which can either be an early symptom of this disorder or may not occur for many years after the diagnosis. Retrograde ejaculation occurs in about 1 percent of diabetic men, but sex drive is usually unaffected. Although the sexual difficulties of diabetic men had been well known for at least two centuries, sexual problems in diabetic women were not identified until the 1970s. It is now known that about one-third of diabetic women have secondary anorgasmia, which usually occurs four to six years after the disease is discovered. In addition, some diabetic women have problems with vaginal lubrication.

The primary cause of these sexual problems in both diabetic men and women is a form of nerve damage that is a complication of diabetes. In a smaller fraction of cases, these dysfunctions are due to circulatory problems. Unfortunately, both types of problems tend to be permanent and untreatable (except for the possibility of using penile implants to overcome erectile dysfunction). In cases where diabetics have sexual difficulties for other reasons, such as anxiety or poor communication, sex therapy can be beneficial.

Disorders of the pituitary, thyroid, or adrenal glands are also commonly associated with sexual difficulties. In each of these conditions, sex drive is likely to be altered along with sexual function. About 40 percent of women with an underactive thyroid or adrenal glands have orgasmic difficulties, and a similar percentage of men have problems with erections. Sexual dysfunction is even higher in people with underactive pituitary glands. Fortunately, these conditions are easily treatable by giving the proper amount of the "missing" hormone in pill form.

Heart Disease

In severe forms of chronic heart disease, the capacity for sexual activity is likely to be greatly limited. In milder types of heart problems, however, there may be no physical limitation on sexual activity. Nevertheless, several studies of men who had heart attacks have shown that six months to a year after recovery, sexual difficulties are common. The primary factor here is not a physical one but a mental one: anxiety, misconception, and avoidance conspire to create sexual difficulties. The man may be worried that his heart attack will cause sexual problems or that sexual excitement will cause another heart attack. This worry is also expressed in stories about men dying from a massive heart attack in the midst of a passionate sexual episode. There is no solid evidence to support this as a meaningful risk for most heart patients, and the cardiac cost of sexual activity — including intercourse and orgasm — seems to be about the same as walking up two flights of steps. Women who have had heart attacks seem to be less likely to develop subsequent sexual problems than men.

Cancer

Until the last few years, most people have automatically assumed that a person with cancer could not possibly have sexual feelings or sexual needs. Now it is clear that this view is incorrect, and increasing attention is being paid this area. We will briefly examine several types of cancer and their impact on sexuality.

An American woman has about a one-in-eleven chance of developing breast cancer, the most frequent female cancer and the leading cause of cancer death in women. More than 100,000 new cases are discovered each year, and late detection seriously lowers the chances of survival.

Breast cancer is usually treated by *mastectomy,* surgical removal of the affected breast. This operation typically creates psychological conflicts for the woman, with concerns about physical attractiveness and possible rejection by her husband or sexual partner ranking high on the list. Perhaps this response is not surprising in light of the degree to which our society emphasizes the breast as a symbol of sexuality and femininity. As a result, the woman who has a mastectomy is likely to feel "incomplete" and unfeminine.

Although surgical removal of a breast does not "cause" sexual problems in a direct, physical sense, the psychological impact of mastectomy can be profound. Women react in a variety of ways ranging from relatively easy adjustment, to depression, loss of sexual desire, or sexual dysfunction. Following mastectomy, a few women avoid all forms of sexual activity. More typically, the frequency of sexual activity declines and women initiate sexual activity less often than they did before.

Nudity may be a matter of self-consciousness for the woman or discomfort for her partner. The couple may shift away from intercourse in the female-on-top position, since the male looks directly at the area of the surgical scar. Women who are without a partner at the time of the mastectomy may be particularly worried about dating and what to tell a prospective sexual partner; they may also be afraid that no man could possibly fall in love with a woman with one breast. For women in ongoing relationships, how their partner responds is an important ingredient of their overall adjustment. In many cases, it is possible to use plastic surgery techniques for breast reconstruction as a way of helping a woman adjust to mastectomy most comfortably. Currently, many physicians favor removal of the tumor without removing the entire breast (a procedure called "lumpectomy") and radiation therapy as treatment options for the breast cancer patient, since these alternative treatments result in less disfigurement and less profound psychological effects and in many cases are just as effective as mastectomy.

Cancer of the cervix accounts for slightly more than 60,000 new cases of cancer annually and cancer of the lining of the uterus is found in another 38,000 women. There are two forms of cervical cancer: carcinoma in situ (CIS) and invasive cancer of the cervix (ICC). Although neither form is likely to cause symptoms, both can be detected by Pap smears during routine pelvic exams. CIS is really a precancerous condition involving cells on the surface of the cervix that do not invade other tissues. On average, it takes eight years or more for CIS to progress to true cancer, or ICC. Treatment of CIS while it is still precancerous results in virtually 100 percent long-term survival. Treatment of ICC depends on whether the cancer cells have spread beyond the cervix: either surgical techniques or radiation therapy can be used. Sexual difficulties are common but depend on the amount of pelvic scarring that occurs. One recent report suggests that surgery is less disruptive to sexual function than radiation therapy.

Cancer of the lining of the uterus is rare before age forty and is usually marked by abnormal bleeding early in its course. Later symptoms include cramping, pelvic discomfort, bleeding after intercourse, or lumps in the groin. Pap smears are *not* a foolproof way of detecting this form of cancer; usually a D&C (dilatation and curettage, scraping the inside of the uterus) is done to establish the diagnosis. Depending on the findings, treatment ranges from surgery (removal of the uterus and ovaries) to radiation therapy, drug therapy, or hormone therapy. If the cancer has not spread beyond the uterus, the patient has an 83 percent chance of being cured — that is, of not having the cancer recur within the next five years. Surgical removal of the uterus, called *hysterectomy*, does not usually have any negative effects on female sexual function. It may actually be beneficial since physical problems such as bleeding or cramping are corrected. However, some women have impaired sexual responsiveness and/or decreased sexual interest after a hysterectomy because they see the operation as a lessening of their femininity.

Prostate cancer accounts for 17 percent of cancers in men, or about 70,000 new cases annually. The disease is rare in men under forty and is generally diagnosed by a rectal examination. Treatment consists of surgical removal of the prostate (*prostatectomy*), radiation therapy, or hormone therapy. Prostatectomy for prostate cancer frequently causes erectile dysfunction because nerves that supply the penis are damaged. When the prostate is removed for reasons other than cancer, erectile dysfunction is a less frequent complication, although a high percentage of men develop retrograde ejaculation.

Cancer of the testis is a relatively rare condition that is most common in the twenties and thirties. Since only about half of these cancers are accompanied by pain, it is important for males to learn a method of testicular self-examination and to seek prompt medical attention if a testicular lump or swelling is noticed.

Treatment of testicular cancer (surgery, drugs, radiation) is sometimes a cause of sexual dysfunction and generally results in infertility. It is common to find that men with this form of cancer develop considerable guilt over their previous sexual practices, often blaming the problem incorrectly on masturbation, venereal disease, or an "overactive" sex life. Other men with this problem feel that the surgical removal of one testis makes them "less than a man." They may also develop fears of sexual performance and spectatoring, leading to difficulty with erection for strictly psychological reasons.

Approximately six million people — pregnant women and their daughters and sons — were exposed to the synthetic estrogen DES (diethylstilbestrol) and other DES-like drugs from 1940 to 1971, when it was widely used to treat so-called high-risk pregnancies — those with a threatened miscarriage or those complicated by a medical problem such as diabetes. In 1971, physicians found evidence of a rare form of cancer of the vagina or cervix (technically called clear cell adenocarcinoma) in daughters whose mothers had taken DES and related drugs during pregnancy. Subsequently, a number of reproductive abnormalities were found in some of the sons of women given DES during pregnancy, including incomplete development of the testes, undescended testes (cryptorchidism), anatomical abnormalities of the penis, and preliminary evidence of infertility. Still more recently, there is some preliminary concern about cancer of the testis in DES-exposed sons, although this finding requires further substantiation.

While the cancer found in DES daughters is relatively infrequent (it is estimated to occur in about one of every 800 exposed women), non-cancerous abnormalities are found in the vagina in one-third of these individuals and abnormalities of the cervix are present in almost every case. It is not certain if a higher rate of vaginal/cervical cancer (or other forms of cancer, such as cancer of the breast or uterus) will emerge as this population of women ages, so that long-term medical follow-up is important. In fact, evidence suggests that there is an increased risk of cancer in the mothers who took the drug during pregnancy. It also appears that there is a higher rate of ectopic pregnancies and miscarriages in DES daughters.

If you were born after 1940, you should ask your mother whether she had any drugs prescribed for her during pregnancy — especially to prevent miscarriage or to treat a problem pregnancy associated with diabetes. If she did (or thinks she might have), go to your physician for evaluation. While it would be helpful to try to find out the dosage of the DES-type drug received, when in the pregnancy it was first administered, and for how long a period, such information is not always available. Nevertheless, a medical examination is in order if you (or your child) had such an exposure or think you may have.

DES daughters should undergo a pelvic examination including a Pap smear and also should have a test using an iodine solution to stain the lining of the vagina temporarily. The physician may also use a special magnifying instrument called a colposcope for this examination, and if there are areas of the vagina that appear abnormal, a biopsy (removal and examination of tissue specimens) may be taken. Usually this type of biopsy causes relatively little discomfort, although a slight amount of bleeding (less than during menstrual flow) may occur for 12 to 24 hours after this procedure is done.

The situation for DES sons is not yet completely understood, but we suggest that until more information is available, they undergo a medical examination on an annual basis.

Generally, treatment is *not* required for DES-related abnormalities, and over 98 percent of exposed individuals will be found to be cancer-free. Nevertheless, follow-up examinations (advisable at least every six months if abnormalities are found, or once a year if not) are important because the success of treating the types of cancer that can develop is partly dependent on detecting them in their earliest stages.

Needless to say, knowledge of DES exposure can lead to anger, guilt, fear of cancer, and concerns about fertility and sex. One twenty-four-year-old nurse summarizes her reactions:

> When I first learned that my mother had used DES while she was pregnant with me, I didn't fully comprehend what it meant. Now that I realize that the risk of cancer is remote, I'm less afraid, but I'd be dishonest if I didn't say that I consider myself abnormal and I constantly worry about what will happen. Statistics mean very little when you're on the line yourself.

Other DES daughters voice dismay at being advised not to use birth control pills (it is possible that the pill might aggravate the DES-induced changes, although whether or not this actually happens is not currently clear) or worry about their future child-bearing risks. Fortunately, there are several groups available to help them deal with such concerns, including DES Action (Long Island Jewish Hospital, New Hyde Park, NY 11040); National Women's Health Network (2025 I Street, N.W., Suite 105, Washington, DC 20006); and the National Cancer Institute (Department DES, Office of Cancer Communications, Bethesda, MD 20205).

Alcoholism

Since many authorities regard alcoholism as an illness, we will discuss its sexual effects here. The alcoholic man or woman has a high chance of having sexual problems. Not all of these problems result from the alcoholism, however: in more than a few cases pre-existing sexual difficulties may have played an important role in starting a person on a path toward heavy drinking. Here, the use of alcohol may help people cope with feelings of sexual inadequacy by making them less interested in sex or less critical of their own performance. Given the popular notion that drinking is "manly" and that liquor can be used to seduce a woman by lowering her resistance (both "facts" being subtly reinforced by the advertising industry), it is easy to see how drinking seems to be "beneficial" from a man's sexual viewpoint. Women who feel guilty or inhibited about sex may find that drinking loosens these restraints and lets them feel more comfortable.

We have found that about 40 percent of alcoholic men have problems with erections and 5 to 10 percent have retarded ejaculation. Thirty to 40 percent of alcoholic women have difficulties in sexual arousal, and 15 percent have problems being orgasmic. Sexual desire is also apt to be low in alcoholics.

There are several reasons behind these problems. Alcoholism directly affects hormone production and lowers testosterone in men and estrogen in women. Shrunken testes and breast enlargement are common in alcoholic men. Other medical complications of alcoholism include liver damage, nerve damage, lowered resistance to infection, and poor nutrition, all of which may provide a biological basis for sexual impairment. Relevant psychosocial factors include fears of performance and spectatoring, marital conflicts, low self-esteem, guilt, and depression.

Even when an alcoholic has stopped drinking completely, there is no guarantee that his or her sexual difficulties will disappear. About half of the time these problems continue, requiring professional counseling to be resolved.

DRUGS AND SEX

For many centuries, there has been an avid search for *aphrodisiacs* — substances that could increase a person's sexual powers or desire. The long list of substances that have been claimed to have such an effect includes oysters, ginseng root, powdered rhinoceros horn, animal testicles, and turtles' eggs, but there is no evidence that an actual aphrodisiac response occurs. "Spanish fly," the most famous supposed aphrodisiac, is made from beetles found in southern Europe. The beetles are ground into a powder that, when taken internally, irritates the bladder and urethra and can also cause ulcers, diarrhea, and even death. The burning sensation in the penis due to irritation of the urethra has been interpreted by some men as a sign of lust.

Aphrodisiacs aside, both prescription drugs and drugs used recreationally (or illicitly, depending on your viewpoint) have some specific effects on sexuality.

Prescription Drugs

Many of the medications used to treat high blood pressure cause sexual difficulties for men and women. For example, Aldomet (alpha-methyldopa), the drug most commonly used to treat this condition, causes erectile dysfunction in 10 to 15 percent of men at low doses and in up to half of men in high doses. Decreased libido and impaired sexual arousal is found in similar proportions of women using this drug. A different type of problem is found with Ismelin (guanethidine), which inhibits ejaculation in more than half of the men using it. Usually these difficulties will disappear within a week or two after stopping the medication, but in some cases the sexual dysfunction may persist because of anxiety. Fortunately, there are many medications available to treat high blood pressure. Some have low rates of sexual side effects, and it is almost always possible to find a combination that will leave sexual function intact while simultaneously controlling high blood pressure.

Tranquilizers such as Librium (chlordiazepoxide) and Valium (diazepam) can sometimes cause erectile dysfunction, anorgasmia, or decreased sexual desire but in other instances may be beneficial by reducing sexual anxiety. Barbiturates and related drugs such as Quaalude (methaqualone) have also been reported to cause a variety of sexual problems, although Quaalude has developed a street reputation as an aphrodisiac. Our research has shown that some people using Quaalude temporarily lose some of their sexual inhibitions, which they translate as a sexual "stimulant" effect. The problem, as a number of experienced drug users confirm, is that Quaalude also depresses the functions of the nervous system and can actually impair sexual performance.

Antihistamines, used in allergy pills and sinus medications, can affect sexuality in two ways. Drowsiness is a prime side effect and one not likely to improve the quality of sex. In women, these drugs often cause a reduction in vaginal lubrication so they may sometimes cause painful intercourse.

Nonprescription Drugs

Alcohol

The effects of alcohol on sexuality have fascinated people throughout history. In *Macbeth,* Shakespeare reported that "it provokes the desire but it takes away the performance" (Act 2, scene 3, line 34), and recent research has shown that this view is fairly accurate. In one study, college men were given alcohol in three different doses while watching erotic movies. Amounts of alcohol well below the legal levels of intoxication suppressed erections. Similar studies in women showed that alcohol had a negative impact on physiologic signs of sexual arousal. Alcohol has also been shown to weaken male masturbatory effectiveness and to decrease the pleasure and intensity of male orgasm, and alcohol, even in moderate amounts, makes it more difficult for women to reach orgasm.

Despite the *physical* inhibition of even two or three drinks of an alcoholic beverage (due to a depressant effect on the nervous system), most people believe that alcohol *increases* their sexual responsiveness. This is partly because alcohol has a "disinhibiting" effect — it lowers the sexual inhibitions a person may ordinarily have, thus making it possible for sexual desire to emerge. The belief that alcohol enhances sex also stems from advertising and cultural myths.

Narcotics

Addictive drugs such as heroin and morphine produce many sexual problems. One large survey found that in 162 male addicts, erectile dysfunction occurred in 48 percent, retarded ejaculation in 59 percent, and low sexual interest in 66 percent; in 85 female addicts, 27 percent had orgasmic dysfunction and 57 percent had low sexual interest. This is a complex area to evaluate, however, because drug addiction may be a means of trying to escape from pre-existing sexual difficulties or may be a substitute for sex. Factors such as hormone problems, infections, and poor nutrition, which occur as a result of addiction, also play a role in causing sexual difficulties.

Narcotic addicts are also likely to have other problems that complicate their sex lives. Rosenbaum (1981) has noted that: (1)

female addicts usually have partners who are also addicted; (2) many addicts find that the "hit" of mainlining heroin is far more pleasurable, intense, and easy to get than an orgasm (in fact, many ex-addicts say the feeling is like dozens of orgasms rolled up into one); and (3) the sensuality and sharing that accompany narcotic use become a replacement for the sharing and sensuality of sex. Furthermore, since most female addicts must turn to prostitution to raise money for their habit, it is not surprising that sex becomes less appealing to them.

Amphetamines and Cocaine

Amphetamines ("speed," "uppers," "pep pills") reportedly increase sexual responsiveness when used in low doses but have the opposite effect in high doses or when used on a long-term basis.

Cocaine ("coke," "snow") has a street reputation as a strong sexual stimulant, but there are also reports of sexual dysfunction with its use. Kolodny found in 1983 that 17 percent of 168 male cocaine users had episodes of erectile failure when they used this drug, and 4 percent had experienced priapism (painful, persistent erections) at least once during or immediately after the use of cocaine. Similarly, Wesson (1982) found evidence of male erectile difficulties during the use of cocaine, and Siegel (1982) reported that the dangerous practice of "free-basing" cocaine consistently leads to sexual lack of interest and situational impotence (twenty of twenty-three men were affected in his study).

Cocaine use as a purported sexual stimulant is interesting for several other reasons. For one thing, many users believe that rubbing cocaine on the tip of the clitoris increases female sexual sensitivity and arousal, but how this could occur is difficult to understand since cocaine is used medically as a topical anesthetic: that is, to deaden nerve endings. The continued use of this practice may indicate how powerful expectations are in interpreting our experiences.

A second interesting point is that cocaine clearly acts as a sexual facilitator in a social sense. Because of its status and expense, as well as its reputation as a sexual stimulant, when a man offers cocaine to a woman (or vice versa) there is usually a sexual invitation implied. As Kolodny notes:

Widely available at singles bars and in the economically advantaged "just-got-a-divorce" crowd, cocaine literally opens the doors of sexual access for many males and provides a convenient excuse for many females who otherwise might pass on having "instant sex" with a partner they hardly know.

Marihuana

Marihuana ("pot," "dope," "grass") is generally reported to enhance sexual feelings. In our own research with more than 1,000 men and women aged eighteen to thirty-five who had used this drug as an accompaniment to sex, 83 percent of the men and 81 percent of the women said that marihuana improved their sexual experience. Most users, however, denied that marihuana led to more sexual desire, quicker sexual arousal, or more intense orgasms. Instead, they indicated that marihuana gave them an increased awareness of touch all over their bodies, led to greater relaxation (both mentally and physically), and put them more in tune with their partners. These are highly subjective judgments which cannot be fully verified in experimental research. However, generally similar findings have been reported by others; for example, in 1982 Halikas, Weller, and Morse described firsthand reports of enhanced touch awareness and physical closeness in a majority of both male and female users.

In nonsexual situations, it has been shown that instead of *increasing* touch sensitivity, marihuana actually produces no change or *lessens* touch perception. There is also considerable documentation that marihuana use slows reflex reactions. The fact that marihuana users say that if *they* are "high" but their partner is not, the sexual experience is unpleasant (disjointed?) also indicates that there is a strongly subjective element to the reported effects.

Some other research findings also bear examination. Erectile dysfunction has been found to affect about 20 percent of men using marihuana on a daily basis, although no association between marihuana use and sexual dysfunction in women has been noted. However, some women who use marihuana report that it causes temporary vaginal dryness which can sometimes cause painful intercourse. Furthermore, heavy marihuana use has been reported to lower testosterone production in animals and men and to disturb sperm production as well. Although these

effects are reversible once the drug is stopped, they may sometimes contribute to sexual problems. A study of chronic, frequent marihuana use in women, done by workers at the Masters
& Johnson Institute, showed menstrual cycle abnormalities and
hormone changes but no negative sexual effects.

Some Miscellaneous Drugs

Lysergic acid diethylamide (LSD, "acid") and related psychedelic
drugs have not been studied extensively from a sexual viewpoint. The few bits of research that exist indicate that these
drugs are not used primarily for their sexual effects and that
they generally lead to preoccupation with mental imagery during a "trip."

In contrast, drugs such as *amyl nitrite* ("snappers," "poppers"),
which are inhaled, are widely used to prolong or intensify the
sensation of orgasm, especially among homosexual men. Severe
headaches are a common side effect, and fainting sometimes
occurs.

In concluding, it is important to note that all of these drug
effects vary considerably from one person to another depending
on health, age, body size, and many other factors. We have tried
to describe the *most typical* sexual effects encountered.

INFECTIONS

Certain types of infections that affect the sex organs are not
transmitted chiefly through sexual intercourse, unlike the conditions called sexually transmitted diseases (STDs), which we will
discuss in the next chapter. These infections, which *can* be sexually transmitted, can produce troublesome symptoms that interfere with sexual pleasure or cause considerable emotional
turmoil. Fortunately, the most common of these infections are
easily treatable and have no major health risks. We discuss them
here to avoid the implication that they are always or usually of
sexual origin.

Vaginitis

Vaginitis refers to any vaginal inflammation, whether caused by infection, allergic reaction, estrogen deficiency, or chemical irritation. Vaginitis can create sexual problems by causing tenderness or pain during intercourse or by causing disagreeable odors that embarrass the woman or reduce her partner's enthusiasm for intimacy. Here, we will consider only the most common forms of vaginal infection.

Trichomonas (pronounced "trick-o-moan'ess") *vaginitis* is caused by a one-cell microorganism called *Trichomonas vaginalis,* ordinarily present in small numbers in the vagina. If these organisms multiply rapidly or are transmitted by sexual contact, the resulting infection produces a frothy, thin, greenish-white or yellowish-brown, foul-smelling discharge, which usually causes burning and itching of the vagina and vulva. The diagnosis is made by examining the discharge under a microscope. The most effective treatment is a prescription drug called Flagyl (metronidazole), which should be given to the woman and her male sexual partner. Flagyl should not be used by nursing mothers because it appears in breast milk, and there is currently some concern about its safety because it has been found to produce tumors in mice.

Monilial vaginitis is a type of fungus or yeast infection caused by an overgrowth of *Candida albicans,* a microorganism that is normally found in the vagina. The discharge is usually thick, white, and cheesy, and is accompanied by intense itching. Diabetic women, pregnant women, and women using birth control pills or antibiotics have an increased incidence of this infection. Treatment involves use of special vaginal creams or suppositories, such as these prescription medications: Monistat (miconazole), Mycostatin (nystatin), or Mycelex-G (clotrimazole) for one or two weeks.

Because yeast infections can mask the presence of gonorrhea or syphilis, specific testing should be done to see if any STDs are also present. In two studies of women with STDs, more than 25 percent were found to have genital yeast infections.

Hemophilus vaginalis is a small bacterium that commonly causes another troublesome vaginitis. The brownish-white or grayish discharge usually has a foul odor and is accompanied by burning or itching. Treatment possibilities include ampicillin or tetracycline pills and various vaginal creams or suppositories. Because there is a high likelihood that the male sexual partner of a woman with hemophilus vaginitis has these bacteria in his urethra, he should also undergo treatment.

In considering how to prevent vaginitis, it is important to begin by realizing that the vagina normally contains a number of different microorganisms. Some of these seem to play a specific role in vaginal physiology, such as maintaining the proper degree of acidity, while others can produce symptoms and infection if they multiply disproportionately. An average of seven different species of bacteria are found in the vagina, and other microorganisms such as yeasts and viruses are also present. Why some women develop vaginal pain or itching without having any detectable infection is unclear. It is also uncertain why many women with documented infections don't have any discharge or other symptoms. However, the fact remains that vaginitis is often an annoying condition that women and their partners would like to prevent if they can. Here are several suggestions for minimizing the risk of developing vaginitis:

1. Wear cotton underpants; nylon and synthetic fiber underpants retain heat and moisture, creating a good environment for bacteria to grow.

2. Avoid frequent douching, since this can irritate the vagina and remove important "natural" microorganisms that protect you. (Many medical authorities believe routine douching is unnecessary and only advise it under specific conditions.)

3. After going to the bathroom, always wipe with a front-to-back motion. This way, bacteria from the rectum will not be brought forward to the vagina.

4. Avoid the long-term use of antibiotics, which can reduce the number of bacteria normally present in the vagina, allowing yeast forms to overgrow.

5. Maintain good habits of personal hygiene, including regular washing of the genital and anal regions with mild soap and water. Avoid so-called feminine hygiene sprays, which can be irritating to the skin.

6. If your partner has an infection of the genitals, avoid sexual

contact. (Using a condom may be of some help in this situation.)

7. Do not put the penis in or near the vagina after anal intercourse, since this can directly introduce "foreign" bacteria into the vagina.

8. Avoid forms of sexual activity that produce any vaginal discomfort.

Cystitis

Cystitis, or infection of the bladder, is closely related to sexual activity in women. Sexual intercourse leads to an increase in bacteria in the urine, presumably because of inward pressure on the urethra during coital thrusting. Because the female urethra is short (about 2.5 cms, or 1 inch) compared to the male urethra (usually more than 15 cms, or 6 inches), cystitis is far more common in women than men (the bacteria have a shorter distance to travel).

The symptoms of cystitis include burning during urination, frequent urination, cloudy or bloody urine, and lower abdominal pain. The diagnosis can be made by examining a urine sample under the microscope and by taking a culture to identify the specific bacteria involved. Broad-spectrum antibiotics such as tetracycline or ampicillin are often prescribed.

One special variety of this infection is the so-called honeymoon cystitis that can occur either when a woman first becomes coitally active (not always on her honeymoon) or when coital activity is resumed after a prolonged period of inactivity.

Toxic Shock Syndrome

Toxic shock syndrome (TSS) first came to public attention in 1980 when it was widely reported as a serious, sometimes fatal illness suddenly striking healthy menstruating women who used tampons. Although TSS was named in 1978 by Todd and coworkers, who reported on a small number of cases appearing in children, it now is clear that it is actually a rare form of scarlet fever that was initially described in 1927 (Stevens, 1927; Reingold, 1983).

TSS is marked by high fever, vomiting, diarrhea, muscle pain, and skin rash that resembles a severe sunburn. Fainting spells, low blood pressure, and dizziness are other common symptoms. TSS is caused by a toxin, or poison, produced by a bacteria called *Staphylococcus aureus,* and it occurs primarily in menstruating women who use tampons. Most of the cases encountered in 1980 were associated with the use of Rely tampons, which were subsequently found to have a relative risk for developing TSS eleven times greater than Playtex tampons, twenty-eight times higher than OB tampons, thirty-eight times higher than Kotex tampons, and seventy-seven times higher than Tampax. After Rely tampons were withdrawn from the market, the number of cases of TSS linked with tampon use dropped, and by 1983, approximately 15 percent of the cases being reported were unrelated to menstruation. It is now clear that TSS can affect men as well as women and all age groups — including infants and the elderly — but the group at highest risk still seems to be tampon-using white females aged fifteen to twenty-five.

While women using tampons have the greatest risk of TSS, and a few cases have been linked to use of the contraceptive sponge as well as to use of a diaphragm, the chances of developing this illness are very low. Preventive measures that can be taken to reduce this risk still further include: (1) switching to sanitary napkins or minipads entirely; (2) alternating the use of tampons and minipads or sanitary napkins several times each day; or (3) changing tampons three or four times daily. If you develop symptoms suggestive of TSS while you are menstruating, you should immediately see a physician, since TSS is a rapidly progressive illness that is fatal in about 4 percent of cases. Fortunately, with proper medical management — including hospitalization, treatment of shock, and aggressive antibiotic therapy — it is now clear that TSS is not as frightening as it first seemed to be.

Prostatitis

Prostatitis, or inflammation of the prostate, can be either acute (sudden) or chronic (long-lasting). The infecting organism is usually *E. coli,* a normal inhabitant of the intestines. Acute prostatitis is marked by fever, chills, perineal or rectal pain, painful urination, and urinary frequency and is likely to interfere with sexual function (painful ejaculation is common). Chronic prostatitis may involve no symptoms at all, or low back pain or perineal discomfort may be present. Chronic prostatitis has sometimes been thought to cause premature or bloody ejaculation. Antibiotic treatment usually clears up acute prostatitis but may be ineffective in curing the chronic form of this disease.

CHAPTER TWENTY

Sexually Transmitted Diseases

INFECTIONS that are spread by sexual contact are referred to as *sexually transmitted diseases,* or STDs. STDs include infections that were formerly known as venereal diseases (VD) — those almost always transmitted by sexual contact — as well as various other infections that are sometimes transmitted by nonsexual routes. In addition to describing a broader category of infections, the term STD has not yet acquired the stigmatizing sound that the label VD carries. The following sections will describe the symptoms, diagnosis, and suggested treatment of sexually transmitted diseases.

GONORRHEA

Gonorrhea is the oldest and one of the most common forms of STD. In the Old Testament, Moses spoke about its infectivity (Leviticus 15); it was also mentioned in the ancient writings of Plato, Aristotle, and Hippocrates. Its modern name was coined by Galen, a Greek physician in the second century A.D. In 1879 Albert Neisser discovered the bacterium that causes it, which was named after him (*Neisseria gonorrheae*).

Although the discovery of penicillin as an effective treatment for this disease slowed its spread in the 1940s and 1950s, the incidence of gonorrhea has grown tremendously in the last twenty years and has reached epidemic proportions today. More than a million cases of gonorrhea are reported annually, and these probably represent only a quarter of the actual cases that occur each year.

Gonorrhea is transmitted by any form of sexual contact, ranging from sexual intercourse to fellatio, anal intercourse, and, infrequently, cunnilingus or even kissing. A woman who has intercourse once with an infected man has a 50 percent chance of getting gonorrhea, while a man who has intercourse once with an infected woman has a lower risk, probably around 20 to 25 percent, of becoming infected. The old excuse, "I caught it from a toilet seat," which was previously laughed at by scientists, has now been shown to be at least theoretically possible since the infective bacteria can survive for up to two hours on a toilet seat or on wet toilet paper. However, it is unlikely that this form of transmission is more than a rare occurrence.

Symptoms

Most men with gonorrhea develop a yellowish discharge from the tip of the penis and painful, frequent urination as the first indications of gonorrhea. These symptoms usually appear within two to ten days after infection but may sometimes start as much as a month later. The symptoms are produced by infection in the urethra, which leads to inflammation (urethritis). The puslike discharge (which often stains the underwear) is part of the body's reaction to this infection. In about 10 percent of cases in men, there may be no symptoms from the infection, which means that the man can spread gonorrhea without realizing it.

Men with symptoms from gonorrhea usually seek treatment promptly and are cured. For men who do not receive treatment, the infection may move up the urethra to the prostate, seminal vesicles, and epididymis and can cause severe pain and fever. If untreated, gonorrhea can lead to sterility, but this is a relatively infrequent complication in men.

Since less than half of the women with gonorrhea have any visible symptoms, they are likely to have their infection for a longer time before treatment is begun. This delay exposes women to a greater risk of complications. In addition, symptomless women may unknowingly spread their infections to their sexual partners. Many women do not find out they are infected until their partner's penile discharge or burning appears.

Even when symptoms occur in women, they may be very mild and tend to go unnoticed or are misdiagnosed. The symptoms include increased vaginal discharge, irritation of the external genitals, pain or burning with urination, and abnormal menstrual bleeding. In women, infection is most commonly found in the cervix (90 percent of cases) but can also be present in the urethra (70 percent), the rectum (30 to 40 percent), the throat (10 percent), or any combination of these sites.

Women who are untreated may develop serious complications. Gonorrhea commonly spreads from the cervix to the uterus, Fallopian tubes, and ovaries, causing pelvic inflammatory disease (PID). PID, although not always caused by gonorrhea, is the most common cause of female infertility because it can produce scarring that blocks the Fallopian tubes. The early symptoms of PID are lower abdominal pain, fever, nausea or vomiting, and pain during intercourse.

In both sexes, gonorrhea can spread through the bloodstream to other organs, causing infection and inflammation of the joints (gonococcal arthritis), the heart (gonococcal endocarditis), or the covering of the brain (gonococcal meningitis). Fortunately, these complications are rare and treatable. Eye infections with gonorrhea occur (rarely) in adults, where they are caused by touching the eye with a contaminated hand. Newborn children may develop eye infections during birth if their mother's cervix is infected. Because this can produce blindness, infection-preventing drops, usually silver nitrate, are routinely put in every newborn baby's eyes.

Diagnosis and Treatment

Gonorrhea in men is diagnosed by examining the urethral discharge under a microscope after staining it with a specially colored dye. As this method is only about 90 percent accurate, it may also be necessary to try to grow the infecting bacteria in a laboratory by a culture test that takes several days. Men who have had homosexual contacts should have cultures of the throat and rectum as well as the urethra.

In women, culture tests are the only reliable means of establishing the diagnosis. Swabs should *always* be taken from the mouth of the cervix *and* the rectum, even if the woman has never had anal intercourse, because a vaginal discharge may drip onto the anus and cause infection there. If the woman has experience with fellatio, a throat swab should also be taken. There is no blood test that can identify gonorrhea reliably at the present time.

The most effective treatment for gonorrhea is a large dose of penicillin G divided into two shots, one given in each buttock. It is recommended that a medicine called probenecid be taken in pill form at the same time to block excretion of the penicillin in the urine, keeping high levels of this antibiotic in the body. For people who are allergic to penicillin, tetracycline pills can be used effectively when taken over five days.

Unfortunately, a form of gonorrhea that is resistant to penicillin treatment (because it produces an enzyme that breaks down and neutralizes penicillin) has been encountered with growing frequency since 1976. Although this penicillin-resistant form of gonorrhea can be successfully treated with other antibiotics such as spectinomycin, if a widespread outbreak of this gonococcus occurs, control would be much more difficult.

In *any* case of gonorrhea, it is important to abstain from sexual activity with a partner until you have been rechecked after treatment to be certain you are cured. *It is also extremely important to notify anyone with whom you had sexual contact, including the person you know or suspect gave you the infection, to insist that he or she see a doctor for proper diagnosis and treatment.*

SYPHILIS

Syphilis first came to public attention at the end of the fifteenth century, when it swept across Europe, decimating armies and towns as it traveled. The source of this widespread outbreak is unclear. Some authorities believe that a particularly infectious type of syphilis was imported from America by Columbus and his crew, while others believe that it was already present in Europe. The spiral-shaped microorganism that causes syphilis, *Treponema pallidum,* was identified in 1905.

Syphilis is far less common than gonorrhea today. In 1983

there were about 25,000 new cases reported in America, with a male-female ratio of two to one. Half of the men with syphilis are homosexual or bisexual.

Syphilis is usually transmitted by sexual contacts, but it can also be acquired from a blood transfusion or can be transmitted from a pregnant mother to the fetus.

Symptoms

The earliest sign of syphilis in its *primary stage* is a sore called a *chancre* (pronounced "shanker"). The chancre generally appears two to four weeks after infection. The most common locations for the chancre, which is painless in 75 percent of cases, are the genitals and anus, but chancres can also develop on the lips, in the mouth, on a finger, on a breast, or on any part of the body where the infecting organism entered the skin. The chancre typically begins as a dull-red spot, which develops into a pimple. The pimple ulcerates, forming a round or oval sore usually surrounded by a red rim. The chancre usually heals within four to six weeks, leading to the erroneous belief that the "problem" went away.

Secondary syphilis begins anywhere from one week to six months after the chancre heals if effective treatment was not received. The symptoms include a pale red or pinkish rash (often found on the palms and soles), fever, sore throat, headaches, joint pains, poor appetite, weight loss, and hair loss. Moist sores called *condyloma lata* may appear around the genitals or anus and are highly infectious. Because of the diversity of symptoms, syphilis is sometimes called "the great imitator." The symptoms of the secondary stage of syphilis usually last three to six months but can come and go periodically. After all symptoms disappear, the disease passes into a *latent stage*. During this stage, the disease is no longer contagious, but the infecting microorganisms burrow their way into various tissues, such as the brain, spinal cord, blood vessels, and bones. Fifty to 70 percent of people with untreated syphilis stay in this stage for the rest of their lives, but the remainder pass on to the *tertiary stage*, or late syphilis. Late syphilis involves serious heart problems, eye problems, and brain or spinal cord damage. These complications can cause paralysis, insanity, blindness, and death.

Syphilis can be acquired by an unborn baby from its mother if the infecting microorganisms are in her bloodstream, since they cross the placenta. The resulting infection, called *congenital syphilis*, produces bone and teeth deformities, anemia, kidney problems, and other abnormalities. Congenital syphilis can be prevented if a pregnant woman with syphilis is treated adequately before the sixteenth week of pregnancy.

Diagnosis and Treatment

Syphilis is usually diagnosed by a blood test. Several different tests are available, including some that are most suitable for screening and others that are more time-consuming and expensive but also more accurate. Although none of these tests is completely foolproof in detecting the primary stage of syphilis, secondary syphilis can be diagnosed with 100 percent accuracy. Diagnosis also depends on a carefully performed physical examination looking for signs of primary or secondary syphilis. Chancres of the cervix or vagina may be detected only by a pelvic examination, since they are usually painless. An examination under a special microscope of the fluid taken from a chancre will usually show the characteristic spiral-shaped organisms.

Syphilis can easily be treated with one injection of penicillin in its primary or secondary stages. Latent, tertiary, or congenital syphilis requires larger doses over a period of time, but the treatment is usually successful in these cases, too. Patients who are allergic to penicillin can be given tetracycline or erythromycin.

GENITAL HERPES

The herpes family of viruses and the infections they cause, such as chicken pox, shingles, and cold sores, are widespread today, as they have been for thousands of years. First named by ancient Greek physicians from the word *herpein,* meaning "to creep," because of the appearance of their characteristic skin rashes, and described in some detail by Roman physicians of the first and second centuries A.D., herpes infections have recently been the subject of considerable public attention.

Genital herpes currently affects some 40 million Americans, with an additional 500,000 cases occurring annually. Viewed by some as a relatively minor skin infection with annoying but brief symptoms and by others as a life-threatening disease or even a heaven-sent directive against loose morals, the genital herpes epidemic of the 1980s had received almost as much media coverage as a presidential campaign by 1983, when it began to be outshadowed in the press by coverage of the AIDS epidemic.

Genital herpes is caused by two different but related forms of the herpes simplex virus, known as herpes virus type 1 and herpes virus type 2. In the past, herpes virus type 1 was almost exclusively a cause of cold sores and fever blisters, while genital herpes infections were almost invariably caused by the type 2 virus. Today, this distinction no longer holds true: in the United States, 10 to 20 percent of cases of genital herpes are now caused by the type 1 virus, while in Japan 35 percent of first episodes are due to the type 1 virus. While some researchers have suggested that this crossover phenomenon may be a result of more frequent oral-genital sex in recent years than in the past, it is not clear that this explanation is correct.

Genital herpes is generally transmitted by sexual contact. Direct contact with infected genitals can cause transmission via sexual intercourse, rubbing the genitals together, oral-genital contact, anal intercourse, or oral-anal contact. In addition, normally protected areas of skin can become infected if there is a cut, rash, or sore, so that infections of the fingers, thighs, or other areas of the body are also possible.

The risk of developing genital herpes in a woman exposed to an infected man is estimated to be 80 to 90 percent. A man's risk of developing genital herpes from a single sexual encounter with an infected woman is estimated to be about 50 percent.

In some cases, genital herpes can be spread by less direct means. For example, transmission of the herpes simplex virus can occur by kissing alone, and if herpes of the mouth develops, it can then be spread by auto-inoculation, that is, touching your genitals after putting your fingers in your mouth. Recently, several reports have noted that the herpes virus can live for at least several hours on toilet seats, plastic, and cloth, raising the possibility of genital herpes infection occurring by nonsexual transmission. It is unlikely, however, that this mode of transmission is very common. The herpes virus does not seem able to survive the chemical purifiers typically used in hot tubs.

Recently, new evidence has come to light that genital herpes can be transmitted by an infected, asymptomatic sex partner — that is, a person who has no herpes blisters or sores and no genital burning or itching. This finding is particularly troubling because it means that someone who doesn't realize that he or she is infected with the herpes simplex virus can inadvertently infect another person. While this mode of transmission is probably not too common, even if it occurs at only one-tenth the rate of symptomatic infection, asymptomatic transmission would occur in 1 out of 20 herpes victims.

The complexity of asymptomatic transmission is compounded further by the fact that someone may have large concentrations of herpes virus present in genital secretions on an episodic basis only, so that cultures of these fluids (semen, cervical mucus, vaginal secretions) that are negative for the herpes virus on one day may be positive a week later. One possibility is for persons with genital herpes to have cultures done on three occasions at least a week apart while they are totally asymptomatic — if all three cultures are negative, they can be fairly certain they won't infect a partner in their asymptomatic phase.

Genital herpes is marked by clusters of small, painful blisters on the genitals. After a few days, these blisters burst, leaving small ulcers in their place. In men, the blisters occur most commonly on the penis but they can also appear in the urethra or rectum. In women, blisters appear on the vaginal lips most often, but the cervix or anal area can also be affected.

The first episode of genital herpes is accompanied by fever, headache, and muscle soreness for two or more consecutive days in 39 percent of men and 68 percent of women. Almost all cases are marked by painful burning at the site of blister formation. Other relatively common symptoms include pain or burning during urination, discharge from the urethra or vagina, and tender, swollen lymph nodes in the groin, but these all tend to disappear within one to two weeks. More serious complications of first episodes, which occur more often in women than men, include the following: aseptic meningitis (an inflammation of the covering of the brain), estimated to occur in 8 percent of cases;

eye infections, which occur in 1 percent of cases; and infection of the cervix in 88 percent of women with primary herpes type 2 infection.

A typical first infection of genital herpes involves the appearance of ten to twenty painful blisters on the genitals. If generalized symptoms such as fever or headache appear, they are usually most prominent within the first four days after the blisters occur, and then they diminish gradually over the first week of the infection. After the blisters burst, they may form larger reddish wet sores or ulcers, which usually heal in one or two weeks. Sores on the penis or on the mons become crusted before they heal, whereas those on the vaginal lips do not. Skin lesions last an average of 16.5 days in men and 19.7 days in women during first episodes of genital herpes, but if the sores become secondarily infected with bacteria, healing may be somewhat delayed.

Although the blisters disappear and the ulcers heal spontaneously within one to three weeks, the herpes simplex virus invades nerves in the pelvic region and continues to live in a dormant state near the base of the spinal cord. In about 10 percent of cases there are no further attacks, but many people have recurrent episodes of genital herpes varying in frequency from once a month to once every few years. Repeat attacks are sometimes brought on by emotional stress, illness, sunburn, physical exhaustion, or extreme climates, or they may occur for no apparent reason. Generally, these recurrences are less severe than the original episode because the body is able to mobilize appropriate antibodies to counteract the infecting virus. Fortunately, many who suffer from herpes find that repeat attacks tend to die out after a few years.

Recurrences are sometimes preceded by warning symptoms that occur up to thirty-six hours before blisters appear. These include itching or tingling sensations in or near the genitals, tenderness or aching in the groin area, and burning or pain with urination or defecation. While these symptoms are not invariably followed by active outbreaks, and while some repeat episodes of genital herpes are totally symptom-free, when such symptoms occur they deserve attention. First, they indicate the possibility that a person may be infectious *even before blisters appear.* This is possible because live herpes virus may be carried in semen or cervical or vaginal secretions even when no rashes or blisters can be seen. Second, it may be possible to prevent a flare-up of herpes by taking steps that reduce stress (e.g., getting more sleep, eating well, and avoiding substances like alcohol or drugs that might suppress the body's immunologic response system).

Scientists are still puzzled by many aspects of genital herpes, and they do not understand clearly why some people never have recurrences while others have a number of repeat attacks. Certainly no one leads a completely stress-free life, nor can most people completely avoid the types of physical illnesses — like the flu — that often trigger recurrences. It is possible that there may be different strains of the herpes simplex virus, types 1 and 2, that partly account for this variability, and the individual's resistance (as determined by the body's immunologic defenses) may play a role, too. However, it would be a mistake to assume that people who experience recurrences of genital herpes aren't taking care of themselves properly or are being reinfected by another sexual partner, since such things usually aren't true.

There are two special problems with genital herpes. First, like syphilis, genital herpes in a pregnant woman can cause birth defects in the developing fetus since the virus can cross the placenta. Fortunately, this is a rare occurrence. Spontaneous abortion and premature labor were also thought to be common in pregnant women with genital herpes, but the validity of this finding has been questioned. More worrisome is that the baby can be infected from the cervix or vagina during delivery, with such infections causing death or serious damage to the brain or eyes more than 50 percent of the time in those newborns who are infected. Recent evidence also suggests that the rate of herpes infections in newborns has increased considerably in the last fifteen years, with 11.9 cases occurring per 100,000 live births in 1978 to 1981. While the risk of infecting the baby may be as high as 50 percent with vaginal delivery during a first attack of genital herpes in the mother, the risk is estimated at about 5 percent during recurrent episodes. Compounding the difficulties of this situation is the fact that newborns may be infected even when the mother has no symptoms at the time of delivery. A recent study showed that cultures taken in the four weeks before delivery do not show accurately whether the infant will be exposed to the herpes simplex virus at delivery. For this reason, cesarean section is often recommended to pregnant women with genital herpes, whether active or latent, but this is an individual matter that each woman should discuss with her physician. In addition, infection can occur after birth if the mother or father has oral lesions or if the virus is transmitted in breast milk.

The second serious issue regarding complications of genital herpes is that there appears to be an association between the herpes simplex type 2 virus and both cervical cancer and cancer of the vulva. While only some women who have genital herpes will be so affected, because these forms of cancer are easily treated if detected early, it is advisable for all women who have had genital herpes to have a Pap smear and pelvic exam every six months.

Just as the physical severity of genital herpes varies greatly from person to person, the emotional response to the discovery of herpes varies considerably, too. Most victims first experience a sense of shock, anger, and disbelief: "It can't really be happening to *me*." Their initial anger is usually (and understandably) directed at the partner who caused the infection, but it can be self-directed as well. One man told us, "I suppose this is the price I have to pay for not being more careful about who I slept with."

Not surprisingly, herpes sometimes causes conflict and suspicion in marriages or other long-term relationships. If one partner infects the other, it is taken as a sign of sexual infidelity, but this is not always true; for instance, a person may have reactivation of herpes that has been dormant for years and that preceded the relationship. Nevertheless, the discovery of herpes in a relationship can lead to so much discord and strife that the partners become combatants rather than lovers.

> *A twenty-eight-year-old woman:* When I found out I had herpes, I was scared out of my wits. I knew my husband would be outraged, but I also knew that trying to hide it was hopeless. So I told him about it, hoping that he'd understand that what was done was done. I wasn't really prepared for what happened, though. He suddenly refused to come near me. He questioned me in minute detail about everywhere I went. He was jealous and angry and rejecting, and it led us to a painful divorce.

As might be expected, genital herpes is at least as emotionally troublesome to single people. On college campuses across the country, many students admit that they have altered their patterns of sexual behavior out of fear of contracting herpes. Here are some typical comments we've heard in interviews: "I couldn't stand the thought of having this infection for the rest of my life — having casual sex may be fun, but it's just not worth it." "To me, it's like playing Russian roulette, only it's my peace of mind at stake." "How could I ever find a husband if I had herpes? I'd have to marry someone else who had it too."

Some herpes sufferers have turned to dating services that provide a means of meeting others with this STD. Others feel such profound shame and depression over their affliction that they withdraw from the social scene completely. Those who continue to date face a difficult dilemma: whether and when to tell a date that they have herpes. Some people feel a moral obligation toward candor but are dismayed by how quickly their honesty scares off potential mates. (Not everyone shares this reaction, though. One woman told us: "At least I knew if they were interested in *me* if they stayed around.") Others unfortunately feel it's all right to deceive their partners, perhaps deciding to make up excuses for abstinence whenever an outbreak occurs (which will *not* always prevent transmission) or convincing themselves that there is so much herpes around that it just doesn't matter. One thirty-year-old lawyer told us, "I've probably given this to twenty-five women in the last two years, and it serves them right — because one of them gave it to me."

Those who find themselves in the throes of emotional anguish over having genital herpes — and even well-adjusted, psychologically stable people can find themselves in this position — can often obtain assistance from support groups composed of people with the same disease. Organizations such as HELP (Herpetics Engaged in Living Productively), initiated by the National Herpes Resource Center of the American Social Health Association, which has chapters in over forty cities nationwide, provide valuable information and perspective (as well as many practical pointers) for people with herpes. In a few cases, particularly if sexual problems develop as a result of herpes or if guilt or depression becomes overwhelming, psychotherapy may be helpful. Most people, however, once past the initial anger and shock at discovering they have herpes, manage to adjust fairly easily to the disease and find that it rarely interferes with their lives. Unfortunately, media discussions of genital herpes have often presented such distorted or alarming "facts" about this disease that they have frightened many people rather than reassured or enlightened them.

The diagnosis of an active infection generally can be made with accuracy by a physician on the basis of a physical examination of the genital blisters and/or ulcers. However, other STDs can mimic genital herpes, and sometimes genital blisters or ulcers are a result of inflammation rather than infection. Thus, making a proper diagnosis is not always a simple matter. Various laboratory tests can establish the diagnosis with more certainty. These include (in increasing order of accuracy): (1) Pap smears in women; (2) blood tests to measure antibodies against herpes viruses; and (3) cultures to grow the virus in the laboratory (these are usually taken by touching a cotton swab to a blister or ulcer; the procedure is generally painless).

There is no known cure for genital herpes, although much research is being done. A new drug called acyclovir is of some usefulness in lessening the severity of symptoms, especially in first attacks. When taken by mouth within five or six days after the onset of symptoms in a first episode of genital herpes, acyclovir reduced the period of virus shedding (virus being present in body fluids such as semen or vaginal secretions) by two-thirds compared with treatment with a placebo. In these same studies, acyclovir shortened the healing time of genital herpes lesions by four to nine days. In addition, taken on a long-term, continuous basis, oral acyclovir seems to cut down the number and duration of recurrences — in fact, in several studies, the rate of recurrence was lowered by at least 75 percent. During genital herpes *recurrences,* however, acyclovir doesn't seem to work as well in reducing pain or viral shedding.

Since its safety hasn't been studied yet in persons using this drug for more than six months, no one is certain at this time what side effects may result from long-term use of acyclovir. Because there is also concern that extensive use of acyclovir will lead to the emergence of new strains of herpes simplex virus that are resistant to it, only people whose recurrences are particularly frequent or troublesome should consider using acyclovir on a long-term basis for suppressive therapy. Other antiviral medications are also being studied for treating genital herpes, and a concerted effort is being made to develop a vaccine that would prevent this infection in the first place.

General measures such as taking aspirin (or an aspirin substitute) and using cold, wet compresses to relieve pain are often helpful during an initial herpes episode or recurrent flare-ups. In addition, avoiding tight underwear or clothing can reduce irritation, and keeping the genitals clean and dry by washing with warm water and soap several times a day can also be beneficial. The skin should be dried with clean towels and a patting, rather than vigorous rubbing motion, and hand-to-eye contact should be avoided after touching the genitals. Towels and washcloths should be kept separately, since they may be contagious to others. In fact, it's wisest to use a separate towel for the face to avoid inadvertent spread of the virus from the genital region to the eyes.

Sexual contact should be completely avoided from the time symptoms of genital herpes first begin until ten days after healing is complete (for a first attack) or until two days after complete healing in recurrent episodes. Unfortunately, a few people seem to shed virus all the time — whether or not they have visible skin lesions or symptoms — so it's impossible to guarantee that there is no risk of contagion. While use of a condom can help prevent transmission of genital herpes, it is not a foolproof method (both because it doesn't cover all lesions and because it isn't always worn from the start of genital contact) and it may actually irritate the condition.

AIDS

The newest and most frightening STD yet discovered was first documented in 1981. Now known by the acronym *AIDS* — for *acquired immune deficiency syndrome* — this devastating illness is the result of a viral infection that leads to a breakdown of the immune system, the system that ordinarily protects the body against infections. Because of the collapse of these immune defenses, AIDS victims fall prey to a variety of rare infections that are usually found only in cancer or transplant patients whose resistance is lowered by medications that impair their immune responses.

According to the Centers for Disease Control, 40,000 cases of AIDS had been reported in the United States as of mid-1987. In western Europe, cases of AIDS are far less common, with only 5,000 cases reported to the World Health Organization as of the same date. In Africa, however, AIDS is running rampant, with an estimated 50,000 cases between 1980 and 1986.

These numbers are alarming for a number of reasons, especially since they appear to be only the tip of the iceberg — many diagnosed cases of AIDS are not reported to government centers that are tracking the disease and other cases are never correctly identified. Even more frightening is the fact that 1.5 million Americans and an estimated 5–10 million persons worldwide have been infected with the AIDS virus and carry it in their blood but have not yet developed any symptoms.

Current studies have shown that more than 30 percent of these "carriers" will go on to develop full-blown cases of AIDS within three years after being infected, and some scientists believe that up to 80 percent will develop AIDS within ten years. Since full-blown cases of AIDS are almost invariably fatal, the implications are staggering. In fact, the U.S. Public Health Service has estimated that by the end of 1991 there will be a cumulative total of 270,000 cases of AIDS in America, with 179,000 deaths. A panel of experts including six Nobel Prize winners has predicted that more than a million Americans will have AIDS by the year 2000 (*New York Times*, March 4, 1987, p. A20). Making the picture even bleaker, most authorities believe that an effective vaccine against the AIDS virus won't be available until 1993 or later and the prospects for developing a successful treatment for AIDS are poor at present. Given this background, it is easy to see why AIDS has become a public health threat almost without parallel in this century.

AIDS is caused by the *human immunodeficiency virus* (*HIV*, formerly called HTLV-III), a newly discovered virus so small that 16,000 could fit on the head of a pin. Typically, the virus gains entry to the body by sexual contact or by intravenous drug use with a contaminated needle. (It is believed that in most cases the virus must enter through a break in the skin or other tissue, such as with a cut, a sore, or a tear.) Once inside, the virus selectively attacks certain white blood cells, called T-helper cells, that are the key coordinators of the immune system. T-helper cells send out chemical signals that stimulate the production of antibodies and largely control the development of several other types of cells that make up the immune system.

After attaching itself to the outer surface of the T-helper cell by a biochemical process that is very much like a key fitting in a lock, the virus is drawn inside the cell, where it establishes a permanent infection. As it does this, the virus copies its genetic information, producing new viruses, which are then released from the host T-cell. These new viruses attack not only other T-cells but also other cells of the immune system and the brain.

Of the first 25,000 cases of AIDS in the United States, 65 percent involved homosexual or bisexual men who were not intravenous drug users, 17 percent occurred in heterosexual IV drug users, 8 percent occurred in homosexual or bisexual IV drug users, 7 percent occurred in heterosexual partners of people with AIDS or of people in high-risk groups, and 2 percent involved people who had received transfusions of contaminated blood products, according to the Centers for Disease Control. In the United States, more than 90 percent of AIDS victims are male, but in Africa cases are divided almost equally between men and women. In fact, while AIDS was initially regarded by many as a problem primarily affecting homosexual and bisexual men, it is now clear that heterosexual transmission is common and that no single group can be regarded as "at risk" for developing this dread illness.

Although there is no certainty at present about how AIDS originated, many scientists believe that it began in central Africa. One possibility is that the AIDS virus first infected monkey colonies there and subsequently spread to humans in the mid-1970s. From there, AIDS may have been carried across the Atlantic Ocean by Haitians who once lived in or visited central Africa. From Haiti, AIDS may have spread to the United States by two routes: Haitian immigrants and vacationing American homosexual males, who often traveled to Haiti. If this explanation is correct, the early clustering of AIDS cases in the United States in gay males may have been largely accidental, which could explain why so many cases of AIDS in Africa are heterosexually transmitted, in contrast to the situation in the United States. The implication for the future, if this is true, is that as time goes by far more AIDS cases in the United States will arise from heterosexual transmission.

Several African and Communist-bloc countries have blamed the AIDS epidemic on American biological warfare experiments gone wrong, letting a deadly virus escape into the general population. (The African countries making this claim are understandably sensitive about having a finger pointed at them as the possible origin of the AIDS virus.) In a variation on this theme, some Americans think that the AIDS virus is a result of genetic engineering gone amok. These beliefs have no more credibility than the even more popular idea that AIDS was sent by God to punish humans. (If God did this intentionally, some observers wonder, does He favor lesbians, since He made them least likely to get AIDS?)

In more than 78 percent of all AIDS cases the virus has been sexually transmitted. It is not, however, as highly contagious as some other STDs. Current estimates suggest that the risk of being infected with HIV from a single act of heterosexual vaginal intercourse with an infected person is 1 in 1,000 for a woman and 1 in 2,000 for a man, although we believe these estimates are overly optimistic. The risk from a single episode of anal intercourse with an infected partner is considerably higher — probably on the order of 1 in 50 to 100. Since the lining of the rectum is very delicate and tears easily during anal sex, infected white blood cells and HIV in the ejaculate can easily enter the tissue and bloodstream of the receptive partner (whether male or female). Since the risk of catching gonorrhea from a single heterosexual exposure to an infected partner is 50 percent for women and 25 percent for men, it is clear that HIV is far less contagious. Nevertheless, anyone who engages in risky sexual activity has a small but definite chance, each time, of getting AIDS. As any student of probability can tell you, the more chances taken and the more exposures, the greater the likelihood of infection.

Despite this finding of a low degree of contagion, many people have been so frightened by the AIDS epidemic that they worry that AIDS might be transmitted by casual contact such as shaking hands with an infected person or coming in contact with the virus on a doorknob, toilet seat, or drinking fountain. Researchers agree that such fears are ungrounded: HIV is transmitted almost exclusively by intimate sexual contact involving the exchange of body fluids and by sharing or reusing contaminated hypodermic needles or syringes. While the AIDS virus has been identified in blood, tears, urine, saliva, semen, and vaginal secretions, many studies have shown that people in close daily contact with AIDS patients — such as a parent caring for a child with AIDS and nurses, doctors, and dentists who work closely with AIDS victims — do not ordinarily develop HIV infections from these contacts.

While it is certain that sexual intercourse — either vaginal or anal — is the primary means by which the AIDS virus is transmitted, it is also very likely that HIV can be transmitted by other forms of sexual activity. Since the concentration of HIV is highest in blood and semen, it should be no surprise that any contact with these fluids poses a risk of transmitting the virus.

Oral-genital sex, which is known to transmit every other form of STD, has not been definitively proven to transmit AIDS as yet, but the lack of absolute proof may reflect the problem of finding people who have engaged exclusively in oral-genital contact and never in coitus (or who have had coitus exclusively while being protected by a condom). In one recent study of 45 married couples in which one spouse had AIDS, it was found that the frequency of oral-genital sex correlated with the previously uninfected partner's becoming infected with the AIDS virus.

French kissing, or soul kissing (which involves exchanging saliva with your partner), must also be considered as a potential means of transmitting the AIDS virus. Although the concentration of HIV is much lower in saliva than in blood, semen, or vaginal secretions, there is a distinct possibility of infection because it is very common to have minor cuts or abrasions of the gums, lips, or inner mouth that may provide a portal of entry for the virus. There have been no empirical studies providing absolute proof that the AIDS virus can be transmitted this way, but it is foolhardy, from a scientific viewpoint, to think that this type of transmission cannot occur. The practical problems that this creates may seem insurmountable at first — will we need a medical certificate to kiss? will kissing be reserved only for long-term relationships? — but the risk of transmission in this way is probably quite small, perhaps on the order of 1 in 100,000 exposures. Nevertheless, it is wisest to avoid French kissing completely with a person known to be infected with the AIDS virus.

Aside from sexual transmission, the major route by which the AIDS virus is transmitted is by intravenous drug users sharing needles or syringes that are contaminated by small amounts of infected blood. Once infected in this way, someone can then pass on the infection by sexual contact as well as by additional sharing of needles.

IV drug users may have a higher susceptibility than others to infection with HIV because their general health and nutritional status is often poor and their immune defenses may already be partly compromised by other illnesses. A few public health officials have suggested providing free sterile needles to all drug abusers who want them as a means of controlling the spread of AIDS, but this strategy has led to cries of outrage from community leaders and law enforcement officials who are afraid it will simply encourage drug abuse. A more practical — and politically acceptable — approach has been for some cities to mount extensive educational campaigns to teach drug abusers how to sterilize their needles and to avoid sharing needles entirely.

The AIDS virus can also be transmitted by the transfusion of contaminated blood or blood products. As of 1986, about 2 percent of AIDS cases in the United States have occurred among hemophiliacs (who require frequent injections of concentrated blood-clotting factors to prevent them from bleeding profusely from even minor cuts or accidents) and recipients of blood transfusions. The routine use since 1985 of screening tests to detect HIV antibodies in donated blood and blood products is now believed to lessen substantially the risk of using contaminated blood for transfusions, and it is thought that this form of transmitting AIDS has become relatively rare. However, the blood supply is not completely safe because the screening tests are not foolproof, as we will discuss shortly. For this reason, some people having elective, nonemergency surgery have donated their own blood months or weeks before the surgery and have had it frozen and stored so it can be used if they need a transfusion. This procedure is called *autologous transfusion* and has recently received favorable medical commentary. It should be stressed that there is absolutely no risk to *giving* blood as long as a sterile needle is used.

Possibly the saddest form of transmission of the AIDS virus is from an infected mother to her developing child during pregnancy or childbirth. Current research suggests that 20 to 50 percent of infants born to infected mothers will be infected, although how many of these children will develop the full-blown syndrome of AIDS is not known at present. Because of this problem, authorities agree that women infected with HIV should not become pregnant. Most experts also feel that a woman who is infected with HIV who becomes pregnant should consider having an abortion, although of course this is a complicated personal decision. The best solution is for women who are infected with HIV to avoid pregnancy completely.

HIV can also be transmitted from a sperm donor or an organ transplant donor to an uninfected person, although instances of this sort are relatively rare.

Is "Safe" Sex Possible?

Experts are somewhat divided on exactly what constitutes "safe" sex. Short of complete abstinence, you should realize that two uninfected partners cannot transmit HIV to one another. While there is no way to tell by someone's appearance if he or she is free of HIV, it is possible to be tested to determine if antibodies to the AIDS virus are present in a person's blood. If you and your partner are both tested and are found to be free of AIDS antibodies and you each decide to have sex only with each other, you are effectively guaranteed that you won't get AIDS. If such an approach is either impractical or distasteful to you, the following guidelines may be of some help.

1. Completely safe sex is possible if there is no exchange of body fluids. While this means cutting out oral sex and intercourse as options — because even without ejaculation there is a danger of transmitting the AIDS virus in pre-ejaculatory fluid and vaginal secretions — techniques such as massage, use of vibrators, and mutual masturbation can be used. Remember, though, that deep kissing is another potential

means of transmitting the virus and should be avoided. And
any sexual practices that can cause injury or rips in tissue
should be avoided, too.

2. Use of condoms will greatly reduce the risk of sexual contact. Studies have found that condoms don't have pores that
are large enough to allow viruses to pass through. However,
since improper use of condoms or a tear in the condom can
certainly lead to leakage, this method is *not* a foolproof
means of preventing infection. Remember, condoms are
not perfect contraceptives, and they're not perfect barriers
to AIDS or other STDs, either.

3. Use of spermicides containing nonoxynol-9 may offer additional protection. Nonoxynol-9 kills HIV under laboratory conditions, although it is uncertain whether under real-
life conditions it provides as much protection as a condom.
The best solution seems to be to use *both* a condom and a
spermicide with nonoxynol-9 during intercourse.

4. Be selective in choosing your sex partners. As we've already
noted, it isn't possible to judge by any symptoms who's been
infected by HIV, since it can take years after becoming infected for symptoms to develop. Under current conditions,
it is prudent to realize that people who have had a large
number of sex partners (or people who inject drugs) are
statistically more likely to have been exposed to and infected
by HIV. Selectivity (and careful questioning of a prospective
sex partner) can be lifesaving.

Infection

Once HIV enters the bloodstream, no matter what the route of
infection, the immune system usually responds with the production of anti-HIV antibodies. Most people have no symptoms that
accompany the initial infection or the production of antibodies,
but 10 to 25 percent may have a brief illness that occurs two to
five weeks after the virus enters the body. Symptoms resemble
infectious mononucleosis ("mono"), including fever, chills,
aches, swollen lymph glands, and itchy rashes. Antibodies to
HIV can usually be detected within two months after the initial

infection, but there are a few cases in which antibodies do not appear for six months or longer.

The infected person then passes into a phase called the *asymptomatic carrier state*. In this phase, there are absolutely no symptoms of illness, but the infection is present and antibodies persist. Many asymptomatic carriers have a reduced number of T-helper cells in their blood, but no other effects are usually detectable. The presence of the live virus in the asymptomatic carrier state means that such a person can infect others without realizing that he or she is harboring the AIDS virus, unknowingly providing it with a launching pad for its silent spread to others. It is important to realize that asymptomatic carriers do not have AIDS (the illness), although they are infected with the virus that causes AIDS.

It is uncertain how long infected people can remain asymptomatic, not developing any signs of illness. Many individuals remain in the asymptomatic carrier state for periods of three to five years or longer before developing AIDS or related symptoms. Current evidence suggests that about 30 percent of asymptomatic carriers will develop full-blown cases of AIDS within five years; and some experts feel that with the passage of additional time, as many as 80 percent may eventually develop AIDS.

In some people, HIV infection leads to symptoms that are less serious than AIDS itself. Generally known as *AIDS-related complex (ARC)*, this condition is usually marked by persistently swollen lymph nodes in several locations in the body (for example, the neck, the armpits, just above the collarbones), which can occur alone or with other symptoms. The most common symptoms of ARC are diarrhea, weight loss, fatigue, and fever. In addition, non-life-threatening infections such as shingles (herpes zoster) or thrush (a fungus infection of the mouth caused by *Candida albicans*) occur with increased frequency. Current research shows that after two years of having ARC, 20 percent of people develop AIDS, and after five years, about 30 percent progress to AIDS. It is very likely that after longer periods of time, the great majority of all persons with ARC will ultimately develop AIDS.

Much remains to be done to clarify what determines the number of people infected with HIV who will progress to ARC or

AIDS. Based on present evidence, it appears that different groups may have different rates of progression. For instance, one study showed that 34 percent of infected homosexual men had developed full-scale cases of AIDS after three years of follow-up, whereas only 12 percent of a group of hemophiliacs had gone on to develop AIDS in this same time period. Another 20 to 30 percent of both groups developed ARC. Such differences may be a result of different lengths of infections with HIV or may relate to other factors. For example, repeated infections with HIV may activate the virus when it has been relatively dormant in the body. It is also possible that persons who have current or past infections with various other viruses (for instance, hepatitis B virus or cytomegalovirus) may be most at risk for developing AIDS. Another theory is that use of certain illicit drugs — most particularly, volatile nitrites like amyl nitrite ("poppers") — may set the stage for developing AIDS more easily by lowering natural resistance to HIV. Whether any of these theories will prove to be correct is uncertain at present.

It is also important to realize that there have been only relatively brief follow-up studies done on people infected with HIV. It is possible that by ten to twenty years after infection, unless a cure is found, most people will progress to fulminant cases of AIDS.

Symptoms

No single pattern of symptoms fits all cases of AIDS. The principal findings are progressive, unexplained weight loss, persistent fever (sometimes accompanied by night sweats), swollen lymph nodes, and slightly raised reddish-purple coin-sized spots on the skin. These skin lesions often turn out to be a form of cancer of the small blood vessels, a condition called *Kaposi's sarcoma* that had previously been quite rare in the United States. About one-quarter of people with AIDS in the United States have been found to have Kaposi's sarcoma.

When symptoms first appear they may remain unchanged for months or they may be quickly followed by one or more opportunistic infections, that is, infections that occur when immunity is compromised. One of the most common of these infections is an unusual form of pneumonia caused by *Pneumocystis carinii,* which serves to establish the diagnosis of AIDS in almost two-thirds of cases seen in the United States. Among other common infections in people with AIDS are severe fungal infections (including a type that can spread to infect the covering of the brain, causing meningitis), tuberculosis, and various forms of herpes that are more severe and recur more often than usual. Encephalitis (inflammation of the brain) is another life-threatening condition that occurs with greatly increased frequency in people with AIDS. In fact, because HIV is able to pass through the blood-brain barrier (a natural filter that normally protects the brain) and directly infect brain cells, a variety of neurologic disturbances are seen in an estimated 30 to 65 percent of AIDS patients. These conditions include memory disturbances, psychiatric symptoms, dementia (severe mental confusion), difficulty walking, seizures, and coma. Although treatment can often fend off these infections temporarily, the typical course is for one after another overwhelming infection to occur until the victim finally dies because the depressed condition of the immune system becomes progressively worse.

At present, it appears that AIDS is almost invariably fatal within a matter of two to four years after it is first diagnosed. Despite this, many people with AIDS are able to lead relatively normal lives early in the course of their disease, although coping with the social, economic, and emotional aspects of their illness is especially difficult for many, as we will discuss shortly.

While some people with AIDS continue their employment and usual activities for six months or more after the diagnosis is established, eventually the weight loss, constant fatigue, and multiple infections take such a toll that even ordinary movement becomes a major effort and the person becomes an invalid. (In Africa, AIDS is often called "slim" because it emaciates its victims so drastically in its late stages that they look as if they have been starving.) Wasting away physically, often suffering from mental confusion and inability to concentrate, and usually in severe pain from high fevers, strenuous coughing, bedsores, and infections, people in the last stages of their struggle with AIDS are haunting reminders to us of a disease about which we know far too little.

Diagnosis and Treatment

Several different blood tests can be used to detect antibodies to HIV. The most widely used test is called ELISA (for *enzyme-linked immunoabsorbent assay*), which was developed to screen blood used for transfusions. Like all biomedical tests, ELISA is not infallible. It fails to detect HIV in about 2 percent of samples known to be positive (a mistake that is known as a "false negative," whose practical implication is that the blood supply is not completely safe), and it incorrectly "finds" HIV antibody in 0.2 percent of samples (in other words, 2 out of 1,000 tests will be mistakenly called positive when they are actually not — a situation known as a "false positive"). In fact, in low-risk populations, up to 90 percent of initially positive ELISA tests proved to be false positives.

Because of these inaccuracies, it is important to verify any positive result found by ELISA by using a different, more expensive test after repeating the ELISA test. The *Western blot test* is most often used for this purpose. When used together (ELISA test first, repeated if positive, and then confirmed by the Western blot test), the accuracy of testing for HIV antibody compares favorably with other screening tests used in medical practice, although even this method of screening is not perfect.

It is important to realize that finding HIV antibody in someone's blood is not the same as detecting HIV directly. (For technical reasons, isolating HIV from a blood sample is currently difficult to do, so the antibody screening test is used as a practical alternative.) *Finding HIV antibodies in a person's blood does not, by itself, show that a person has AIDS.* A false positive test can be the result of a technical error in performing the test or a clerical error in identifying the blood sample. A false positive can occur when the reagents used in the test have reacted to a substance in the blood that gave a mistaken reading, registering as though it were HIV antibody when it was not. (False positives are more common in women than men and may also be related to previous pregnancy.) A positive test could also mean that a person has been exposed to the AIDS virus and has successfully fought it off. In such a case, antibodies would remain in the blood, although no virus is present in the person's body. (In such a circumstance, the result would be a true positive, not a false positive, since HIV antibodies are really present.) While only 65 percent of antibody-positive people have had recoverable virus circulating in their blood when tested on a single occasion, this may mean that current laboratory techniques are not sensitive enough to always identify the virus when it is present.

Despite such possibilities, we should stress that a positive test *is* cause for concern and should be followed up by careful long-term medical evaluation. In addition, as pointed out recently by a panel of AIDS experts, "All persons who are antibody positive for HIV, whether they are symptom free or ill, must be considered to be potentially infectious to others by sexual transmission, by sharing of drug injection equipment, by childbearing, or by donation of blood, semen, or organs."

Several large-scale screening studies have found generally low rates of HIV antibody in the U.S. population. For instance, in 1986 the U.S. military tested one million men and women on active duty with the armed forces and found a 0.15 percent positive rate (meaning 15 out of 10,000), which was exactly the same rate found in an earlier test of 300,000 military recruits (*New York Times*, October 18, 1986, p. A6). In an earlier study of one million blood donors by the American Red Cross, it was found that 4 out of each 10,000 units of blood were positive for HIV antibody. In contrast, several studies of prostitutes in Africa have found the prevalence of HIV antibody to be in the 25 to 66 percent range, and a study of 4,936 homosexual men in five U.S. cities found a 38 percent prevalence rate. Various estimates suggest that 70 to 80 percent of IV drug users in New York City and San Francisco have HIV antibodies in their blood as well.

It should be stressed that a positive blood test does not establish the diagnosis of AIDS. The diagnosis of AIDS is made when a major disease such as *Pneumocystis carinii* pneumonia or Kaposi's sarcoma that signals an underlying deficiency in the immune system occurs in the absence of other conditions known to be risk factors for these illnesses. AIDS can also be diagnosed if a person has a positive antibody test, evidence of a suppressed immune system, and at least one disease from a list of certain lesser infections. While many physicians have argued that such a strict definition underestimates the true number of AIDS cases (perhaps by as much as 20 percent), the U.S. Centers for Disease Control has insisted on a strict definition for scientific purposes to assist it in tracking the spread of this epidemic by keeping definitions relatively constant over time.

Although as of early 1987 no successful cure for AIDS had been found or even hinted at, there are some signs of progress in the fight against this dreaded disease. A drug called *AZT*, for azidothymidine, has emerged from early clinical trials with impressive credentials. In one study, only 1 out of 145 patients taking AZT had died over an eight-month period, compared with 16 deaths among 137 patients taking placebos. As a result, AZT is being made more widely available to people with AIDS, although researchers stress that it is not a cure — it may simply keep the disease under control. However, AZT is far from an ideal drug since it has particularly toxic side effects in many people (most notably, suppression of the bone marrow, which causes severe anemia) and since it is quite expensive, with a year's supply currently costing $8,000 to $10,000. Other antiviral drugs, such as ribavirin, cyclosporine A, and alpha interferon, are also being studied in the attempt to find the best treatment strategy. None of these approaches, however, has succeeded in overcoming, on a long-term basis, the underlying deficiency of the immune response system that is the major problem in AIDS.

Social and Emotional Reactions

- Although the U.S. Surgeon General has reported that no cases of AIDS have been transmitted from one child to another in school, day care, or foster care settings, parent groups in many communities have tried to ban children with AIDS from attending public schools.
- Some homosexual fathers are finding that their rights to visit their children after a divorce are being challenged on the grounds that they are members of a "high-risk" group for contracting AIDS.
- In some cities, physicians are leading "safer sex" parties, "which are something like the familiar Tupperware parties," with a focus on ways of reducing the risk of catching AIDS.

- Many gay leaders are worried that mandatory reporting of the results of AIDS antibody blood tests will lead to widespread discriminatory repercussions — "a public witch-hunt," as one person put it.
- Homosexual men have filed several lawsuits against insurance companies, claiming that they are being denied insurance coverage because they are in a high-risk group for AIDS.

The impact of AIDS on our society has already been divisive and emotional. For many, the AIDS epidemic seems to be a ready-made excuse for intensified prejudice against homosexuals, particularly among the misinformed, who think that shaking hands with a gay person or being served by a homosexual waiter could transmit a fatal infection. Such people frequently suggest that everyone infected with the AIDS virus should be quarantined — a concept that is both inhumane and economically infeasible in light of the estimated 1.5 million Americans now harboring the AIDS virus. Others are complacent about this major public health crisis because they mistakenly see AIDS as a problem confined to homosexuals, bisexuals, and drug users rather than a problem for American society as a whole. Still others blame the AIDS crisis on unbridled teenage sex or on blatant sexual themes in movies, books, and rock music. Even among those not caught up in this sort of hysteria, many are willing to give up a number of fundamental civil rights to combat the growing epidemic.

In the gay community, which has clearly been most directly affected, pervasive fear and dismay are rampant. Worry about possible exposure to AIDS has apparently led to widespread changes in sexual behavior in this group: many gay men now avoid the bathhouse scene entirely and a large number have cut back to having sex with only a small number of partners who are well known to them. This is not a universal change, however: some gay men still seem bent on proving how "macho" they are by ignoring warnings from many gay organizations and continuing their life-styles of frequent sex with multiple anonymous partners. At the opposite extreme are homosexual males who are so paralyzed by fear of AIDS that they decide to be completely celibate until a vaccine or cure is found, as well as a small number of male homosexuals and bisexuals who have switched to heterosexual partners, at least temporarily.

Within the gay community itself, some have questioned whether enough has been done by homosexuals themselves to reduce the spread of AIDS by promoting monogamy and safer sex. Larry Kramer, a founder of the Gay Men's Health Crisis, a leading New York group focused on providing a broad range of services to AIDS victims, says, "The movement of the 60's and 70's legitimized promiscuity. To in any way criticize that as wrong was tantamount to heresy, and still is in certain quarters" (*New York Times*, September 22, 1985, p. 56). The Gay Men's Health Crisis has, however, opposed the closing of New York's bathhouses (which are geared toward anonymous sex with multiple partners), arguing that the bathhouses provide a centralized location for education about AIDS. "The building doesn't give you AIDS; people give you AIDS," said Barry Davidson, the Gay Men's Health Crisis director of community information. San Francisco, under court order, has attempted a compromise: the bathhouses are still allowed to operate, but doors have been removed from cubicles and "lifeguards" monitor all sexual contacts to try to ensure that safe sex practices are followed. Many gays disagree with Davidson's opinion and San Francisco's compromise solution. One former participant in the bathhouse scene who has now given up this activity puts it succinctly: "To keep the baths open at this stage of the game, when we know what a killer AIDS is, is to show the rest of the world that gays are crazy."

Unfortunately, in some cities — among them New York, Newark, Chicago, and San Francisco — where there has been extensive infection with the AIDS virus in the gay communities and where most gays have watched friends and lovers die of AIDS, a swing back to freewheeling sex is beginning to occur. At least some of those who test positive for HIV antibodies are so resigned to having AIDS brewing in their bodies that they are throwing caution out the window and are escalating the frequency and variety of their sexual activity, dropping precautions such as using condoms or avoiding ingestion of a partner's semen.

Needless to say, anyone who discovers that one of his or her previous sex partners has AIDS has particular reason for concern. So far, this concern has primarily affected gays and IV drug users, but heterosexuals are increasingly going to have to confront the same possibility as the AIDS epidemic spreads into the broad population. Just how does it feel to be in such a situation?

> *A twenty-eight-year-old heterosexual man:* I heard from a friend that a girl I used to date had come down with AIDS. At first I thought it was just a crazy story, something to try to scare me with, but when I checked it out I was horrified to discover it was true. God, I thought, we had sex together a couple of dozen times, and we never used a condom because she was on the pill. My doctor sent me for a blood test, which much to my relief came back okay, but in the twenty-four hours before I got that report I can only say I had visions of a slow, agonizing death right before my eyes, and I was plenty worried. You can be sure that I'll remember this episode for a long time.

Some homosexual men have developed such a degree of AIDS anxiety that they have become preoccupied with every minor physical ailment they experience. To these gay men, a sore throat, a skin rash, or a fever is a sign of impending doom, and they may even mistakenly tell their partners that they have AIDS before it is diagnosed. Among the very fearful it is common to see a sharp decline in sexual interest and sexual activity.

In heterosexuals, concerns about AIDS are mounting rapidly as awareness is rising that AIDS is not just a homosexual disease. For instance, it is not unusual now for a woman to ask a man she's dating to have a blood test for HIV antibody before they begin sexual relations (some men have also made the same request of women they're going out with). As one college-age woman told us, "It's not that I'm totally nervous about AIDS, but considering the stakes are so high, what's the purpose in pretending that everyone you'll meet will be honest? If someone doesn't care enough about me to have the test, the relationship isn't going to go anywhere anyway." At the same time, it's important to be cautious about bogus "blood test certificates" that some unscrupulous entrepreneurs are selling in an effort to make a quick buck from people's legitimate concerns about AIDS.

People who are diagnosed as having AIDS are usually young adults who have been generally healthy, so they have relatively little preparation for facing a life-threatening disease with a very bleak prognosis. As if this were not difficult enough, AIDS victims, whether gay or straight, must also face a hostile and fearful society. Many are ostracized at work, rejected by their family and friends, and put in a position of social isolation. In addition, they must face the prospect of numerous medical complications and hospitalizations, deal with concerns about the economic cost of their illness, and come to terms with their own identity and emotions as part of the ultimate task of preparing for death.

Immediately after learning that they have AIDS, most people react with shock, anger, and denial. These are perfectly normal defense mechanisms that can initially cushion the blow of receiving such a diagnosis. In the weeks or months after the diagnosis, this initial reaction typically undergoes a transition into guilt, sadness, and resignation to one's fate — a sort of unhappy acceptance of the situation. Of course, different people react in very different ways. Some people with AIDS have been anticipating the illness with dread for months or years as they've watched friends and lovers become stricken. (One such person told us, "After attending the funerals of eight of my friends who'd died of AIDS and sitting in hospital rooms dozens of times with them, I knew that my time was coming soon — it was just a matter of when.") Others begin to obsessively review their past to discover what they might have done to "deserve" AIDS or who might have given them the AIDS virus. While it is not unusual for people with AIDS to withdraw from contact with family and friends, others with AIDS continue to work actively in their jobs or professions and also become actively involved in efforts to educate the general public about this illness.

Ethical Concerns Related to AIDS

Discrimination against AIDS victims has occurred on many different levels, and it is not directed only at those who are gay. Children with AIDS have been banned from their classrooms; landlords have evicted people with AIDS from their apartments or refused to renew their leases; many employers have fired people with AIDS or forbidden them to come to work. Some health care personnel and members of the clergy have even refused to provide appropriate services to AIDS patients because of their fears of possible contagion.

As a result of such clear-cut discrimination, many issues of ethics and confidentiality are involved in the AIDS epidemic. For example, does an employer have a right to know the results of an employee's blood test for HIV antibody? (Does it make any difference if the employee is a food handler, a nurse, or a doctor?) Should insurance companies be permitted to refuse to provide coverage for people with positive antibody tests or for people who are members of high-risk groups? Should mandatory premarital blood testing for AIDS antibody be required — and should those who have positive tests be prohibited from marrying or having children? At the other end of this spectrum are concerns for individual rights of another sort. Do parents have a right to know if their children are in a classroom with another child with AIDS? If, as happened in San Francisco, a policeman is bitten by a man who claimed he had AIDS, does the policeman have a right to confidential medical information about this man, and does the state have the right to such information in deciding whether to charge the man with attempted murder?

In addition to these broad ethical concerns, many leaders of the gay community are worried about the effects AIDS publicity is having on the general public in terms of increasing homophobia. Having made considerable progress in the 1970s in attaining civil rights for homosexuals in employment and housing situations and in legal matters (for example, the right to adopt children), gays are distressed that a disease that is not understood very well — but is clearly not only a gay disease — may be pushing heterosexuals toward intolerance of homosexuals once again, this time out of fear for their personal safety. Those who proclaim that AIDS is an expression of the wrath of God against homosexuality in a sort of latter-day Sodom and Gomorrah further fuel this type of reaction — and it is likely that as the number of AIDS cases escalates, as it inevitably will, such reasoning will be accepted by more and more people.

There has been marked resentment in the gay community that not enough money was spent on AIDS research initially. Donald Currie, manager of the San Francisco Kaposi's sarcoma hot line, was quoted in *Time* as saying, "If the same number of Boy Scouts had been dying of this, there would have been a hell of a lot more money for research" (March 28, 1983, p. 55). While many scientists and politicians alike now admit that this was true, there have been remarkable increases in federal spending on AIDS: from $5.5 million in 1981 to $411 million in 1986. Still, a prestigious panel cosponsored by the Institute of Medicine and the National Academy of Science in 1986 indicated that even this amount of funding was inadequate — billions of dollars must be mobilized if we are to have a reasonable chance of successfully putting this epidemic to an end over the next ten years.

The long-term implications of the current AIDS epidemic probably will not be fully apparent for some years. The cost of the epidemic, both in young lives and billions of dollars, will undoubtedly be astronomical before it is all over. If AIDS education is brought into elementary schools in an unskilled, haphazard manner, we are liable to give an entire generation of young people the message that sex kills before they learn about the loving, responsible, pleasurable aspects of sex. Conceivably, if the AIDS epidemic continues, many adolescent males may be dissuaded from becoming actively homosexual and others may be kept from "coming out." Premarital intercourse and extramarital sex among heterosexuals may also become, statistically, things of the past, and within both the straight and gay communities there may be an increasing trend toward close-coupled, monogamous, long-term relationships. It now seems almost inevitable that AIDS will spread increasingly to the heterosexual community — with some experts suggesting that in five years 20 to 25 percent of AIDS cases in the United States may be among non-drug-using heterosexuals.

In short, there is no way to simplify the problems and challenges posed by AIDS. While research continues in the effort to develop both a preventive vaccine and a true cure, we must also try to keep the broadest possible perspective on the issues that this crisis has brought forth. Confronting difficult questions and making tough decisions with fairness and compassion will be needed as our society struggles with this devastating disease.

CHLAMYDIAL INFECTIONS

Infections caused by the bacteria *Chlamydia trachomatis* have been surprisingly overlooked until very recently, although they are now known to be the most common bacterial STD in the United States, with an estimated 3 to 4 million cases annually. Part of the reason for this relative neglect is that many people have never heard of chlamydial infections (unlike more familiar STDs such as gonorrhea, syphilis, or AIDS). Furthermore, because it has been difficult to grow *C. trachomatis* in the laboratory, the disease often went undetected as a cause of nonspecific medical symptoms. In addition, because chlamydial infections are not reportable diseases nationally, physicians may have been lulled into thinking that they are less important or less transmissible STDs than others. Unfortunately, however, the health problems caused by chlamydial infections are serious, so being properly informed about this group of STDs is important.

The number of new cases of chlamydial infections has been growing rapidly over the last decade and has now reached an alarming incidence. Data from several different surveys show the extent of this problem. Among several thousand sexually active women attending family planning clinics, 9 percent were found to have chlamydial infections. This figure is similar to the prevalence rate of 10 percent found in men in military settings. Some college health services are reporting rates of chlamydial infection as high as 17 percent in undergraduate women. At STD clinics, an average of 20 to 40 percent of all people tested are found to have chlamydial infections.

Many diseases and medical complications are caused by *C. trachomatis*. In men, about half of all cases of nongonococcal urethritis (infection of the urethra not due to gonorrhea) is caused by this organism. This conditon alone — *chlamydial urethritis* — has an incidence about 2.5 times that of urethritis caused by gonorrhea. In addition, *Chlamydia trachomatis* causes about half of the 500,000 annual cases of *acute epididymitis* (infection of the epididymis) in the United States. Both of these conditions are more common in younger males: chlamydial urethritis is most prevalent in males aged fifteen to twenty-four, while chlamydial epididymitis is predominantly found in males under thirty-five years of age. Chlamydial urethritis is only one-third as common in homosexual males as in heterosexuals, but 6 to 8 percent of homosexual males have evidence of chlamydial infections of the rectum. Based on current knowledge, there do not seem to be major long-term consequences to males from chlamydial infections, even if they recur or are chronic. However, relatively little research has been done on this topic, so it should not be presumed that it is safe for men to ignore such an infection. Furthermore, men who are not properly treated will almost inevitably transmit their infections to their sex partners, with potentially dire consequences for females.

In females, *C. trachomatis* causes a number of different problems at various levels of the reproductive tract. The female version of chlamydial urethritis is a condition called the *urethral syndrome*. In addition, chlamydia causes *cervicitis* (infection of the cervix) and accounts for an estimated 250,000 to 500,000 cases of *PID* (pelvic inflammatory disease) annually. If you recall that many cases of PID eventually progress to the point where they cause scarring of the Fallopian tubes, you will see why this complication is a major cause of female infertility as well as a major factor in some ectopic pregnancies. In many cases chlamydial infection also involves the endometrium (inner lining) of the uterus. In one recent study, it was found that 41 percent of women infected with *C. trachomatis* had chlamydial endometritis.

Another major problem in females is that maternal chlamydial infections during pregnancy are commonly passed to the newborn during childbirth, presumably from the infant coming in contact with infected secretions in the birth canal. Up to 50 percent of infants born to infected mothers develop conjunctivitis (a type of eye infection), and 3 to 18 percent develop chlamydial pneumonia (a lung infection) before they are four months old. While chlamydial pneumonia is not usually serious, chlamydial conjunctivitis can sometimes cause chronic eye disease. (This is not the same type of eye infection as trachoma, another form of chlamydial disease and a major cause of blindness in developing countries. Trachoma is spread primarily by flies, not by sexual transmission.)

A number of studies also link chlamydial infections with various complications of pregnancy, such as premature rupture of the fetal membranes, premature delivery, and postpartum endometritis. However, the data are not conclusive.

A different strain of *C. trachomatis* also causes an STD called *lymphogranuloma venereum (LGV),* which is rare in North America and Europe but common in South America, Africa, and Asia. LGV is about four times as common in males as in females. Like syphilis, this disease has three stages. The primary lesion is a small, hardly noticeable ulcer or pimple on the genitals that appears after an incubation period of 3 to 12 days and then heals rapidly. While this lesion is usually painless, if it occurs in the urethra it may cause pain or burning.

The secondary stage of LGV, which occurs several months after the primary lesion, is marked by painful swelling of lymph nodes in the groin (usually on one side of the body), with accompanying fever, chills, and generalized aching. While almost all males with LGV have such a reaction in the secondary stage, only 20 to 30 percent of females do. Another one-third of women have lower abdominal and back pain. In addition, both men and women may experience symptoms of rectal infection that include a mucuslike discharge, rectal bleeding, and abscess formation. While the great majority of people recover from LGV after the secondary stage, even if untreated, late complications of this STD are serious. They include genital elephantiasis (large swellings of the genitals) and severe scarring of the rectum, sometimes leading to partial bowel obstruction.

Chlamydial infections (except for trachoma) are most commonly sexually transmitted. As an STD, chlamydial infections follow a pattern that is quite similar to that of gonorrhea, although with a lower degree of transmissibility. In one study, male sexual partners of women who had either gonorrhea or chlamydial cervicitis were found to have chlamydial infections less often than gonorrhea: 28 percent versus 81 percent. In the same study, male partners of women who had dual infections of gonorrhea and chlamydia also got chlamydial infections much less often than they got gonorrhea (28 percent versus 77 percent).

Chlamydial infections can be transmitted by vaginal or anal intercouse and (less often) by oral-genital contact. Chlamydial infections, not surprisingly, are also most common in persons with multiple sex partners. As with gonorrhea, women seem more susceptible to chlamydial infections. About 70 percent of the female sex partners of men with documented chlamydial infections are found to be infected with chlamydia, too, while 25 to 50 percent of the male sex partners of women with chlamydia have been found to be infected.

One of the difficult problems with chlamydial infections is

that, much like gonorrhea, they often do not produce symptoms in infected females. According to some estimates, half of all women with chlamydial infections are symptom-free. A smaller number of men — estimated to be about 15 to 30 percent — do not have symptoms, either, but since the long-term consequences for males are not as serious as for females, this is a far less serious medical problem.

In males, *C. trachomatis* most commonly causes symptoms of urethritis after one to three weeks' incubation. The two primary symptoms of urethritis are burning with urination and a whitish or clear urethral discharge. While these two symptoms typically occur together, either can occur without the other. In most cases, these symptoms are milder than in gonorrhea.

Epididymitis in men under age 35 is presumed to be due to chlamydia unless proved otherwise and is marked by swelling, pain and tenderness in the scrotum (usually on one side only), and, often, fever. There may also be an associated urethritis. The pain of epididymitis can be so severe as to interfere with walking or running, but it also can be low-grade, feeling more like an aching sensation than a searing pain. Epididymitis usually makes sexual activity quite unpleasant for the affected male, partly because during sexual arousal the scrotum pulls up toward the body. Along with the vasocongestion in the testes, this creates pressure on the infected epididymis that is usually painful.

In females, chlamydial cervicitis does not generally produce symptoms. A cloudy mucous discharge from the cervix is sometimes seen but one study found this in only 37 percent of women who had *Chlamydia trachomatis* isolated from the cervix. There may be itchiness or mild discomfort in the genitals, but in most instances there are no symptoms, so the female has no way of knowing that she has an active chlamydial infection that requires treatment. The lack of symptoms is a problem because of the likelihood that the infection will move higher in the female reproductive tract, where it is apt to do more damage.

Females with chlamydial urethritis are in a similar situation. Only about one-third of these women develop pain, burning with urination, or frequent urination, and only a small percentage develop a urethral discharge or soreness at the opening of the urethra. Thus, few seek treatment. Although women with chlamydial endometritis may have intermittent vaginal bleeding, this occurs only in a minority; this type of chlamydial infection, too, is typically without any noticeable symptoms.

In contrast, chlamydial PID is more likely to be accompanied by symptoms, including the sudden onset of lower abdominal pain, vaginal bleeding, fever, and dyspareunia (painful intercourse). Another key symptom is that pressure on the lower abdomen during a pelvic exam is painful. However, as many as half of all cases of chlamydial PID may not have any symptoms, which is especially true for those cases that represent low-grade, smoldering infections that persist over time.

Because *Chlamydia trachomatis* grows only intracellularly (inside living cells), it is technically difficult and expensive to culture. In addition, because it usually takes at least three days to get the result of a culture for chlamydial infections, treatment is often started on a presumptive basis. Nevertheless, a culture is the most accurate means of establishing the diagnosis.

To avoid these difficulties, several new types of antibody testing have recently become available. These tests, which can detect chlamydial infections in genital secretions with a fairly high degree of accuracy, can generally be done within a few hours and are less expensive than cultures.

A number of antibiotics are very effective in treating chlamydial infections. Tetracycline or doxycycline is currently the drug of choice except in pregnant women. Erythromycin and sulfamethoxazole are alternative possibilities that are also highly effective. (Erythromycin is the drug of choice for pregnant women.)

It should be noted that penicillin is *not* effective against chlamydial infections. Because it is common to get chlamydial and gonorrheal infections at the same time, people are often treated for gonorrhea without realizing that they have a coexistent chlamydial infection — in part because gonorrhea has a shorter incubation time and in part because gonorrhea is easier to detect in the laboratory. This creates a situation in which the gonorrhea is cured by the penicillin, but the chlamydial infection is untouched. In some instances, this leads to a new flare-up of symptoms (chlamydial urethritis is common after gonorrhea in men), but if no symptoms occur or if they are ignored and go away spontaneously, the person remains infected and contagious to others and runs the risk of additional health complications.

It is particularly important that people who are diagnosed as having a chlamydial infection (or a suspected chlamydial infection) notify any partner with whom they had sex in the thirty days prior to the appearance of symptoms so that these individuals can be evaluated and treated.

PUBIC LICE

Pubic lice, or "crabs," are parasites that invade the pubic region. Although crabs are usually transmitted by sexual contact, they may also be inadvertently picked up from sheets, towels, or clothing used by an infested person. The lice attach themselves to pubic hair and require fresh blood at least twice a day to survive. Eggs laid by female lice are cemented onto the pubic hairs and cannot be washed off.

The crab louse (known officially as *Phthirus pubis* and called "papillon d'amour," or "butterfly of love," by the French) causes intense itching, which is mainly felt at night. A few people have no real symptoms; others develop an allergic rash that can be infected by bacteria after a lot of scratching. The lice can be killed by gamma benzene hexachloride, marketed in cream, lotion, or shampoo form under the trade name Kwell. Although pubic lice can survive for only twenty-four hours once they leave the human body, eggs that fall off into sheets or onto clothing can survive for six days. For this reason, fresh bedding and clean clothing should always be used to avoid reinfestation.

NONSPECIFIC URETHRITIS

Nonspecific urethritis (NSU) is any inflammation of the male urethra that is not caused by gonorrhea. Among male college students, 80 to 90 percent of cases of urethritis are due to causes other than gonorrhea. The incidence of NSU has been increasing at a faster rate than gonorrhea in the last ten years and may now be twice as common, but the exact cause cannot always be found. The organisms most frequently isolated are *Chlamydia trachomatis* and *T. mycoplasma;* however, the inflammation can also be of chemical or allergic origin.

Chlamydia trachomatis is found in about 30 to 50 percent of women who are sex partners of men with chlamydia-positive NSU, and it is also frequently found in women with the so-called acute urethral syndrome marked by painful, frequent urination. In addition, 30 to 60 percent of women with gonorrhea also have chlamydia infections.

The symptoms of NSU are similar to gonorrhea but are usually milder. The urethral discharge is generally thin and clear. Treatment of NSU with tetracycline will usually clear it up, but other antibiotics may be required. Penicillin is usually *not* effective in treating NSU.

VIRAL HEPATITIS

Viral hepatitis is an infection of the liver that can vary in severity from a completely symptomless state to mild gastrointestinal symptoms (poor appetite, indigestion, diarrhea) to an acute debilitating illness with fever, jaundice (yellowish appearance of the skin), vomiting, abdominal pain, and — occasionally — more serious medical complications. There are three main types of viral hepatitis: hepatitis A, hepatitis B, and non-A, non-B hepatitis.

Hepatitis A (formerly known as infectious hepatitis) has an incubation period of fifteen to forty-five days. The hepatitis A virus is spread mainly by the fecal-oral route, with person-to-person transmission, food-borne epidemics (usually caused by infected food handlers), and the consumption of raw or poorly cooked shellfish from contaminated waters accounting for most cases in industrialized nations. Recent data show that homosexual males have a higher incidence of hepatitis A than heterosexuals. It appears that oral-anal contact is the primary explanation for this finding, so that at least within the homosexual population hepatitis A may sometimes be sexually transmitted. (Oral-anal contact by heterosexual partners also can transmit this infection, but this is relatively uncommon.)

Hepatitis B, previously called serum hepatitis, is usually

spread by blood or blood products but it can also be transmitted by saliva, seminal fluid, vaginal secretions, and other biologic fluids. Many of the approximately 200,000 annual cases of hepatitis B in the United States are sexually transmitted, and recent evidence suggests that homosexual men have the highest rates of previous or current infection with this disorder. It is thought that trauma to the rectal mucosa from anal intercourse, manual stimulation of the rectum, or frequent use of enemas may predispose to the spread of this infection. Whether transmission occurs only when infected saliva is introduced into the rectum (either from oral-anal stimulation or from the use of saliva as a lubricant for anal intercourse) or whether seminal fluid can transmit the infection rectally is not clear at present. Although male homosexuals may have the highest past and current prevalence of this infection, many more cases of hepatitis B are sexually transmitted between heterosexual partners because heterosexuals outnumber homosexuals by approximately ten to one.

Studies of the prevalence of hepatitis B infection among various groups show that overall prevalence in homosexual males is 40 to 60 percent, whereas in heterosexual males it ranges from 4 to 18 percent. In the same study, it was found among heterosexual university students that students with three or more sexual partners in the four months before the study were much more likely to have been infected with hepatitis B virus than students with fewer than three partners in this same time period (14 percent versus 1.5 percent respectively). This result might mean that persons whose life-style exposes them to numerous sex partners are actually exposed to a pool of people with much higher rates of hepatitis B infection than are the group of people with few sex partners. If this view is correct, it means that a person having many sex partners has a risk of infection that is not just numerically proportional to the number of partners but is actually considerably higher since each of the partners is "riskier" from a health viewpoint — that is, more likely to be infected with the hepatitis B virus and thus more likely to be a carrier.

Hepatitis B persists in a carrier state in approximately 5 to 10 percent of infected adults, with an estimated 400,000 to 800,000 carriers in the United States and 150 to 200 million worldwide. This condition, in which the person is still infectious although not generally ill, can persist for months, years, or a lifetime. Carriers of hepatitis B have an increased risk of developing liver cancer in their lifetime.

Non-A, non-B hepatitis, which is the most common form after blood transfusions, is not thought to be sexually transmitted.

Hepatitis is diagnosed on the basis of laboratory tests (for instance, blood tests show abnormalities in liver enzymes), and the specific type of hepatitis is determined by immunologic testing of blood samples. Treatment is generally symptomatic, with hospitalization required only in the more severe cases.

Persons in intimate contact with someone with hepatitis A can obtain partial immunity by getting a shot of immune serum globulin, but as a practical matter they have often been exposed for weeks before the diagnosis has been made. Viral hepatitis B vaccines are now available that are safe and effective.

OTHER SEXUALLY TRANSMITTED DISEASES

Genital warts (condylomata acuminata) are dry, usually painless warts that grow on or near the genitals and around the anus. They are caused by a sexually transmitted virus called the *human papilloma virus* (HPV) and are usually pink or grayish-white with a cauliflower-like surface. Genital warts were previously thought to be more of a nuisance than a health problem, but recent evidence suggests that HPV may be associated on a long-term basis with the development of cervical cancer in women and other genital cancers in both sexes. Since genital warts commonly coexist with other STDs, anyone with such growths should undergo thorough medical evaluation to check for the possibility of an asymptomatic infection with such organisms as *N. gonorrhea* or *C. trachomatis*. Genital warts are most effectively treated by removal with a carbon dioxide laser, which is a simple, painless office technique. The use of liquid nitrogen is also effective, but the previously used treatment involving the application of a liquid or ointment containing podophyllin is now recognized as less than fully effective in preventing the recurrence of such lesions.

Molluscum contagiosum is caused by a pox virus that typically produces raised lesions on the external genitals or on the thighs, buttocks, or lower abdomen. The painless lesions, which usually appear three to six weeks after exposure, vary from one millimeter to one centimeter in diameter and have a pinkish-orange color with a pearly top. If the lesion is squeezed, the cheesy plug of material within it can be expressed, much like a blackhead. Since the infection usually causes little trouble and often disappears spontaneously in about six months, treatment is not always necessary. Local applications of liquid nitrogen or frozen carbon dioxide have been used to remove the skin lesions.

Chancroid and *granuloma inguinale* are other STDs that are rare in the United States but more common in tropical climates. If you have traveled to these areas or had sexual contact with a partner who has, you may be at some risk from these infections.

PSYCHOSOCIAL ASPECTS OF STDs

There is little question that in the 1980s our society has been overwhelmed by new developments regarding STDs. First there was a tremendous amount of attention given to the still-raging genital herpes epidemic. Although the media focus on this disease was initially informative and useful, it gradually took on overtones of emotional hysteria in some quarters. This reaction may have peaked in mid-1982, but by early 1983, the public's intense fears about genital herpes were put dramatically in perspective by news coverage about AIDS. While one is a disease that in most instances simply causes painful skin blisters for a few days, the other is a disease that kills.

Increased concern over AIDS has removed the spotlight from the millions of adults with genital herpes, but it has also led to a climate of fear surrounding AIDS and its contagiousness. This fear has mushroomed as it has become increasingly clear that AIDS can be heterosexually transmitted. That AIDS can and does strike people who are not homosexuals or IV drug users has clearly had an impact on sexual behavior patterns in a variety of ways.

The fear that governs many people's reactions to the AIDS epidemic is in some ways similar to the fear experienced by people when they discover that they have symptoms that *might* indicate the presence of an STD. Fear, along with a disbelief that they might have an STD, leads many people to delay a visit to a physician. They deny the reality of the situation as if pretending an infection isn't there will make it go away. Unfortunately, since most symptoms of STDs disappear within a matter of a few weeks, many individuals fool themselves into thinking "it was nothing after all." By gaining a temporary sense of psychological relief, people like this are doubly foolish: they continue to harbor the infection in their bodies and expose their sex partners to the risk of infection, as well. In the case of sexually active people who have been infected with the AIDS virus, this denial and delay are deplorable because they involve a risk to someone else's life.

Some people are reluctant to go to a physician when they have symptoms of a possible STD because they're worried about getting a lecture or they're worried about the confidentiality that will be extended to them. Although physicians are legally bound to keep information they learn about their patients confidential, there are circumstances in which physicians are obligated to act differently. For example, in some states a physician's records can be subpoenaed by a court order. In all localities, a physician is supposed to report instances of certain communicable diseases to state public health departments (although as of mid-1987, no state had as yet required the mandatory reporting of AIDS cases). Physicians' records are often accessible to insurance companies, since patients consent to having their records examined whenever they file a health insurance claim or when they apply for a new health or life policy. For this reason, anyone who is particularly concerned about this issue needs to discuss with his or her physician the limits to absolute confidentiality that might apply to his or her case. Since it is conceivable that new legislation in some states may require physicians to report all cases of people who test positive for HIV antibodies, those who are particularly concerned about this issue should be tested only at centers that offer testing anonymously (which is usually done using a code number, rather than a name, to identify the people who are tested).

Just a few years ago, people were dismayed if they received a lecture from their physician that advocated traditional sexual values and pointed out the health risks of casual sex with multiple partners. Now it is not so clear that physicians who give advice about safer sexual practices are old-fashioned or trying to impose their moral values on their patients; with what our society has learned about STD epidemics in this decade, such physicians may simply be preaching about sensible public health practices. Nevertheless, if you encounter a physician who makes you uncomfortable by giving you unwanted moralistic-sounding advice, don't be intimidated; it will generally be in your best interests to switch to another doctor with whom you feel more at ease.

What is the impact of an STD on a person's sexuality? In most cases other than AIDS, assuming the infection is properly detected and treated, there will be little if any physical effect. During the acute phase of many STDs some people have little interest in sex because it's painful, while others continue functioning sexually without noticing that anything is wrong. Even when an STD has been untreated and has become chronic, it is not likely to depress libido or interfere with sexual functioning (genital herpes is the major exception, since sex can be quite painful during flare-ups). But even though STDs don't typically interfere with the *physical* component of sexual functioning, some people find themselves having sexual difficulties due to the psychological effects of finding out they have an STD. Often, these individuals are guilty and embarrassed about what has happened. They sometimes decide that their infection was God's way of warning or punishing them for sexual transgressions. Since they equate sex with sin, it is no wonder that such people occasionally develop subsequent sexual inhibitions.

Others develop obsessive concerns about sexual cleanliness and worry constantly about the possibility of being reinfected or catching another form of STD. Some men who fall in this category find themselves having erectile problems, while women who are exceptionally fearful about STDs may develop vaginismus as an unconscious way of protecting themselves from infection. Needless to say, such major concerns about sexual cleanliness may lead people to change their patterns of sexual behavior, too — for instance, not participating in oral-genital sex.

When one partner in an intimate relationship develops an STD that the other person doesn't have, it immediately implies that the infected person has been sexually active outside the relationship. While this isn't always true, and some types of STD (such as hepatitis B) are also commonly transmitted by nonsexual means, doubt and suspicion may affect even a stable, loving relationship. If one partner infects the other with an STD, there may be even more conflict and hostility. This hostility may be expressed in sexual terms (that is, by rejecting sexual activity with the other person) and may spill over into the entire relationship. This type of reaction can infect a relationship far more destructively than an STD and, fortunately, most people soon return to a better state of relating to their partners.

Except for AIDS, STDs are no longer as frightening as they once were, but for many people they remain somehow "different" from other infections because they are transmitted by sexual contact and generally affect the sex organs. Until people are able to think about sex as naturally as they think about breathing or eating, it is likely that this form of stigmatization will continue as a fact of life.

PREVENTING SEXUALLY TRANSMITTED DISEASES

The People's Republic of China apparently has managed to practically wipe out syphilis and many other STDs by enforcing rigid codes of sexual behavior and stopping prostitution, but most other countries must contend with the STDs as a price for social and sexual freedom. Even so, some practical guidelines can be offered to help minimize the chances of contracting an STD or spreading it once you've caught it.

1. *Be informed.* Knowing about the symptoms of STDs can help protect you against exposing yourself to the risk of infection from a partner and help you know when to seek treatment.
2. *Be observant.* Knowledge alone is not enough. Looking is the

best way of discovering if you or your partner has a genital discharge, sore, rash, or other sign of sexual infection. (This can't be done with the lights out or in a moonlit back seat of a car). If you see a suspicious sore or blister, don't be a hero about it — refrain from sexual contact and insist that your partner be examined. While "looking" may seem clinical and crass, you don't have to announce *why* you're looking, and often a good, close look can be obtained in the sexual preliminaries (getting undressed, giving your partner a massage). A step beyond looking is commonly used by prostitutes (and doctors) to check for gonorrhea and NSU in men. The penis is gently but firmly "milked" from its base to its head to see if any discharge is present; the "exam" is sometimes called a short-arm inspection.

3. *Be selective.* Having numerous sex partners greatly increases the risk of developing an STD. Likewise, anonymous sex is risky, too: you don't know if you can trust your partner or who he or she has been with in the recent past. Being selective in your choice of sex partners will improve your chances of avoiding STDs.

4. *Be honest.* If you have (or think you *may* have) an STD, tell your partner (or partners). This can avoid spreading the infection and will alert your partner to watch for her or his own symptoms, or to be examined or tested. Similarly, if you are worried about your partner's status, don't hesitate to ask. It's foolish to jeopardize your health to protect someone else's feelings.

5. *Be cautious.* Use of a condom will significantly lower the chances of getting or spreading STDs. Using an intravaginal chemical contraceptive (foams, jellies, creams) reduces the woman's risk of getting gonorrhea. Urinating soon after sexual activity helps flush invading organisms out of the urethra and so limits risks to a small degree. *If you think you've been exposed, promptly contact your doctor for advice.* If you *know* that you've been exposed, also abstain from sexual activity until your tests show that everything is okay.

6. *Be promptly tested and treated.* A quick diagnosis and an effective course of treatment will help you prevent some of the serious complications of STDs. Treatment can be obtained

from your private doctor, hospital clinics, or public health service clinics. After treatment, you must be rechecked to be certain that the disease has been eradicated. In addition, be sure to urge your partner to be tested (and treated if necessary) so as to avoid reinfection.

EPILOGUE

The Future of Sexuality

IT IS QUITE EDUCATIONAL in certain respects to look
back on the changes that have occurred in sexual attitudes
and behavior in American society over the past quarter-century.
While the postwar 1940s and 1950s were in the main steeped in
sexual repression and decorum as a lingering legacy of the Victorian era, the sixties and seventies became, for many, a time for
sexual experimentation and discovery. With the advent of birth
control pills as a powerful catalyst to this transformation —
aided, to be sure, by a multitude of social factors that contributed to such a broad-based change — sex came out of the closet.
Today there can be no doubt that changes in female sexual
behavior were statistically the most pronounced, as a definite
trend emerged wherein females were becoming sexually experienced at far younger ages than their mothers and grandmoth-
ers had been. In addition, women of all ages were also
considerably more willing to experiment with various types of
sexual self-pleasuring, erotic fantasies, and non-monogamous
sexual relationships. Concomitantly, many women became interested in learning about their bodies, developing awareness of
the workings of their sexual and reproductive apparatus and
also adopting sex-positive attitudes that allowed them to become
more sexually assertive. For females and males alike, sex took
on a new aura of spontaneity and freedom from worry over
reproductive consequences, although new sexual worries —
most notably, performance anxieties and a sense of pressure to
be "with it" sexually — arose to cloud the horizon. Thus, some
women worried about not being multiorgasmic (or not being

orgasmic at all), while more males than ever began to question the adequacy of their sexual prowess.

In this climate, it is perhaps no surprise that along with the zero population growth movement and the millions of adults who proudly chose permanent sterilization as a contraceptive method, we saw the rise and fall of Sandstone and Plato's Retreat. Neither is it surprising that divorce rates soared, premarital cohabition became widely accepted, media portrayals of sexual themes became candid, if not downright clinical, and transsexual surgery was matter-of-factly accepted as a psychologically beneficial — and humane — act.

By the end of the seventies, however, a number of forces were set in motion that we believe have begun a definite shift of our sexual attitudes and behavior away from the all-out hedonistic path of the "me" decade. These forces include, in no particular order, a growing dissatisfaction with one-night stands and mechanical, impersonal sex as a substitute for intimacy or commitment; the profound, as yet uncharted, impact that biotechnological advances of relatively recent vintage will have on reproductive choices and behavior; increasing public awareness of the risks posed by currently raging epidemics of sexually transmitted diseases; increasing progress in the move toward egalitarianism between the sexes; and the new social conscience our society seems to be slowly developing in relation to sex victimology. To be certain, these forces will not operate independently, without regard for or interaction with important vectors of change from the past two decades, but we are convinced that their long-range repercussions have not been widely considered from a sexual viewpoint. Our intent in this chapter is to outline the ways in which we see contemporary sexuality being transformed by these and other factors over the next twenty-five years. While we claim neither omniscience nor a particularly powerful crystal ball for future-gazing, we have some confidence that our outlook is on target, as we have been generally accurate in similar assessments over the past quarter-century.

THE SEXUAL IMPACT
OF BIOTECHNOLOGY

Advances in reproductive biotechnology have suddenly thrust us into a world previously seen only in science fiction novels and other futuristic writings. With the methods of in vitro fertilization now widely available, at least to those with sufficient funds, the conception and birth of large numbers of so-called test-tube babies is already fact, having moved beyond the experimental stage in a matter of just a few years after the 1978 announcement by Patrick Steptoe and Robert Edwards of the birth of Louise Brown. Despite lingering uncertainties about the ethical implications of certain aspects of in vitro fertilization, frozen embryo transfer, and surrogate mothers — women who, for a fee, agree to be artificially inseminated with the sperm of the husband of an infertile woman and carry the pregnancy to term, giving the baby up to adoption by the couple who has hired her — it is almost a certainty that such practices will become commonplace.

What are the implications of this statement? For one thing, there is a strong possibility that a substantial number of women will opt to have children without experiencing a pregnancy. By using in vitro fertilization — that is, having a woman's ovum fertilized outside her body by her husband's (or some other man's) sperm — followed by implanting the developing embryo at an early stage in the uterus of a mother-for-hire, the woman is assured of having a child who is genetically "hers" but is sheltered and sustained prior to birth by someone else. In fact, it is quite possible that within the next decade the surrogate mother can be circumvented entirely by the development of an artificial uterus that maintains the developing embryo-fetus in an incubator-like environment, hooked up to an artificial placenta that functions not unlike a kidney dialysis machine. While some may be horrified at first by this notion (they'll miss the emotional experience of being pregnant and the joy of delivering their own child, critics will undoubtedly object), several surveys conducted to date have found that as many as 40 percent of women in their twenties would choose to have children only in this manner.

And, not too surprisingly, one out of three women surveyed indicated that they would choose this method after first having a pregnancy of their own. Good-bye, morning sickness. Who needs varicose veins? Why risk the medical complications of pregnancy if I can hire someone to take these risks for me or if I can avoid them by using a computer-controlled artificial uterus?

Another implication that is not as farfetched as it might at first sound is that the availability and acceptance of the new reproductive biotechnology, as we envision it, will eventually lead to a reduction in the marriage rate in our society. We are already seeing evidence of never-married women using artificial insemination to become mothers. This is true both among women in their mid-thirties who have given up on being able to find a mate, although they may well have found one or more men whom they consider desirable biological fathers, and among lesbians who have no interest in a heterosexual relationship of any sort. In fact, in at least one instance a lesbian successfully obtained a court order forcing a medical school's artificial insemination program to accept her as a client although she was unmarried. The flip side of this coin is that we have also already encountered two instances wherein homosexual men hired surrogate mothers so they could have their own children to raise. We expect this too will become far more widespread as surrogate mothers come to be regarded in a fashion not unlike that extended to women who serve as wet nurses in other cultures. Here, too, court precedents in several jurisdictions establishing the right of homosexual men to be adoptive fathers have already been handed down and may presage changing social attitudes in this area.

Reproductive practices will also be altered, we suspect, by the future availability of methods for preselecting a baby's sex. While reliable sex preselection methods have not yet been adequately proven, we believe that within the next decade such techniques will not only become available, but will also be widely used. Although there has already been a great deal of conjecture that this will lead to a disproportionate number of male births, we believe that the change in sex ratio of births will principally involve firstborn children, where the number of males preselected by parents may raise the proportion of males to a figure as high as 75 percent. If this occurs, the proportion of second-born females will probably rise to the 60 to 65 percent level as many American couples strive for an "ideal" family of one boy and one girl. The socioeconomic and psychological consequences of such a change in patterns of childbirth are difficult to predict at present, but many experts believe that it would inevitably lead a generation of females into second-class status both in terms of occupational achievement and self-esteem.

Biotechnological advances over the next twenty-five years will also result in the development of nearly perfect reversible contraceptive methods for both sexes. Although, as we have previously noted, research on male contraception has lagged far behind research on female contraception for reasons that probably reflect social and political factors more than scientific concerns alone, economic incentives presently in place are likely to ensure that much progress will be made in this quest. As a result of these new contraceptive methods, we envision seeing two notable effects: a sizable reduction in the number of unplanned teenage pregnancies and an increasing emphasis on sex in a relational or recreational context, as opposed to a procreative act. The development of a new generation of highly effective contraceptive methods will also result in a substantial reduction in the number of abortions performed in America each year although this reduction will be partly offset by a marked increase in the number of first-trimester abortions performed as a result of improved prenatal diagnostic techniques that will allow the early identification of a surprisingly wide number of potential birth defects and genetically transmitted diseases. On balance, we expect that the abortion rate, expressed in terms of numbers of women of reproductive age, will be some 50 percent lower in a quarter-century than it is today.

FUTURE PATTERNS
OF SEXUAL BEHAVIOR

Over the next twenty-five years there will be a gradual but definite increase in recognizing the importance of sex education for children. This will eventually lead to the routine inclusion of sex education in public schools from the earliest grade levels on, with this material generally integrated into everyday curricula rather than taught as a separate, distinct course. As a result, we expect that children will become better prepared to cope with situations of potential sexual abuse and that teenagers will become considerably more likely to use contraception when they are sexually active. It is also possible, but somewhat less likely, that such a widespread prevalence of childhood and adolescent sex education will in time contribute to a reduced rate of adult sexual problems, particularly those that stem from misinformation and inhibitions about sex.

At the same time, we do not anticipate much further acceleration from present levels of sexual activity for adolescents. Twenty-five years from now, we expect that approximately half of all teenagers will no longer be virgins at age sixteen, and at age eighteen, three-quarters of all teenagers will be coitally experienced. Thus, we do not believe that early sex education will result in a great deal of sexual experimentation that otherwise would not have occurred. We base these projections on our view that even in late adolescence, there is a substantial group of teens who will be sexually abstinent for reasons such as religious beliefs, personal immaturity, the inability to attract a partner, or simple personal preference. In addition, some teens will be abstinent because of their own sexual confusion or fear of sexually transmitted disease.

We also expect to see a continuation and accentuation of the present trend of young adults to postpone marriage until later ages, so that by 2010 the average age of first marriage in America will be closer to twenty-five than to twenty-one. At the same time, there will be more young adult cohabitors than there are today, including many who choose this relationship style as an acceptable substitute for marriage. Accompanying this change, there will also be a modest but discernible drop in the rate of marriage. These projections are based in part on anticipating that the divorce rate over the next two decades will continue to climb from today's levels — although at a slower pace than during the last twenty years. The relevance of this observation is that many children of divorced parents (when the divorce occurred during childhood or adolescence) indicate antipathy toward attempting marriage themselves and — at least according to some preliminary demographic data we have gathered — are less likely to marry during their twenties than age-matched, socioeconomically matched children of intact marriages.

What implications do these projections hold for future trends in adult sexual behavior? In our view, it is unlikely that there will be a return to the frenzied sexual gropings and multiple partners of the seventies in the near future. While we do not anticipate a total return to monogamy, we do think that increasing value will be placed on long-term sexual relationships, with less emphasis on sex for the sake of sex. This pattern will be true in both homosexual and heterosexual populations, although homosexual males as a group will continue to have a lifetime number of sexual partners that is several times higher than that for either lesbians or heterosexuals. One factor in reaching this conclusion is our belief that public concern will continue to mount over the consequences of sexually transmitted diseases. Unfortunately, the current situation with AIDS is far from over, and even if a means of preventing the transmission of AIDS were to be found today, it is likely that at least half a million cases are already incubating, with perhaps one-quarter of these cases involving heterosexuals. In addition to the AIDS epidemic, which has already led to considerable change in patterns of sexual behavior among gay males across the nation, there are also alarming rates of other STDs such as chlamydia and genital herpes that presently are of considerable concern from a public health viewpoint. Since it is quite plausible that even the "older" STDs, such as gonorrhea and syphilis, could break out of their current containment patterns as a result of developing resistance to current antibiotics used successfully in their treatment, and since it is also possible that other "new," as yet undiscovered, STDs may emerge from the shadows, we believe that public awareness of STDs will increase over the foreseeable future and will be a factor in steering many people toward a relative degree of sexual selectivity. On the positive side, however, is the likelihood that effective vaccines will be developed to prevent a number of the STDs with which we now contend.

To be sure, people won't simply be frightened into restricting their sexual activity largely to long-term relationships. There are other, more compelling, reasons why this pattern will become widespread. Chief among them is the fundamental fact that more people find greater personal comfort and satisfaction with sex in the context of an intimate, caring relationship than find these qualities in the relatively impersonal context of sex without commitment or caring. To the extent that sex is a shared experience of emotions and meanings that transcends the purely physical aspects of two bodies coupling, the most intense erotic gratifications — the pleasure bond between lovers — will prove far more rewarding as a total experience than passion devoid of its interpersonal dimension. The synergy of sex born of intimacy and caring, sex in which the physical action and the private, inner scenario of the psyche are merged with a partner's feelings and desires, is not easily matched by an earth-shattering orgasm disconnected from the fabric of our being.

Before we are criticized for putting forth a somewhat utopian view, let us hasten to add that we do not envision a move toward total relationship fidelity. Extramarital (or extra-relationship) sex will continue to occur much as it does today, although a greater attitude of tolerance toward such behavior will probably develop, particularly in the middle and upper classes. Prostitution in its various forms will also continue, although we ascribe only a slim chance to the possibility that prostitution will become widely legalized (and regulated) in this country. At the same time, there will probably be a small segment of the adult male population, both hetero- and homosexual, that maintains a pattern of numerous sexual contacts with relatively anonymous multiple partners.

What other changes can be anticipated for the next twenty-five years? For one thing, there will probably be a marked change in attitudes toward sexuality and aging as an increasing percentage of the population reaches its geriatric years, due to both current demographic patterns and advances in health care that result in further longevity than we enjoy today. We envision that sexual activity among older adults will become more acceptable and that this changing attitude will lead to some liberalization of sexual restrictions on even the institutionalized geriatric population — those in nursing homes, for example. Another change that we expect to see is an increasing tolerance of homosexuality. While this trend will be forestalled for the next few years while the AIDS epidemic continues, eventually the combined effects of a mounting body of psychological research on homosexuality that will continue to document that it is not an abnormality and an increasing public and even legislative perception of homosexuality as a civil rights issue will push the pendulum in the direction of increased tolerance. Still another change that we envision is that sex will begin to be accepted more matter-of-factly — in a sense, it will be somewhat demystified and deglamorized — rather than being seen as a category of human activity totally separate and apart from the rest of our lives.

It is likely that the next five years will witness the passage in various locales of legislation that restricts access to certain types of sexually explicit materials. While there is some present consensus among professionals in strong opposition to child pornography today, there are more mixed reactions in singling out violent pornography or pornography that degrades or objectifies women, with substantial concern over eroding First Amendment rights on the basis of as yet unproven presumptions about the behavioral effects of viewing such materials. We are of the opinion that such restrictive legislation will quickly prove unworkable and undesirable so that the longer-term trend toward free access to adult sexually explicit materials will remain intact.

In a society that can become more open and honest about sex, as ours has certainly done over the last twenty-five years, it is also possible to envision a continued lessening of the strictures imposed on sexual attitudes and behaviors by traditional gender-role stereotypes and expectations. A gradual move in this direction has already begun. We expect that even more of a sense of flexibility and freedom will develop as both sexes are able to drop the unwanted legacies they have previously been hobbled with — the sexual double standard, the notion that men must make sexual conquests to "prove" their masculinity, the myth of the sexless female who offers her body to buy love and financial security. Relieved of such burdens of the past, males and females will be more apt to communicate effectively with one another, sexually and otherwise. And in the final analysis, it is effective communication that is really what puts us in touch with one another.

Selected Bibliography

Adams, D. B.; Gold, A. R.; and Burt, A. D. "Rise in Female-Initiated Sexual Activity at Ovulation and Its Suppression by Oral Contraceptives." *New England Journal of Medicine* 299:1145–50, 1978.

Addiego, F., et al. "Female Ejaculation: A Case Study." *Journal of Sex Research* 17(1):13–21, 1981.

Alan Guttmacher Institute. *Safe and Legal: 10 Years' Experience with Legal Abortion in New York State.* New York: Alan Guttmacher Institute, 1980.

———. *Teenage Pregnancy: The Problem That Hasn't Gone Away.* New York: Alan Guttmacher Institute, 1981.

Alington-MacKinnon, D., and Troll, L. E. "The Adaptive Function of the Menopause: A Devil's Advocate Position." *Journal of the American Geriatrics Society* 29:349–353, 1981.

Altman, Dennis. *The Homosexualization of America.* Boston: Beacon Press, 1982.

Alzate, H., and Londono, M. L. "Vaginal Erotic Sensitivity." *Journal of Sex & Marital Therapy* 10:49–56, 1984.

American Journal of Psychiatry. "Historical Notes: A Letter from Freud." 107:786–787, 1951.

American Psychiatric Association. *Diagnostic and Statistical Manual of Mental Disorders,* 3rd ed. (DSM-III). Washington, D.C.: American Psychiatric Association, 1980.

Anderson, T. P., and Cole, T. M. "Sexual Counseling of the Physically Disabled." *Postgraduate Medicine* 58:117–123, 1975.

Annon, J. S. *The Behavoral Treatment of Sexual Problems: Brief Therapy.* New York: Harper & Row, 1976.

Apfelbaum, B. "Why We Should *Not* Accept Sexual Fantasies." In Apfelbaum, B. (ed.), *Expanding the Boundaries of Sex Therapy,* rev. ed., pp. 101–108. Berkeley, Calif.: Berkeley Sex Therapy Group, 1980.

———. (ed.). *Expanding the Boundaries of Sex Therapy,* rev. ed. Berkeley, Calif.: Berkeley Sex Therapy Group, 1980.

———. *Expanding the Boundaries of Sex Therapy,* 2nd ed. Berkeley, Calif.: Berkeley Sex Therapy Group, 1983.

Arafat, I., and Cotton, W. L. "Masturbation Practices of Males and Females." *Journal of Sex Research* 10:293–307, 1974.

Arentewicz, G., and Schmidt, G. (eds.). *The Treatment of Sexual Disorders.* New York: Basic Books, 1983.

Athanasiou, R. "Pornography: A Review of Research." In Wolman, B. B., and Money, J. (eds.), *Handbook of Human Sexuality*, pp. 251–265. Englewood Cliffs, N.J.: Prentice-Hall, 1980.

Backhouse, C., and Cohen, L. *Sexual Harassment on the Job.* Englewood Cliffs, N. J.: Prentice-Hall, 1981.

Baker, C. D. "Preying on Playgrounds: The Sexploitation of Children in Pornography and Prostitution." In Schultz, L. (ed.), *The Sexual Victimology of Youth.* Springfield, Ill.: Charles C. Thomas, 1980.

Bakwin, H. "Erotic Feelings in Infants and Young Children." *Medical Aspects of Human Sexuality* 8(10):200–215, 1974.

Ballinger, C. B. "The Menopause and Its Syndromes." In Howells, J. G. (ed.), *Modern Perspectives in the Psychiatry of Middle Age*, pp. 279–303. New York: Brunner/Mazel, 1981.

Barbach, L. G. *For Yourself: The Fulfillment of Female Sexuality.* New York: Doubleday, 1975.

———. *Women Discover Orgasm.* New York: Free Press, 1980.

Barclay, A. M. "Sexual Fantasies in Men and Women." *Medical Aspects of Human Sexuality* 7(5):205–216, 1973.

Bart, P. B., and Jozsa, M. "Dirty Books, Dirty Films, Dirty Data." In Lederer, L. (ed.), *Take Back the Night*, pp. 204–217. New York: William Morrow, 1980.

Bart, P. B., and Grossman, M. "Menopause." In Notman, M. T., and Nadelson, C. C. (eds.), *The Woman Patient*, pp. 337–354. New York: Plenum Press, 1978.

Bayer, R. *Homosexuality and American Psychiatry: The Politics of Diagnosis.* New York: Basic Books, 1981.

Beach, F. *Human Sexuality in Four Perspectives.* Baltimore: Johns Hopkins University Press, 1977.

Becker, J. V., et al. "Incidence and Types of Sexual Dysfunctions in Rape and Incest Victims." *Journal of Sex & Marital Therapy* 8:65–74, 1983.

Bell, A. P., and Weinberg, M. S. *Homosexualities.* New York: Simon & Schuster, 1978.

Bell, A. P.; Weinberg, M. S.; and Hammersmith, S. K. *Sexual Preference — Its Development in Men and Women.* Bloomington, Ind.: Indiana University Press, 1981.

Belzer, E. "Orgasmic Expulsions of Women: A Review and Heuristic Inquiry." *Journal of Sex Research* 17:1–12, 1981.

Bem, S. L. "Sex Role Adaptability: One Consequence of Psychological Androgyny." *Journal of Personality and Social Psychology* 31(4):634–643, 1975.

Bene, E. "On the Genesis of Male Homosexuality: An Attempt at Clarifying the Role of the Parents." *British Journal of Psychiatry* 111:803–813, 1965.

Berger, Raymond M. *Gay and Gray: The Older Homosexual Male.* Urbana, Ill.: University of Illinois Press, 1982.

Berlin, F. S., and Meinecke, C. F. "Treatment of Sex Offenders with Antiandrogenic Medication." *American Journal of Psychiatry* 138:601–607, 1981.

Berne, E. *Sex in Human Loving.* New York: Pocket Books, 1971.

Bieber, I., et al. *Homosexuality: A Psychoanalytic Study.* New York: Basic Books, 1962.

Binkin, N. J., and Alexander, E. R. "Neonatal Herpes: How Can It Be Prevented?" *Journal of the American Medical Association* 250:3094–95, 1983.

Blumstein, P. W., and Schwartz, P. "Bisexuality: Some Social Psychological Issues." *Journal of Social Issues* 33(2):30–45, 1977.

———. *American Couples.* New York: William Morrow, 1983.

Bohlen, J. G. " 'Female Ejaculation' and Urinary Stress Incontinence." *Journal of Sex Research* 18:360–363, 1982.

Borneman, Ernest. "Progress in Empirical Research on Childhood Sexuality." Presented at the 6th World Congress of Sexology, Washington D.C., May 24, 1983.

Boston Women's Health Book Collective. *Our Bodies, Ourselves,* 2nd ed. New York: Simon & Schuster, 1976.

Boswell, J. *Christianity, Social Tolerance, and Homosexuality.* Chicago: University of Chicago Press, 1980.

Branden, N. *The Psychology of Romantic Love.* Los Angeles: J. P. Tarcher, 1980.

Brown, G. *The New Celibacy.* New York: McGraw-Hill, 1980.

Brownmiller, S. *Against Our Will.* New York: Simon & Schuster, 1975.

Bullard, D. G., and Knight, S. E. (eds.). *Sexuality and Physical Disability.* St. Louis, Mo.: C. V. Mosby, 1981.

Bullough, V. L. *Sexual Variance in Society and History.* New York: Wiley, 1976.

Bullough, M., and Bullough, B. *Sin, Sickness, and Sanity.* New York: New American Library, 1977.

Burgess, A. W., et al. *Sexual Assault of Children and Adolescents.* Lexington, Mass.: D. C. Heath, 1978.

———. "Response Patterns in Children and Adolescents Exploited Through Sex Rings and Pornography." *American Journal of Psychiatry* 141:656–662, 1984.

Burgess, A. W., and Holmstrom, L. L. "Coping Behavior of the Rape Victim." *American Journal of Psychiatry* 133:413–418, 1976.

Bush, P. *Drugs, Alcohol and Sex.* New York: Richard Marek Publishers, 1980.

Butler, R. M., and Lewis, M. I. *Sex after Sixty.* New York: Harper & Row, 1976.

Cass, V. C. "Homosexual Identity Formation: A Theoretical Model." *Journal of Homosexuality* 4(3):219–235, 1979.

Catania, J. A., and White, C. B. "Sexuality in an Aged Sample: Cognitive Determinants of Masturbation." *Archives of Sexual Behavior* 11:237–245, 1982.

Centers for Disease Control. *Abortion Surveillance 1978.* Atlanta, Ga.: U.S. Department of Health and Human Services/Public Health Service, November 1980.

Cherlin, A. J. *Marriage, Divorce, Remarriage.* Cambridge, Mass.: Harvard University Press, 1981.

Chilman, C. *Adolescent Sexuality in a Changing American Society* (no. NIH 79–1426). Bethesda, Md.: U.S. Department of Health, Education, and Welfare, 1979.

Chipouras, S., et al. *Who Cares? A Handbook of Sex Education and Counseling Services for Disabled People.* Washington, D.C.: George Washington University, 1979.

Chiriboga, D. A. "The Developmental Psychology of Middle Age." In Howells, J. G. (ed.), *Modern Perspectives in the Psychiatry of Middle Age,* pp. 3–25. New York: Brunner/Mazel, 1981.

Cohen, M. L.; Seghorn, T.; and Calmas, W. "Sociometric Study of the Sex Offender." *Journal of Abnormal Psychology* 74:249–255, 1969.

Coleman, E. "Developmental Stages of the Coming Out Process." *Journal of Homosexuality* 7(2/3):31–43, 1981/82.

Comfort, A. *The Joy of Sex.* New York: Crown, 1972.

Constantine, L. L., and Constantine, J. M. *Group Marriage: A Study of Contemporary Multilateral Marriage.* New York: Macmillan, 1973.

Constantine, L. L., and Martinson, F. M. (eds.). *Children and Sex.* Boston: Little, Brown, 1981.

Corey, L., et al. "Genital Herpes Simplex Virus Infections: Clinical Manifestations, Course, and Complications." *Annals of Internal Medicine* 98:958–972, 1983.

Corey, L., and Holmes, K. K. "Genital Herpes Simplex Virus Infections: Current Concepts in Diagnosis, Therapy, and Prevention." *Annals of Internal Medicine* 98:973–983, 1983.

Council on Scientific Affairs. "Estrogen Replacement in the Menopause." *Journal of the American Medical Association* 249:359–361, 1983.

Crepault, C., and Couture, M. "Men's Erotic Fantasies." *Archives of Sexual Behavior* 9:565–581, 1980.

Crepault, C., et al. "Erotic Imagery in Women." In Gemme, R., and Wheeler, C. C. (eds.), *Progress in Sexology,* pp. 267–283. New York: Plenum Press, 1977.

Culp, R. E.; Cook, A. S.; and Housley, P. C. "A Comparison of Observed and Reported Adult-Infant Interactions: Effect of Perceived Sex." *Sex Roles* 9:475–479, 1983.

Cunningham, Susan. "Violent Pornography Said to Spur Aggression." *APA Monitor,* p. 30, March, 1983.

D'Augelli, J. F., and D'Augelli, A. R. "Moral Reasoning and Premarital Sexual Behavior: Toward Reasoning about Relationships." *Journal of Social Issues* 33(2):46–66, 1977.

Davis, K. B. *Factors in the Sex Life of Twenty-Two Hundred Women.* New York: Harper, 1929.

DeLamater, J., and MacCorquodale, P. *Premarital Sexuality: Attitudes, Relationships, Behavior.* Madison, Wis.: University of Wisconsin Press, 1979.

Delaney, J.; Lupton, M. J.; and Toth, E. *The Curse: A Cultural History of Menstruation.* New York: New American Library, 1977.

DeMartino, M. F. (ed.). *Human Autoerotic Practices.* New York: Human Sciences Press, 1979.

Dion, K. K.; Berscheid, E.; and Walster, E. "What Is Beautiful Is Good." *Journal of Personality and Social Psychology* 24:285–290, 1972.

Dion, K. K., and Dion, K. L. "Self-Esteem and Romantic Love." *Journal of Personality* 43:39–57, 1975.

Djerassi, C. *The Politics of Contraception.* New York: Norton, 1979.

Dodson, B. *Liberating Masturbation.* New York: Bodysex Designs, 1974.

Donnerstein, E. "Massive Exposure to Sexual Violence and Desensitization to Violence and Rape." Presented at the 26th Annual Meeting of the Society for the Scientific Study of Sex, Chicago, Ill.: November 20, 1983.

Duldt, B. W. "Sexual Harassment in Nursing." *Nursing Outlook,* pp. 336–343, June 1982.

Dumm, J. J.; Piotrow, P. T.; and Dalsimer, I. A. "The Moder: Quality Product for Effective Contraception." *Population* H(2):May 1974.

Ellis, A. *The American Sexual Tragedy*. New York: Twayne Publishers, 1959.

Erikson, E. *Childhood and Society*, 2nd ed. New York: Norton, 1963.

———. *Identity: Youth and Crisis*. New York: Norton, 1968.

Fast, J. *Body Language*. New York: M. Evans, 1972.

Fauci, A. S. "The Acquired Immune Deficiency Syndrome: The Ever-Broadening Clinical Spectrum." *Journal of the American Medical Association* 249:2375–76, 1983.

Faust, B. *Women, Sex and Pornography*. New York: Macmillan, 1980.

Feldman, M. P., and MacCulloch, M. J. *Homosexual Behavior: Therapy and Assessment*. Oxford: Pergamon Press, 1971.

Feldman-Summers, S.; Gordon, P.; and Meagher, J. R. "The Impact of Rape on Sexual Satisfaction." *Journal of Abnormal Psychology* 88:101–105, 1979.

Finkelhor, D. *Sexually Victimized Children*. New York: Free Press, 1979.

———. "Sex among Siblings: A Survey on Prevalence, Variety, and Effects." *Archives of Sexual Behavior* 9:171–194, 1980.

———. "Sex between Siblings." In Constantine, L. L., and Martinson, F. M. (eds.), *Children and Sex: New Findings, New Perspectives*. Boston: Little, Brown, 1981, pp. 129–149

———. *Child Sexual Abuse*. New York: Free Press, 1984.

Fiscella, K. "Relationship of Weight Change to Required Size of Vaginal Diaphragm." *Nurse Practitioner* 7(7):21, 25, Jul.–Aug. 1982.

Fisher, S. *The Female Orgasm*. New York: Basic Books, 1973.

Fisher, W. A.; Branscombe, N. R.; and Lemery, C. R. "The Bigger the Better? Arousal and Attributional Responses to Erotic Stimuli That Depict Different Size Penises." *Journal of Sex Research* 19:337–396, 1983.

Fisher, W. A., and Byrne, D. "Sex Differences in Response to Erotica? Love vs. Lust." *Journal of Personality and Social Psychology* 36:117–125, 1978.

FitzGerald, M., and FitzGerald, D. "Deaf People Are Sexual Too!" *SIECUS Report* 6(2):1, 13–15, 1977.

Forbes, G. B., and King, S. "Fear of Success and Sex-Role: There Are Reliable Relationships." *Psychological Reports* 53:735–738, 1983.

Ford, C. S., and Beach, F. A. *Patterns of Sexual Behavior*. New York, Harper & Brothers, 1951.

Frank, E; Anderson, C.; and Rubinstein, D. "Frequency of Sexual Dysfunction in 'Normal' Couples." *New England Journal of Medicine* 299:111–115, 1978.

Freud, S. *A General Introduction to Psychoanalysis*. Garden City, N.Y.: Garden City Publishing, 1943.

Friday, N. *My Secret Garden*. New York: Trident, 1973.

———. *Forbidden Flowers*. New York: Pocket Books, 1975.

———. *Men In Love*. New York: Delacorte, 1980.

Frieze, I. "Investigating the Causes and Consequences of Marital Rape." *Signs* 8:532–553, 1983.

Frisch, R. E., and McArthur, J. W. "Menstrual Cycles: Fatness as a Determinant of Minimum Weight for Height Necessary for Their Maintenance or Onset." *Science* 185:949–951, 1974.

Frisch, R. E.; Wyshak, G.; and Vincent, L. "Delayed Menarche and Amenorrhea in Ballet Dancers." *New England Journal of Medicine* 303:17–19, 1980.

Fromm, E. *The Art of Loving*. New York: Harper & Row, 1956.

Furstenberg, F., Jr.; Mencken, J.; and Lincoln, R. *Teenage Sexuality, Pregnancy, and Childbearing*. Philadelphia: University of Pennsylvania Press, 1981.

Gadpaille, W. J. *The Cycles of Sex*. New York: Scribner, 1975.

Gager, N., and Schurr, C. *Sexual Assault: Confronting Rape in America*. New York: Grosset & Dunlap, 1976

Gay Liberation v. University of Missouri. 1977[416 F. Supp. 1350(W. D. Mo. 1976)].

Gebhard, P. H. "Factors in Marital Orgasm." *Journal of Social Issues* 22(4):88–95, 1966.

General Accounting Office. *Sexual Exploitation of Children — A Problem of Unknown Magnitude* (HRD–82–64). Washington, D.C.: U.S. General Accounting Office, 1982.

George, L. K., and Weiler, S. J. "Sexuality in Middle and Later Life." *Archives of General Psychiatry* 38:919–923, 1981.

Gilbaugh, J. H., Jr., and Fuchs, P. C. "The Gonococcus and the Toilet Seat." *The New England Journal of Medicine* 301:91–93, 1979.

Goldberg, D. C., et al. "The Grafenberg Spot and Female Ejaculation: A Review of Initial Hypotheses." *Journal of Sex & Marital Therapy* 9:27–37, 1983.

Goldman, R., and Goldman, J. *Children's Sexual Thinking*. Boston: Routledge and Kegan Paul, 1982.

———. "Children's Sexual Thinking: Report of a Cross-national Study." *SIECUS Report* 10(3):3–7, January 1982a.

Goleman, D., and Bush, S. "The Liberation of Sexual Fantasy." *Psychology Today* 11:48–53, 104–107, October 1977.

Gordon, S. Speech at the Annual Meeting of the American Association for Sex Educators, Counselors, and Therapists, New York, N.Y., March 13, 1982. (Printed in *Impact 1982/83*, Institute for Family Research and Education, Syracuse University.)

Gottlieb, B. "Incest: Therapeutic Intervention in a Unique Form of Sexual Abuse." In Warner, C. (ed.), *Rape and Sexual Assault*, pp. 121–140. Germantown, Md.: Aspen Systems Corp., 1980.

Gove, W. R. "Sex Differences in the Epidemiology of Mental Disorder: Evidence and Explanations." In Gomberg, E. S., and Franks, V. (eds.), *Gender and Disordered Behavior*, pp. 23–68. New York: Brunner/Mazel, 1979.

Grafenberg, E. "The Role of the Urethra in Female Orgasm." *The International Journal of Sexology* 3:145–148, 1950.

Graham, S., et al. "Sex Patterns and Herpes Simplex Virus Type 2 in the Epidemiology of Cancer of the Cervix." *American Journal of Epidemiology* 115:729–735, 1982.

Greenblatt, D. R. "Semantic Differential Analysis of the 'Triangular System' Hypothesis in 'Adjusted' Overt Male Homosexuals." Ph.D. dissertation, University of California, 1966.

Greer, D. M., et al. "A Technique for Foreskin Reconstruction and Some Preliminary Results." *Journal of Sex Research* 18:324–330, 1982.

Greer, G. *The Female Eunuch*. New York: Bantam Books, 1972.

Grosskopf, D. *Sex and the Married Woman*. New York: Simon & Schuster, 1983.

Grossman, R., and Sutherland, J. (eds.). *Surviving Sexual Assault*. New York: Congdon & Weed, 1982/83.

Groth, A. N. *Men Who Rape*. New York: Plenum Press, 1979.

Groth, A. N., and Burgess, A. W. "Male Rape: Offenders and Victims." *American Journal of Psychiatry* 137:806–810, 1980.

———. "Sexual Dysfunction during Rape." *The New England Journal of Medicine* 297:764–766, 1977.

Groth, A. N.; Burgess, A. W.; and Holmstrom, L. "Rape: Power, Anger, and Sexuality." *American Journal of Psychiatry* 134:1239–43, 1977.

Grumbach, M. "The Neuroendocrinology of Puberty." *Hospital Practice*, pp. 51–60, March 1980.

Hacker, H. M. "Blabbermouths and Clams: Sex Differences in Self-Disclosure in Same-Sex and Cross-Sex Friendship Dyads." *Psychology of Women Quarterly* 5:385–401, 1981.

Halikas, J.; Weller, R.; and Morse, C. "Effects of Regular Marihuana Use on Sexual Performance." *Journal of Psychoactive Drugs* 14:59–70, 1982.

Haller, J. S., and Haller, R. M. *The Physician and Sexuality in Victorian America.* New York: Norton, 1977.

Hallstrom, T. "Sexuality of Women in Middle Age: The Göteborg Study." *Journal of Biosocial Sciences (Suppl.)* 6:165–175, 1979.

Hariton, E. B., and Singer, J. L. "Women's Fantasies during Marital Intercourse: Normative and Theoretical Implications." *Journal of Consulting and Clinical Psychology* 42(3):313–322, 1974.

Harlow, H. F. "The Nature of Love." *American Psychologist* 13:673–685, 1958.

Harrison, F. *The Dark Angel — Aspects of Victorian Sexuality.* New York: Universe Books, 1977.

Hass, A. *Teenage Sexuality.* New York: Macmillan, 1979.

Hatcher, R. A., et al. *Contraceptive Technology 1980–1981.* New York: Irvington Publishers, 1980.

Hatfield, E. "Passionate Love, Companionate Love, and Intimacy." In Fisher, M. and Stricker, G. (eds.), *Intimacy*, pp. 267–292. New York: Plenum Press, 1982.

Hatterer, L. J. *Changing Homosexuality in the Male.* New York: McGraw-Hill, 1970.

Heinlein, R. A. *Stranger in a Strange Land.* New York: Putnam, 1961.

Hier, D. B., and Crowley, W. F., Jr. "Spatial Ability in Androgen-deficient Men." *New England Journal of Medicine* 306:1202–5, 1982.

Higham, E. "Sexuality in the Infant and Neonate: Birth to Two Years." In Wolman, B. B., and Money, J. (eds.), *Handbook of Human Sexuality*, pp. 16–27. Englewood Cliffs, N.J.: Prentice-Hall, 1980.

Hilberman, E. "Rape: The Ultimate Violation of the Self." *American Journal of Psychiatry* 133:436, 1976.

Hill, C. T.; Rubin, Z.; and Peplau, L. A. "Breakups before Marriage: The End of 103 Affairs." *Journal of Social Issues* 32:147–168, 1976.

Hite, S. *The Hite Report.* New York: Dell, 1977.

———. *The Hite Report on Male Sexuality.* New York: Alfred A. Knopf, 1981.

Hollender, M. H. "Women's Wish to Be Held: Sexual and Nonsexual Aspects." *Medical Aspects of Human Sexuality* 5(10):12–26, 1971.

Hooker, E. "The Adjustment of the Male Overt Homosexual." *Journal of Projective Techniques* 21:18–31, 1957.

Horner, M. "Toward an Understanding of Achievement Related Conflicts in Women." *Journal of Social Issues* 28: 157–175, 1972.

Howells, J. G. (ed.). *Modern Perspectives in the Psychiatry of Middle Age.* New York: Brunner/Mazel, 1981.

Hunt, M. *The Natural History of Love.* New York: Funk & Wagnalls, Minerva Press, 1967.

———. *Sexual Behavior in the 1970s.* New York: Dell, 1975.

Hunt, M. "The Future of Marriage." In DeBurger, J. E. (ed.), *Marriage Today*, pp. 638–698. New York: Wiley, 1977.

Huston, T. L., and Levinger, G. "Interpersonal Attraction and Relationships." *Annual Review of Psychology* 29:115–156, 1978.

Jaffe, H. W., et al. "National Case-Control Study of Kaposi's Sarcoma and *Pneumocystitis Carinii* Pneumonia in Homosexual Men." *Annals of Internal Medicine* 99:145–151, 1983.

Jessor, S. L., and Jessor, R. "Transition from Virginity to Nonvirginity among Youth: A Social-Psychological Study over Time." *Developmental Psychology* 11:473–484, 1975.

———. *Problem Behavior and Psychosocial Development: A Longitudinal Study of Youth*. New York: Academic Press, 1977.

Jong, E. *Fear of Flying*. New York: Signet, 1974.

Kallman, F. J. "Comparative Twin Study on the Genetic Aspects of Male Homosexuality." *Journal of Nervous and Mental Disease* 115:283–298, 1952.

Kanin, E. "Selected Dyadic Aspects of Male Sex Aggression." *Journal of Sex Research* 5:12–28, 1969.

———. "Date Rapists: Differential Sexual Socialization and Relative Deprivation." *Archives of Sexual Behavior* 14:219–231, 1985.

Kaplan, A., and Sedney, M. A. *Psychology and Sex Roles: An Androgynous Perspective*. Boston: Little, Brown, 1980.

Kaplan, H. *The Evaluation of Sexual Disorders*. New York: Brunner/Mazel, 1983.

Kaplan, H. S. *The New Sex Therapy*. New York: Brunner/Mazel, 1974.

———. *Disorders of Sexual Desire*. New York: Brunner/Mazel, 1979.

Karlen, A. *Sexuality and Homosexuality: A New View*. New York: Norton, 1971.

———. "Homosexuality: The Scene and Its Students." In Henslin, J. M., and Sagarin, E., *The Sociology of Sex*, pp. 223–248. New York: Schocken Books, 1978.

Kegel, A. "Sexual Functions of the Pubococcygeus Muscle." *Western Journal of Surgery, Obstetrics, and Gynecology* 60:521–524, 1952.

Kempton, W. "Sex Education for the Mentally Handicapped." *Sex and Disability* 1(2):137–146, 1978.

Kern, S. "Freud and the Discovery of Child Sexuality." *History of Childhood Quarterly* 1:117–141, 1973.

Kinsey, A. C., et al. *Sexual Behavior in the Human Female*. Philadelphia: Saunders, 1953.

Kinsey, A. C.; Pomeroy, W. B.; and Martin, C. E. *Sexual Behavior in the Human Male*. Philadelphia: Saunders, 1948.

Kirby, D.; Alter, J.; and Scales, P. *An Analysis of U.S. Sex Education Programs and Evaluation Methods*. Atlanta, Ga.: U.S. Department of Health, Education, and Welfare, Public Health Service, Centers for Disease Control, Bureau of Health Education, Report # CDC-2021-79-DK-FR, 1979.

Knutson, D. C. (ed.). "Introduction." *Journal of Homosexuality* 5(1,2):5–23, 1979/1980.

Kolodny, R. C. "Sexual Dysfunction in Diabetic Females." *Diabetes* 208(8):557–559, 1971.

———. "Evaluating Sex Therapy: Process and Outcome at the Masters & Johnson Institute." *Journal of Sex Research* 17:301–318, 1981.

———. "The Clinical Management of Sexual Problems in Substance Abusers."

In Bratter, T. E., and Forrest, G. (eds.), *Current Management of Alcoholism and Substance Abuse*. New York: Free Press, 1985.

———. "Depression of Plasma Testosterone levels after Chronic Intensive Marihuana Use." *New England Journal of Medicine* 290:872–874, 1974.

Kolodny, R. C.; Masters, W. H.; and Johnson, V. E. *Textbook of Sexual Medicine*. Boston: Little, Brown, 1979.

Kols, A., et al. "Oral Contraceptives in the 1980s." *Population Reports*, Series A, Number 6, May-June 1982.

Kutchinsky, B. "The Effect of Easy Availability of Pornography on the Incidence of Sex Crimes." *Journal of Social Issues* 229:163–182, 1973.

Ladas, A. K.; Whipple, B.; and Perry, J. D. *The G Spot and Other Recent Discoveries About Human Sexuality*. New York: Holt, Rinehart & Winston, 1982.

Lawrence, D. H. *Lady Chatterley's Lover*. New York: Grove Press, Black Cat Edition, 1962.

LeBolt, S. A.; Grimes, D. A.; and Cates, W., Jr. "Mortality from Abortion and Childbirth." *Journal of the American Medical Association* 248:188–191, 1982.

Lederer, L. (ed.). *Take Back the Night*. New York: William Morrow, 1980.

Lee, A. L., and Scheurer, V. L. "Psychological Androgyny and Aspects of Self-Image in Women and Men." *Sex Roles* 9:289–306, 1983.

Lee, J. A. *The Colours of Love*. Toronto: New Press, 1973.

Leiblum S., et al. "Vaginal Atrophy in the Postmenopausal Woman: The Importance of Sexual Activity and Hormones." *Journal of the American Medical Association* 249:2195–98, 1983.

Levin, R. J., and Levin, A. "Sexual Pleasure: The Surprising Preferences of 100,000 Women." *Redbook*, pp. 51–58, September 1975.

Levinger, G., and Raush, H. L. (eds.). *Close Relationships: Perspectives on the Meaning of Intimacy*. Amherst, Mass.: University of Massachusetts Press, 1977.

Levinson, D. J., et al. *The Seasons of a Man's Life*. New York: Ballantine, 1978.

Levy, N. L. "The Middle-aged Male and Female Homosexual." In Howells, J. G. (ed.), *Modern Perspectives in the Psychiatry of Middle Age*, pp. 116–131. New York: Brunner/Mazel, 1981.

Liebowitz, M. R. *The Chemistry of Love*. Boston: Little, Brown, 1983.

Long Laws, J. *The Second X*. New York: Elsevier, 1979.

LoPiccolo, J. "Direct Treatment of Sexual Dysfunction in the Couple." In Money, J., and Musaph, H. (eds.), *Handbook of Sexology*, pp. 1227–44. Amsterdam: Elsevier/North-Holland, 1977.

Lowry, T. P. "The Volatile Nitrites as Sexual Drugs: A User Survey." *Journal of Sex Education and Therapy* 1:8–10, 1979.

Lowry, T. P., and Williams, G. R. "Brachioproctic Eroticism." *Journal of Sex Education and Therapy* 9(1):50–52, 1983.

McCandlish, B. M. "Therapeutic Issues with Lesbian Couples." *Journal of Homosexuality* 7(2-3):71–78, 1981/82.

McCarthy, B. W. "Sexual Dysfunctions and Dissatisfactions among Middle-Years Couples." *Journal of Sex Education and Therapy* 8(2):9–12, 1982.

Maccoby, E., and Jacklin, C. *The Psychology of Sex Differences*. Stanford: Stanford University Press, 1974.

McGee, E. A. *Too Little, Too Late: Services for Teenage Parents*. New York: Ford Foundation, 1982.

McGhee, P. E., and Frueh, T. "Television Viewing and the Learning of Sex-role Stereotypes." *Sex Roles* 6:179–188, 1980.

McGuinness, D., and Pribram, K. "The Origins of Sensory Bias in the Development of Gender Differences in Perception and Cognition." In Bortner, M. (ed.), *Cognitive Growth and Development: Essays in Honor of Herbert G. Birch,* pp. 3–56. New York: Brunner/Mazel, 1978.

MacKinnon, C. A. *Sexual Harassment of Working Women.* New Haven: Yale University Press, 1979.

Macklin, E. D. "Unmarried Heterosexual Cohabitation on the University Campus." In Wiseman, J. P. (ed.), *The Social Psychology of Sex,* pp. 108–142. New York: Harper & Row, 1976.

———. "Review of Research on Nonmarital Cohabitation in the United States." In Murstein, B. I. (ed.), *Exploring Intimate Life Styles,* pp. 197–243, New York: Springer, 1978.

———. "Nontraditional Family Forms: A Decade of Research." *Journal of Marriage and the Family* 42:905–922, 1980.

MacNamara, D. E., and Sagarin, E. *Sex, Crime, and the Law.* New York: Free Press, 1977.

McWhirter, D. P., and Mattison, A. M. *The Male Couple: How Relationships Develop.* Englewood Cliffs, N.J.: Prentice-Hall, 1984.

Malamuth, N., and Donnerstein, E. (eds.). *Pornography and Sexual Aggression.* New York: Academic Press, 1984.

Malatesta, V. J. "Alcohol Effects on the Orgasmic-Ejaculatory Response in Human Males." *Journal of Sex Research* 15:101–107, 1979.

Malatesta, V. J., et al. "Acute Alcohol Intoxication and Female Orgasmic Response." *Journal of Sex Research* 18:1–17, 1982.

Malinowski, B. *The Sexual Life of Savages.* New York: Harcourt, Brace & World, 1929.

Marcus, S. *The Other Victorians.* New York: Bantam Books, 1967.

Marin, Peter. "A Revolution's Broken Promises." *Psychology Today,* pp. 50–57, July 1983.

Marmor, J. (ed.). *Homosexual Behavior.* New York: Basic Books, 1980.

Marshall, D. "Sexual Behavior on Mangaia." In Marshall, D., and Suggs, R. (eds.), *Human Sexual Behavior,* pp. 103–162. New York: Basic Books, 1971.

Martinson, F. M. "Eroticism in Infancy and Childhood." In Constantine, L. L., and Martinson, F. M. (eds.), *Children and Sex: New Findings, New Perspectives,* pp. 23–35. Boston: Little, Brown, 1981.

Masters, W. H. "Outcome Studies at the Masters & Johnson Institute." Presented at the 6th World Congress of Sexology, Washington, D.C.: May 26, 1983.

Masters, W. H., and Johnson, V. E. *Human Sexual Response.* Boston: Little, Brown, 1966.

———. *Human Sexual Inadequacy.* Boston: Little, Brown, 1970.

———. *The Pleasure Bond.* New York: Bantam Books, 1976.

———. *Homosexuality in Perspective.* Boston: Little, Brown, 1979.

Meiselman, K. C. *Incest.* San Francisco: Jossey-Bass, 1978.

Metzger, D. "It Is Always the Woman Who Is Raped." *American Journal of Psychiatry* 133:405–408, 1976.

Meyer, J. D., and Reter, D. J. "Sex Reassignment." *Archives of General Psychiatry* 36:1010–15, 1979.

Meyers, Robert. *D.E.S. — The Bitter Pill.* New York: Seaview/Putnam, 1983.

Miller, H. *Tropic of Cancer.* New York: Grove Press, 1961.

Millett, K. *Sexual Politics*. New York: Doubleday, 1970.

Mohr, J. C. *Abortion in America: The Origins and Evolution of National Policy, 1800–1900*. New York: Oxford University Press, 1978.

Monat, R. K. *Sexuality and the Mentally Retarded*. San Diego, Calif.: College-Hill Press, 1982.

Money, J. *Love and Love Sickness*. Baltimore: Johns Hopkins University Press, 1980.

Mooney, T. O.; Cole, T. M.; and Chilgren, R. A. *Sexual Options for Paraplegics and Quadraplegics*. Boston: Little, Brown, 1975.

Murstein, B. I. (ed.). *Exploring Intimate Life Styles*. New York: Springer, 1978.

Nadelson, C. C. "The Emotional Impact of Abortion." In Notman, M. T., and Nadelson, C. C. (eds.), *The Woman Patient*, vol. 1, pp. 173–179. New York: Plenum Press, 1978.

Nadelson, C. C., et al. "A Follow-up Study of Rape Victims." *American Journal of Psychiatry* 139:1266–70, 1982.

Offit, A. *The Sexual Self*. New York: Ballantine, 1977.

O'Neill, N. *The Marriage Premise*. New York: Bantam Books, 1978.

O'Neill, N., and O'Neill, G. *Open Marriage: A New Life Style for Couples*. New York: M. Evans and Company, 1972.

Ory, H. W.; Rosenfeld, A.; and Landman, L. C. "The Pill at 20: An Assessment." *Family Planning Perspectives* 12:278–283, 1980.

Paludi, M. A., and Bauer, W. D. "Goldberg Revisited: What's in an Author's Name?" *Sex Roles* 9:387–390, 1983.

Peele, S., with Brodsky, A. *Love and Addiction*. New York: New American Library, 1976.

Peplau, L. A.; Padesky, C.; and Hamilton, M. "Satisfaction in Lesbian Relationships." *Journal of Homosexuality* 8(2):23–35, 1982.

Perelman, M. A. "Treatment of Premature Ejaculation." In Lieblum, S. R., and Pervin, L. A. (eds.), *Principles and Practice of Sex Therapy*, pp. 199–233. New York: Guilford Press, 1980.

Perkins, R. P. "Sexual Behavior and Response in Relation to Complications of Pregnancy." *American Journal of Obstetrics and Gynecology* 134:498–505, 1979.

Perry, J. D., and Whipple, B. "Pelvic Muscle Strength of Female Ejaculators: Evidence in Support of a New Theory of Orgasm." *Journal of Sex Research* 17(1):22–39, 1981.

Peters, J. B.; Bryson, Y.; and Lovett, M. A. "Genital Herpes: Urgent Questions, Elusive Answers." *Diagnostic Medicine*, pp. 71–74, 76–88, March/April 1982.

Petersen, J. R. "Desire," *Playboy*, p. 180, December 1980.

Pivar, D. J. *Purity Crusade, Sexual Morality, and Social Control, 1868–1900*. Westport, Conn.: Greenwood Press, 1973.

Pogrebin, L. C. *Growing Up Free: Raising Your Child in the 80's*. New York: McGraw-Hill Book Co., 1980.

Rachman, S. J., and Wilson, G. T. *The Effects of Psychological Therapy*, 2nd ed. New York: Pergamon Press, 1980.

Rada, R. T. (ed.). *Clinical Aspects of the Rapist*. New York: Grune & Stratton, 1978.

Ramcharan, S. *The Walnut Creek Contraceptive Drug Study: A Prospective Study of the Side Effects of Oral Contraceptives* (no. NIH 81–564). Bethesda, Md.: U.S. Department of Health and Human Services, National Institutes of Health, 1981.

Reingold, A. L. "Nonmenstrual Toxic Shock Syndrome: The Growing Picture." *Journal of the American Medical Association* 249:932, 1983.

Reingold, A. L., et al. "Toxic Shock Syndrome Surveillance in the United States, 1980 to 1981." *Annals of Internal Medicine* 96(Part 2):875–880, 1982.

Reiss, I. L. *Family Systems in America,* 3rd ed. New York: Holt, Rinehart and Winston, 1980.

Rheingold, H. L. and Cook, K. V. "The Contents of Boys' and Girls' Rooms as an Index of Parents' Behavior." *Child Development* 46:459–463, 1975.

Robbins, M., and Jensen, G. D. "Multiple Orgasm in Males." *Journal of Sex Research* 14:21–26, 1978.

Robboy, S. J., et al. *Prenatal Diethylstilbestrol (DES) Exposure: Recommendations of the Diethylstilbestrol Adenosis (DESAD) Project for the Identification and Management of Exposed Individuals* (no. NIH 81–2049). Bethesda, Md.: U.S. Department of Health and Human Services, 1981.

Roberts, C. L., and Lewis, R. A. "The Empty Nest Syndrome." In Howells, J. G. (ed.), *Modern Perspectives in the Psychiatry of Middle Age,* pp. 328–336. New York: Brunner/Mazel, 1981.

Rogers, C. R. *Becoming Partners.* New York: Delacorte Press, 1972.

Rommel, E. "Grade School Blues." *Ms. Magazine,* pp. 32–35, January 1984.

Rosenbaum, M. "When Drugs Come into the Picture, Love Flies out the Window: Women Addicts' Love Relationships." *International Journal of the Addictions* 16:1197–1206, 1981.

Ross, M. W. "Retrospective Distortion in Homosexual Research." *Archives of Sexual Behavior* 9:523–532, 1980.

Rotheram, M. J., and Weiner, N. "Androgyny, Stress, and Satisfaction." *Sex Roles* 9:151–158, 1983.

Rubenstein, C., and Shaver, P. *In Search of Intimacy.* New York: Random House, 1982.

Rubin, L. "Sex and Sexuality: Women at Midlife." In Kirkpatrick, M. (ed.), *Women's Sexual Experiences — Explorations of the Dark Continent,* pp. 61–82. New York: Plenum Press, 1982.

———. *Intimate Strangers: Men and Women Together.* New York: Harper & Row, 1983.

Rubin, Z. et al. "Self-Disclosure in Dating Couples: Sex-Roles and the Ethic of Openness." *Journal of Marriage and the Family* 42:305–317, 1980.

Rubenstein, C. "The Modern Art of Courtly Love." *Psychology Today,* pp. 43–49, July 1983.

Rush, F. "The Sexual Abuse of Children: A Feminist Point of View." In Connell, N., and Wilson, C. (eds.), *Rape: The First Sourcebook for Women,* pp. 65–75. New York: New American Library, 1974.

Russell, Diana E. H. *Rape in Marriage.* New York: Macmillan, 1982.

Sadoff, R. L. "Other Sexual Deviations." In Freedman, A. M.; Kaplan, H. I.; and Sadock, B. J. (eds.), *Comprehensive Textbook of Psychiatry/II,* pp. 1539–44. Baltimore: Williams & Wilkins, 1975.

Safran, C. "What Men Do to Women on the Job: A Shocking Look at Sexual Harassment." *Redbook,* p. 148, November 1976.

————. "Sexual Harassment: The View from the Top." *Redbook*, pp. 47–51, March 1981.

Saghir, M. T., and Robins, E. *Male and Female Homosexuality*. Baltimore: Williams & Wilkins, 1973.

Sanford, L. T. *Come Tell Me Right Away*. Fayetteville, N.Y.: Ed-U Press, 1982.

Schachter, S. "The Interaction of Cognitive and Physiological Determinants of Emotional State." In Berkowitz, L. (ed.). *Advances in Experimental Social Psychology*, vol. 1, pp. 49–80. New York: Academic Press, 1964.

Schlech, W. F., III, et al. "Risk Factors for Development of Toxic Shock Syndrome." *Journal of the American Medical Association* 248:835–839, 1982.

Scholl, T.O., et al. "Effects of Vaginal Spermicides on Pregnancy Outcome." *Family Planning Perspectives* 15:244–250, 1983.

Schover, L. R., and LoPiccolo, J. "Treatment Effectiveness for Dysfunctions of Sexual Desire." *Journal of Sex & Marital Therapy* 8:179–197, 1982.

Schover, L. R., et al. "The Multi-axial Problem-oriented Diagnostic System for the Sexual Dysfunctions: An Alternative to DSM-III." *Archives of General Psychiatry* 39:614–619, 1982.

Schreiner-Engel, P., et al. "Sexual Arousability and the Menstrual Cycle." *Psychosomatic Medicine* 43:199–214, 1981.

Schultz, L. G. (ed.). *The Sexual Victimology of Youth*. Springfield, Ill.: Charles C. Thomas, 1980.

Schwartz, M. F., and Masters, W. H. "Conceptual Factors in the Treatment of Paraphilias: A Preliminary Report." *Journal of Sex & Marital Therapy* 9:3–18, 1983.

Scott, J. F. *The Sexual Instinct: Its Use and Dangers as Affecting Heredity and Morals*, 3rd ed. Chicago: Login Brothers, 1930.

Sevely, J., and Bennett, J. "Concerning Female Ejaculation and the Female Prostate." *Journal of Sex Research* 14:1–20, 1978.

Shainess, N. "How 'Sex Experts' Debase Sex." *World* 2(1):21–25, 1973.

Shainess, N., and Greenwald, H. "Debate: Are Fantasies during Sexual Relations a Sign of Difficulty?" *Sexual Behavior* 1:38–54, 1971.

Shanor, K. *The Fantasy Files*. New York: Dial Press, 1977.

Sheehy, G. *Passages: Predictable Crises of Adult Life*. New York: E. P. Dutton, 1976.

Sherfey, M. J. *The Nature and Evolution of Female Sexuality*. New York: Random House, 1972.

Shostak, A.; McLouth, G.; and Seng, L. *Men and Abortions: Lessons, Losses, and Love*. New York: Praeger, 1984.

Siegel, R. K. "Cocaine and Sexual Dysfunction." *Journal of Psychoactive Drugs* 14:71–74, 1982.

Siegelman, M. "Parental Background of Male Homosexuals and Heterosexuals." *Archives of Sexual Behavior* 3:3–18, 1974.

Simenauer, J., and Carroll, D. *Singles: The New Americans*. New York: Simon and Schuster, 1982.

Singer, J. L. "Romantic Fantasy in Personality Development." In Pope, K. S. (ed.), *On Love and Loving*, pp. 172–194. San Francisco: Jossey-Bass, 1980.

Singer, J., and Singer, I. "Types of Female Orgasm." *Journal of Sex Research* 8:255–267, 1972.

Socarides, C. W. "Homosexuality and Medicine." *Journal of the American Medical Association* 212:1199–1202, 1970.

Somers, A. "Sexual Harassment in Academe: Legal Issues and Definitions." *Journal of Social Issues* 38(4):23–32, 1982.

Sorenson, R. C. *Adolescent Sexuality in Contemporary America.* New York: World Publishing Co., 1973.

Sorokin, P. A. *The American Sex Revolution.* Boston: Porter Sargent, 1956.

Spanier, G. B., and Furstenberg, F. F., Jr. "Remarriage after Divorce: A Longitudinal Analysis of Well-being." *Journal of Marriage and the family* 44:709–720, 1982.

Spence, J. T., and Helmreich, R. L. *Masculinity & Feminity: Their Psychological Dimensions, Correlates, and Antecedents.* Austin, Tex.: University of Texas Press, 1978.

Starr, B. D., and Weiner, M. B. *The Starr-Weiner Report on Sex & Sexuality in the Mature Years.* New York: Stein & Day, 1981.

Steinam, G. "Erotica vs. Pornography." In Steinam, G., *Outrageous Acts and Everyday Rebellions.* New York: Holt, Rinehart and Winston, 1983.

Stevens, F. A. "The Occurrence of *Staphyloccocus aureus* Infection with a Scarlatiniform Rash." *Journal of the American Medical Association* 88:1957–58, 1927.

Stoller, R. J. "Sexual Deviations." In Beach, F. (ed.), *Human Sexuality in Four Perspectives,* pp. 190–214. Baltimore: Johns Hopkins University Press, 1977.

———. *Sexual Excitement.* New York: Pantheon Books, 1979.

Sue, D. "Erotic Fantasies of College Students during Coitus." *Journal of Sex Research* 15:299–305, 1979.

Sullivan-Bolyai, J., et al. "Neonatal Herpes Simplex Virus Infection in King County, Washington." *Journal of the American Medical Association* 250:3059–62, 1983.

Summit, R., and Kryso, J. "Sexual Abuse of Children: A Clinical Spectrum." *American Journal of Orthopsychiatry* 48:237–251, 1978.

Szasz, T. *Sex by Prescription.* New York: Anchor Press/Doubleday, 1980.

Tangri, S. S.; Burt, M. R.; and Johnson, L. B. "Sexual Harassment at Work: Three Explanatory Models." *Journal of Social Issues* 38(4):33–54, 1982.

Tannahill, R. *Sex in History.* New York: Stein & Day, 1980.

Tanner, J. M. "Sequence and Tempo in the Somatic Changes in Puberty." In Grumbach, M. M.; Grave, G. D.; and Mayer, F. E. (eds.), *Control of the Onset of Puberty.* New York: Wiley, 1974.

Tavris, C. *Anger — The Misunderstood Emotion.* New York: Simon and Schuster, 1982.

Tavris, C., and Offir, C. *The Longest War: Sex Differences in Perspective.* New York: Harcourt Brace Jovanovich, 1977.

Taylor, G. R. *Sex in History.* New York: Vanguard, 1954.

Tennov, D. *Love and Limerence.* New York: Stein & Day, 1979.

Tripp, C. A. *The Homosexual Matrix.* New York: McGraw-Hill, 1975.

Trudel, G., and Saint Laurent, S. "A Comparison Between the Effects of Kegel's Exercises and a Combination of Sexual Awareness Relaxation and Breathing on Situational Orgasmic Dysfunction in Women." *Journal of Sex & Marital Therapy* 9:204–209, 1983.

Udry, J. R., and Cliquet, R. L. "A Cross-cultural Examination of the Relationship between Ages at Menarche, Marriage, and First Birth." *Demography* 19:53–63, 1982.

Vontver, L. A., et al. "Recurrent Genital Herpes Simplex Virus Infection in Pregnancy: Infant Outcome and Frequency of Asymptomatic Recurrences." *American Journal of Obstetrics and Gynecology* 143:75–81, 1982.

Walster, E., and Walster, G. W. *A New Look at Love.* Reading, Mass.: Addison-Wesley, 1978.

Walum, L. R. *The Dynamics of Sex and Gender: A Sociological Perspective.* Chicago: Rand McNally College Publishing Co., 1977.

Washburn, S. *Partners: How to Have a Loving Relationship after Women's Liberation.* New York: Atheneum, 1981.

Waterman, C. D., and Chiauzzi, E. J. "The Role of Orgasm in Male and Female Sexual Enjoyment." *Journal of Sex Research* 18:146–159, 1982.

Weitzman, L. J. "Sex-Role Socialization." In Freeman, J. (ed.), *Women: A Feminist Perspective.* Palo Alto, Calif.: Mayfield, 1975.

Weitzman, L. J., et al. "Sex Role Socialization in Picture Books for Pre-School Children." *American Journal of Sociology* 77:1125–50, 1972

Wilcox, D., and Hager, R. "Toward Realistic Expectation for Orgasmic Response in Women." *Journal of Sex Research* 16:162–179, 1980.

Wilson, G. T., and Lawson, D. M. "Effects of Alcohol on Sexual Arousal in Women." *Journal of Abnormal Psychology* 85:489–497, 1976.

———. "Effects of Alcohol on Sexual Arousal in Male Alcoholics." *Journal of Abnormal Psychology* 87:609–616, 1978.

Wolfe, L. "The Sexual Profile of That Cosmopolitan Girl." *Cosmopolitan,* pp. 254–265, September 1980.

Wolff, C. *Love between Women.* New York: Harper & Row, 1971.

Wolpe, J. *Psychotherapy by Reciprocal Inhibition.* Stanford, Calif.: Stanford University Press, 1958.

Wong, H. "Typologies of Intimacy." *Psychology of Women Quarterly* 5:435–443, 1981.

World Health Organization. "A Prospective Multicentre Trial of the Ovulation Method of Natural Family Planning. II. The Effectiveness Phase." *Fertility & Sterility* 36:591–598, 1981.

Zacharias, L., and Wurtman, R. J. "Age at Menarche: Genetic and Environmental Influences." *New England Journal of Medicine* 280:868–875, 1969.

Zamichow, N. "Is It Something in the Food?" *Ms. Magazine,* pp. 92–93, 141–143, October 1983.

Zeiss, A.M. "Expectation for the Effects of Aging on Sexuality in Parents and Average Married Couples." *Journal of Sex Research* 18:47–57, 1982.

Zelnik, M., and Kantner, J. F. "Sexual Activity, Contraceptive Use, and Pregnancy among Metropolitan-Area Teenagers: 1971–1979." *Family Planning Perspectives* 12:230–237, 1980.

Zelnik, M.; Kantner, J. F.; and Ford, K. *Sex and Pregnancy in Adolescence.* Beverly Hills, Calif.: Sage Publications, 1981.

Zelnik, M., and Kim, Y. J. "Sex Education and Its Association with Teenage Sexual Activity, Pregnancy and Contraceptive Use." *Family Planning Perspectives* 14:117–126, 1982.

Zilbergeld, B. *Male Sexuality: A Guide to Sexual Fulfillment.* Boston: Little, Brown, 1978.

Zilbergeld, B., and Evans, M. "The Inadequacy of Masters and Johnson." *Psychology Today* 14:29–43, 1980.

Zimmer, D.; Borchardt, E.; and Fischle, C. "Sexual Fantasies of Sexually Distressed and Nondistressed Men and Women: An Empirical Comparison." *Journal of Sex & Marital Therapy* 9:38–50, 1983.

Zwerner, J. "Sexual Issues of Women with Spinal Cord Injury." *Sexuality and Disability* 5:158–171, 1982.

Index